AP® CHEMISTRY

PREMIUM PREP

25th Edition

The Staff of The Princeton Review

PrincetonReview.com

Penguin
Random
House

The Princeton Review
110 East 42nd St, 7th Floor
New York, NY 10017

Published in the United States by Penguin Random House LLC, New York.

ISBN: 978-0-593-51676-8
ISSN: 2690-5329

Editor: Meave Shelton
Production Artist: Jason Ullmeyer
Production Editors: Chris Stobart, Becky Radway
Content Contributor: Nick Leonardi

Printed in the United States of America.

10 9 8 7 6 5 4 3 2 1

25th Edition

The Princeton Review Publishing Team
Rob Franek, Editor-in-Chief
David Soto, Senior Director, Data Operations
Stephen Koch, Senior Manager, Data Operations
Deborah Weber, Director of Production
Jason Ullmeyer, Production Design Manager
Jennifer Chapman, Senior Production Artist
Selena Coppock, Director of Editorial
Orion McBean, Senior Editor
Aaron Riccio, Senior Editor
Meave Shelton, Senior Editor
Chris Chimera, Editor
Patricia Murphy, Editor
Laura Rose, Editor

Penguin Random House Publishing Team
Tom Russell, VP, Publisher
Alison Stoltzfus, Senior Director, Publishing
Brett Wright, Senior Editor
Emily Hoffman, Associate Managing Editor
Ellen Reed, Production Manager
Suzanne Lee, Designer
Eugenia Lo, Publishing Assistant

For customer service, please contact **editorialsupport@review.com**, and be sure to include:

- full title of the book
- ISBN
- page number

Acknowledgments

The Princeton Review would like to give special thanks to Nick Leonardi and Brad Hutnick for their work on this edition, and for their dedication to the ongoing improvement of this book.

We would also like to thank the production staff—Jason Ullmeyer, Chris Stobart, and Becky Radway—for their hard work and careful attention to every page.

Contents

Get More (Free) Content
at PrincetonReview.com/prep

As easy as 1·2·3

1 Go to PrincetonReview.com/prep or scan the **QR code** and enter the following ISBN for your book: **9780593516768**

2 Answer a few simple questions to set up an exclusive Princeton Review account. *(If you already have one, you can just log in.)*

3 Enjoy access to your **FREE** content!

Once you've registered, you can...

- Get our take on any recent or pending updates to the AP Chemistry Exam.

- Access your 6th and 7th AP Chemistry practice tests (there are 5 in your book and 2 online), plus answer keys and complete answers and explanations.

- Take a full-length practice SAT and ACT.

- Get valuable advice about the college application process, including tips for writing a great essay and where to apply for financial aid.

- Use our searchable rankings of *The Best 389 Colleges* to find out more information about your dream school.

- Access comprehensive study guides and a variety of printable resources, including: answer bubble sheets, the periodic table, and a list of equations to review.

- Check to see if there have been any corrections or updates to this edition.

Need to report a potential **content** issue?

Contact **EditorialSupport@review.com** and include:

- full title of the book
- ISBN
- page number

Need to report a **technical** issue?

Contact **TPRStudentTech@review.com** and provide:

- your full name
- email address used to register the book
- full book title and ISBN
- Operating system (Mac/PC) and browser (Chrome, Firefox, Safari, etc.)

Look for These Icons Throughout the Book

 PROVEN TECHNIQUES

 APPLIED STRATEGIES

 MORE GREAT BOOKS

 ONLINE ARTICLES

 ONLINE PRACTICE TESTS

 ONLINE VIDEO TUTORIALS

Part I
Using This Book to Improve Your AP Score

PREVIEW ACTIVITY: YOUR KNOWLEDGE, YOUR EXPECTATIONS

Your route to a high score on the AP Chemistry Exam depends a lot on how you plan to use this book. Please respond to the following questions.

1. Rate your level of confidence about your knowledge of the content tested by the AP Chemistry Exam:

    A. Very confident—I know it all
    B. I'm pretty confident, but there are topics for which I could use help
    C. Not confident—I need quite a bit of support
    D. I'm not sure

2. Circle your goal score for the AP Chemistry Exam.

    5   4   3   2   1   I'm not sure yet

3. What do you expect to learn from this book? Circle all that apply to you.

    A. A general overview of the test and what to expect
    B. Strategies for how to approach the test
    C. The content tested by this exam
    D. I'm not sure yet

YOUR GUIDE TO USING THIS BOOK

This book is organized to provide as much—or as little—support as you need, so you can use this book in whatever way will be most helpful to improving your score on the AP Chemistry Exam.

-   The remainder of **Part I** will provide guidance on how to use this book and help you determine your strengths and weaknesses.
-   **Part II** of this book contains Practice Test 1, its Diagnostic Answer Key, and its answers and explanations. (Bubble sheets can be found in the very back of the book for easy tear-out.) We strongly recommend that you take this test before going any further, in order to realistically determine:

    o   your starting point right now
    o   which question types you're ready for and which you might need to practice
    o   which content topics you are familiar with and which you will want to carefully review

Once you have nailed down your strengths and weaknesses with regard to this exam, you can focus your test preparation, build a study plan, and be efficient with your time. Our Diagnostic Answer Key will assist you with this process.

- **Part III** of this book will:

  - provide information about the structure, scoring, and content of the AP Chemistry Exam
  - help you to make a study plan
  - point you toward additional resources

- **Part IV** of this book will explore:

  - how to tackle multiple-choice questions
  - how to write high-scoring free-response answers
  - how to manage your time to maximize the number of points available to you

- **Part V** of this book covers the content you need for your exam.

- **Parts VI** and **VII** of this book contain Practice Tests 2 through 5, and their answers and explanations. (Bubble sheets can be found in the very back of the book for easy tear-out.) If you skipped Practice Test 1, we recommend that you do both Practice Tests 1 and 2 (with at least a day or two between them) so that you can compare your progress between the two. Additionally, this will help to identify any external issues: if you get a certain type of question wrong both times, you probably need to review it. If you only got it wrong once, you may have run out of time or been distracted by something. In either case, this will allow you to focus on the factors that caused the discrepancy in scores and to be as prepared as possible on the day of the test.

You may choose to use some parts of this book over others, or you may work through the entire book. This will depend on your needs and how much time you have. Let's now look at how to make this determination.

Don't Forget!
To take your 6th and 7th Practice Tests, be sure to register your book online following the instructions on pages x–xi. You'll also gain access to a wealth of other helpful free Student Tools, including study guides and printable bubble sheets.

HOW TO BEGIN

1. Take a Test

Before you can decide how to use this book, you need to take a practice test. Doing so will give you insight into your strengths and weaknesses, and the test will also help you make an effective study plan. If you're feeling test-phobic, remind yourself that a practice test is a tool for diagnosing yourself—it's not how well you do that matters but how you use information gleaned from your performance to guide your preparation.

So, before you read further, take AP Chemistry Practice Test 1 starting at page 9 of this book. Be sure to do so in one sitting, following the instructions that appear before the test.

2. Check Your Answers

Using the Diagnostic Answer Key on page 36, follow our three-step process to identify your strengths and weaknesses with regard to the tested topics. This will help you determine which content review chapters to prioritize when studying this book. Don't worry about the explanations for now, and don't worry about why you missed questions. We'll get to that soon.

3. **Reflect on the Test**

After you take your first test, respond to the following questions:

- How much time did you spend on the multiple-choice questions?

- How much time did you spend on each long form free-response question? What about each short form free-response question?

- How many multiple-choice questions did you miss?

- Do you feel you had the knowledge to address the subject matter of the free-response questions?

- Do you feel you wrote well-organized, thoughtful answers to the free-response questions?

4. **Read Part III of this Book and Complete the Self-Evaluation**

As discussed previously, Part III will provide information on how the test is structured and scored. As you read Part III, re-evaluate your answers to the questions above. You will then be able to make a study plan, based on your needs and time available, that will allow you to use this book most effectively.

5. **Engage with Parts IV and V as Needed**

Notice the word *engage*. You'll get more out of this book if you use it intentionally than if you read it passively, hoping for an improved score through osmosis.

The strategy chapters will help you think about your approach to the question types on this exam. Part IV will open with a reminder to think about how you approach questions now and then close with a reflection section asking you to think about how/whether you will change your approach in the future.

The content chapters are designed to provide a review of the content tested on the AP Chemistry Exam, including the level of detail you need to know and how the content is tested. You will have the opportunity to assess your grasp of the content of each chapter through test-appropriate questions.

6. **Take Test 2 and Assess Your Performance**

Once you feel you have developed the strategies you need and gained the knowledge you lacked, you should take Test 2, which starts at page 333 of this book. You should do so in one sitting, following the instructions at the beginning of the test.

When you are done, check your answers to the multiple-choice sections on page 361. See if a teacher will read your answers to the free-response questions and provide feedback.

Once you have taken Practice Test 2, reflect on the areas you still need to work on, and revisit the units in this book that address those deficiencies. Once you feel confident, take the additional tests and repeat this process. Through this type of reflection and engagement, you will continue to improve.

Score Yourself
To calculate your approximate scaled score on each of your practice tests, visit your Student Tools online and download our handy Score Converter worksheet.

7.   **Keep Working**

After you have revisited certain topics in this book, continue the process of testing, reflecting, and engaging with the remaining practice tests. Each time, consider what additional work you need to do and how you will change your strategic approach to different parts of the test.

Bonus Tips and Tricks
Check us out on YouTube for additional test-taking tips and must-know strategies at www.youtube.com/ThePrincetonReview

Part II
Practice Test 1

Practice Test 1

The Exam

AP® Chemistry Exam

SECTION I: Multiple-Choice Questions

DO NOT OPEN THIS BOOKLET UNTIL YOU ARE TOLD TO DO SO.

At a Glance

Total Time
1 hour and 30 minutes
Number of Questions
60
Percent of Total Grade
50%
Writing Instrument
Pencil required

Instructions

Section I of this examination contains 60 multiple-choice questions. Fill in only the ovals for numbers 1 through 60 on your answer sheet.

Indicate all of your answers to the multiple-choice questions on the answer sheet. No credit will be given for anything written in this exam booklet, but you may use the booklet for notes or scratch work. After you have decided which of the suggested answers is best, completely fill in the corresponding oval on the answer sheet. Give only one answer to each question, If you change an answer, be sure that the previous mark is erased completely. Here is a sample question and answer.

Sample Question   Sample Answer

Chicago is a   Ⓐ ● Ⓒ Ⓓ
(A) state
(B) city
(C) country
(D) continent

Use your time effectively, working as quickly as you can without losing accuracy. Do not spend too much time on any one question. Go on to other questions and come back to the ones you have not answered if you have time. It is not expected that everyone will know the answers to all the multiple-choice questions.

About Guessing

Many candidates wonder whether or not to guess the answers to questions about which they are not certain. Multiple-choice scores are based on the number of questions answered correctly. Points are not deducted for incorrect answers, and no points are awarded for unanswered questions. Because points are not deducted for incorrect answers, you are encouraged to answer all multiple-choice questions. On any questions you do not know the answer to, you should eliminate as many choices as you can, and then select the best answer among the remaining choices.

CHEMISTRY
SECTION I
Time—1 hour and 30 minutes

INFORMATION IN THE TABLE BELOW AND ON THE FOLLOWING PAGES MAY BE USEFUL IN ANSWERING THE QUESTIONS IN THIS SECTION OF THE EXAMINATION

PERIODIC TABLE OF THE ELEMENTS

DO NOT DETACH FROM BOOK.

1	2	3	4	5	6	7	8	9	10	11	12	13	14	15	16	17	18
1 H 1.008																	2 He 4.00
3 Li 6.94	4 Be 9.01											5 B 10.81	6 C 12.01	7 N 14.01	8 O 16.00	9 F 19.00	10 Ne 20.18
11 Na 22.99	12 Mg 24.30											13 Al 26.98	14 Si 28.09	15 P 30.97	16 S 32.06	17 Cl 35.45	18 Ar 39.95
19 K 39.10	20 Ca 40.08	21 Sc 44.69	22 Ti 47.87	23 V 50.94	24 Cr 52.00	25 Mn 54.94	26 Fe 55.85	27 Co 58.93	28 Ni 58.69	29 Cu 63.55	30 Zn 65.38	31 Ga 69.72	32 Ge 72.63	33 As 74.92	34 Se 78.97	35 Br 79.90	36 Kr 83.80
37 Rb 85.47	38 Sr 87.62	39 Y 88.91	40 Zr 91.22	41 Nb 92.91	42 Mo 95.95	43 Tc	44 Ru 101.07	45 Rh 102.91	46 Pd 106.42	47 Ag 107.87	48 Cd 112.41	49 In 114.82	50 Sn 118.71	51 Sb 121.76	52 Te 127.60	53 I 126.90	54 Xe 131.29
55 Cs 132.91	56 Ba 137.33	57-71 *	72 Hf 178.49	73 Ta 180.95	74 W 183.94	75 Re 186.21	76 Os 190.23	77 Ir 192.22	78 Pt 195.08	79 Au 196.97	80 Hg 200.59	81 Tl 204.38	82 Pb 207.2	83 Bi 208.98	84 Po	85 At	86 Rn
87 Fr	88 Ra	89-103 †	104 Rf	105 Db	106 Sg	107 Bh	108 Hs	109 Mt	110 Ds	111 Rg	112 Cn	113 Nh	114 Fl	115 Mc	116 Lv	117 Ts	118 Og

*Lanthanoids	57 La 138.91	58 Ce 140.12	59 Pr 140.91	60 Nd 144.24	61 Pm	62 Sm 150.36	63 Eu 151.97	64 Gd 157.25	65 Tb 158.93	66 Dy 162.50	67 Ho 164.93	68 Er 167.26	69 Tm 168.93	70 Yb 173.05	71 Lu 174.97
†Actinoids	89 Ac	90 Th 232.04	91 Pa 231.04	92 U 238.03	93 Np	94 Pu	95 Am	96 Cm	97 Bk	98 Cf	99 Es	100 Fm	101 Md	102 No	103 Lr

GO ON TO THE NEXT PAGE.

AP® CHEMISTRY EQUATIONS & CONSTANTS

Throughout the exam the following symbols have the definitions specified unless otherwise noted.

L, mL	=	liter(s), milliliter(s)	mm Hg	= millimeters of mercury
g	=	gram(s)	J, kJ	= joule(s), kilojoule(s)
nm	=	nanometer(s)	V	= volt(s)
atm	=	atmosphere(s)	mol	= mole(s)

ATOMIC STRUCTURE

$E = h\nu$

$c = \lambda\nu$

E = energy
ν = frequency
λ = wavelength

Planck's constant, $h = 6.626 \times 10^{-34}$ J s

Speed of light, $c = 2.998 \times 10^8$ m s^{-1}

Avogadro's number $= 6.022 \times 10^{23}$ mol^{-1}

Electron charge, $e = -1.602 \times 10^{-19}$ coulomb

EQUILIBRIUM

$K_c = \dfrac{[C]^c[D]^d}{[A]^a[B]^b}$, where $a\,A + b\,B \rightleftarrows c\,C + d\,D$

$K_p = \dfrac{(P_C)^c(P_D)^d}{(P_A)^a(P_B)^b}$

$K_a = \dfrac{[H^+][A^-]}{[HA]}$

$K_b = \dfrac{[OH^-][HB^+]}{[B]}$

$K_w = [H^+][OH^-] = 1.0 \times 10^{-14}$ at 25°C

$\quad = K_a \times K_b$

$pH = -\log[H^+]$, $pOH = -\log[OH^-]$

$14 = pH + pOH$

$pH = pK_a + \log\dfrac{[A^-]}{[HA]}$

$pK_a = -\log K_a$, $pK_b = -\log K_b$

Equilibrium Constants

K_c (molar concentrations)
K_p (gas pressures)
K_a (weak acid)
K_b (weak base)
K_w (water)

KINETICS

$[A]_t - [A]_0 = -kt$

$\ln[A]_t - \ln[A]_0 = -kt$

$\dfrac{1}{[A]_t} - \dfrac{1}{[A]_0} = kt$

$t_{1/2} = \dfrac{0.693}{k}$

k = rate constant
t = time
$t_{1/2}$ = half-life

GO ON TO THE NEXT PAGE.

GASES, LIQUIDS, AND SOLUTIONS

$$PV = nRT$$

$$P_A = P_{total} \times X_A, \text{ where } X_A = \frac{\text{moles A}}{\text{total moles}}$$

$$P_{total} = P_A + P_B + P_C + \dots$$

$$n = \frac{m}{M}$$

$$K = {}^{\circ}C + 273$$

$$D = \frac{m}{V}$$

$$KE_{molecule} = \frac{1}{2}mv^2$$

Molarity, M = moles of solute per liter of solution

$$A = \varepsilon bc$$

P = pressure
V = volume
T = temperature
n = number of moles
m = mass
M = molar mass
D = density
KE = kinetic energy
v = velocity
A = absorbance
ε = molar absorptivity
b = path length
c = concentration

Gas constant, R = 8.314 J mol^{-1} K^{-1}
$\qquad$ = 0.08206 L atm mol^{-1} K^{-1}
$\qquad$ = 62.36 L torr mol^{-1} K^{-1}
1 atm = 760 mm Hg = 760 torr
STP = 273.15 K and 1.0 atm
Ideal gas at STP = 22.4 L mol^{-1}

THERMOCHEMISTRY/ ELECTROCHEMISTRY

$$q = mc\Delta T$$

$$\Delta S^{\circ} = \sum S^{\circ} \text{ products} - \sum S^{\circ} \text{ reactants}$$

$$\Delta H^{\circ} = \sum \Delta H_f^{\circ} \text{ products} - \sum \Delta H_f^{\circ} \text{ reactants}$$

$$\Delta G^{\circ} = \sum \Delta G_f^{\circ} \text{ products} - \sum \Delta G_f^{\circ} \text{ reactants}$$

$$\Delta G^{\circ} = \Delta H^{\circ} - T\Delta S^{\circ}$$

$$= -RT \ln K$$

$$= -nFE^{\circ}$$

$$I = \frac{q}{t}$$

$$E_{cell} = E_{cell}^{\circ} - \frac{RT}{nF} \ln Q$$

q = heat
m = mass
c = specific heat capacity
T = temperature
S° = standard entropy
H° = standard enthalpy
G° = standard Gibbs free energy
n = number of moles
E° = standard reduction potential
I = current (amperes)
q = charge (coulombs)
t = time (seconds)
Q = reaction quotient

Faraday's constant, F = 96,485 coulombs per mole
$\qquad$ of electrons

$$1 \text{ volt} = \frac{1 \text{ joule}}{1 \text{ coulomb}}$$

GO ON TO THE NEXT PAGE.

Step 1: $H_2(g) + ICl(g) \rightarrow HCl(g) + HI(g)$
Step 2: $HI(g) + ICl(g) \rightarrow HCl(g) + I_2(g)$

1. The reaction between H_2 and ICl can be broken down into two elementary steps. If the first step is known to be the rate determining one, what is the correct rate law for the overall reaction?

   (A) Rate = $k[H_2][ICl]$
   (B) Rate = $k[HI][ICl]$
   (C) Rate = $k[H_2][ICl]^2$
   (D) Rate = $k[H_2][HI]$

$$2\,H_2O_2(aq) \rightarrow 2\,H_2O(l) + O_2(g)$$

2. The decomposition of hydrogen peroxide is a first-order process which can be catalyzed in the presence of the iodide ion, I^-. What would happen to the concentration of the iodide ion as the reaction progresses?

   (A) It would decrease at the same rate as the H_2O_2.
   (B) It would decrease half as quickly as the H_2O_2.
   (C) It would decrease twice as quickly as the H_2O_2.
   (D) It would remain unchanged.

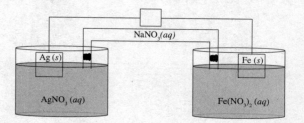

3. Based on its Lewis diagram, what would be the empirical formula for a molecule of caffeine?

   (A) $C_8H_{10}N_4O_2$
   (B) $C_4H_5N_2O$
   (C) $C_{16}H_{20}N_8O_4$
   (D) $C_{12}H_{15}N_6O_3$

Use the following information to answer questions 4–8.

Two half cells are connected to form a voltaic cell, as shown in the diagram. The initial concentration of both solutions is 1.0 M, and the temperature is 25°C.

The standard reduction potentials for both half-cells can be found in the table.

Half-Reaction	E^o (V)
$Ag^+(aq) + e^- \rightarrow Ag(s)$	0.80
$Fe^{2+}(aq) + 2e^- \rightarrow Fe(s)$	−0.44

The net ionic reaction which occurs in the cell is
$2\,Ag^+(aq) + Fe(s) \rightarrow Fe^{2+}(aq) + 2\,Ag(s)$

4. What is the standard reduction potential for the full reaction?

   (A) 1.68 V
   (B) 1.24 V
   (C) 1.16 V
   (D) 0.36 V

GO ON TO THE NEXT PAGE.

5. Which diagram below correctly shows the pathway and direction of the electron flow in the cell?

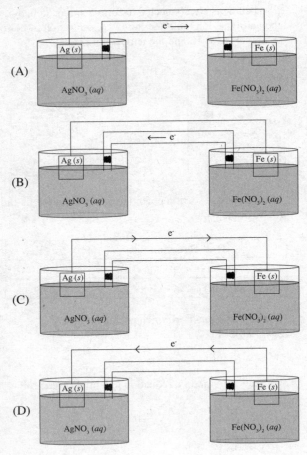

(A)

(B)

(C)

(D)

6. Which of the following starting concentration values for each solution would produce the highest cell potential?

(A) $[Ag^+] = 2.0\ M$, $[Fe^{2+}] = 1.0\ M$
(B) $[Ag^+] = 2.0\ M$, $[Fe^{2+}] = 2.0\ M$
(C) $[Ag^+] = 0.5\ M$, $[Fe^{2+}] = 1.0\ M$
(D) $[Ag^+] = 0.5\ M$, $[Fe^{2+}] = 2.0\ M$

7. What is true about the values for K_{eq} and ΔG for this cell?

	K_{eq}	ΔG
(A)	<0	>1
(B)	<1	<0
(C)	>0	<1
(D)	>1	<0

8. What is the function of the salt bridge in the cell?

(A) It prevents the mass of the electrodes from changing.
(B) It donates ions to keep the solutions at the electrodes electrically neutral.
(C) It allows the solutions from both electrodes to mix freely.
(D) It serves as a return pathway for the electrons to move between electrodes.

9. Stainless steel is primarily made up of three elements: iron, carbon, and chromium. Which of the below diagrams is an accurate representation of stainless steel on a particular level?

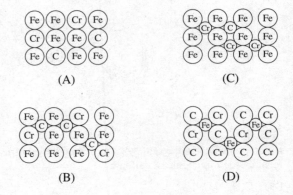

(A)

(B)

(C)

(D)

10. Light in which region of the electromagnetic spectrum supports the hypothesis that the N-O bonds in a nitrate (NO_3^-) ion are longer than those in a nitrite (NO_2^-) ion?

(A) Gamma Rays
(B) Ultraviolet
(C) Infrared
(D) Microwave

GO ON TO THE NEXT PAGE.

Use the following information to answer questions 11–13.

The photoelectron spectrum for an element in its standard state is shown below.

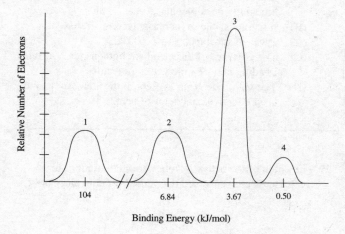

Binding Energy (kJ/mol)

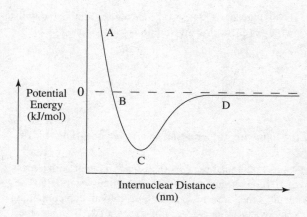

11. Which element does the spectrum above belong to?

(A)   Boron
(B)   Fluorine
(C)   Sodium
(D)   Aluminum

12. Electrons from which peak would have originally been closest to the nucleus?

(A)   Peak 1
(B)   Peak 2
(C)   Peak 3
(D)   Peak 4

13. Electrons from which peak would have the highest velocity after being ejected?

(A)   Peak 1
(B)   Peak 2
(C)   Peak 3
(D)   Peak 4

14. At which point in the graph are the interactions between the two nuclei the strongest?

(A)   Point A
(B)   Point B
(C)   Point C
(D)   Point D

15. Acrylic acid, $C_3H_7NO_2$, has a Lewis structure as shown above. Which hydrogen is the most likely to be transferred when acrylic acid is mixed with water?

(A)   Hydrogen 1
(B)   Hydrogen 2
(C)   Hydrogen 3
(D)   Hydrogen 4

$$PCl_5(g) \rightleftharpoons PCl_3(g) + Cl_2(g) \quad K_c = 2.0 \times 10^{-3}$$

16. The equilibrium constant for the above reaction is 2.0×10^{-3} at a certain temperature. What would the equilibrium constant for the reaction $2\,Cl_2(g) + 2\,PCl_3(g) \rightleftharpoons 2\,PCl_5(g)$ be at the same temperature?

(A) 4.0×10^{-3}
(B) 5.0×10^2
(C) 1.0×10^3
(D) 2.5×10^5

GO ON TO THE NEXT PAGE.

17. Liquid hexane, C_6H_{14}, would be the most miscible with which of the following solutions?

    (A) $NaCl(aq)$
    (B) $NF_3(l)$
    (C) $H_2O(l)$
    (D) $CCl_4(l)$

Use the following information to answer questions 18–21.

A piece of vanadium ($c = 0.50$ J/g·°C) metal with a mass of 10.0 g is heated to 100°C in a boiling water bath. The metal is then transferred into a Styrofoam cup containing some water initially at a temperature of 22.0°C, and the solution is stirred with a stir bar until the temperature reaches a maximum value of 23.0°C.

18. Approximately how much energy was transferred from the vanadium to the water?

    (A) 150 J
    (B) 265 J
    (C) 310 J
    (D) 385 J

19. Why is the specific heat of vanadium significantly lower than the specific heat of the water ($c = 4.18$ J/g·°C)?

    (A) The strength of the interactions between the vanadium atoms is lower than the strength of the intermolecular forces in water.
    (B) Water has a significantly lower molar mass than vanadium.
    (C) The strength of the interactions between the vanadium atoms exceeds the strength of the intermolecular forces in water.
    (D) It is easier to break the bonds between vanadium atoms than it is to break the bonds between the hydrogen and oxygen atoms in water.

20. If the vanadium sample were replaced with an aluminum ($c = 0.900$ J/g·°C) sample of the same mass and the experiment were repeated, how would that affect the final temperature of the water, compared with the final temperature in the vanadium experiment?

    (A) It would be the same.
    (B) It would be lower.
    (C) It would be higher.
    (D) The aluminum will dissolve, so it is impossible to say.

21. The student observes that the maximum temperature of the boiling water bath never increases above 100°C. Why is this true?

    (A) The energy being added to the water goes into breaking intermolecular forces.
    (B) It is impossible to increase the temperature of a covalent substance above 100°C.
    (C) The presence of the metal strengthens the hydrogen bonding between the water molecules.
    (D) The sea of electrons present in the water molecules is an effective conductor of heat.

22. In which of the below isotopes does the number of valence electrons exceed the number of core electrons?

    (A) $^{13}_{7}N$
    (B) $^{33}_{16}S$
    (C) $^{22}_{11}Na$
    (D) $^{4}_{2}Be$

$$4\ Fe(s) + 3\ O_2(g) \rightarrow 2\ Fe_2O_3(s)$$

23. When 112 g of iron (MM = 56 g/mol) reacts with 64 g of oxygen (MM = 32 g/mol) via the above reaction, how many grams of excess reactant are left after the reaction goes to completion?

    (A) 16 g
    (B) 32 g
    (C) 48 g
    (D) 64 g

GO ON TO THE NEXT PAGE.

$$2\,N_2O_5(g) \rightleftharpoons O_2(g) + 4\,NO_2(g) \quad K_p = 0.62 \text{ at } 400 \text{ K}$$

Substance	Initial Pressure (atm)
N_2O_5	0.20
O_2	0.20
NO_2	0.10

24. Three different gases are introduced into an empty flask at the pressures given above with the temperature held at 400 K. As the reaction moves towards equilibrium while the temperature remains constant, what will happen to the partial pressure of each species?

	N_2O_5	O_2	NO_2
(A)	Increase	Decrease	Decrease
(B)	Decrease	Increase	Increase
(C)	No change	Increase	Increase
(D)	No change	No change	No change

$$CH_3CHO(g) \rightarrow CH_4(g) + CO(g)$$

25. The decomposition of acetaldehyde is known to be a second-order process, as shown in the reaction above. Which graph below correctly demonstrates the relationship of concentration of acetaldehyde vs. time during its decomposition?

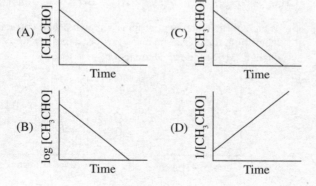

Substance	Melting Point (°C)
NaF	993
KCl	770
RbBr	693

26. The melting point trend above is best explained by which of the following?

   (A) The nuclear charge in the anion
   (B) The electronegativity difference between the ions
   (C) The atomic radii of both ions
   (D) The conductivity of each ion

Use the following information to answer questions 27–30.

Ethylamine, $CH_3CH_2NH_2$, is a weak base that can accept a single proton. A 20.0 mL sample of 1.0 M ethylamine is titrated with a strong acid, and the pH is monitored as the titration progresses.

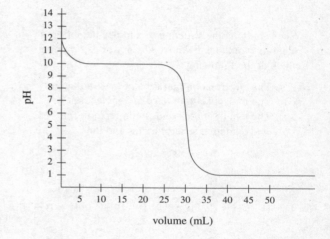

27. What is the concentration of the HCl?

   (A) 0.67 M
   (B) 1.0 M
   (C) 1.33 M
   (D) 1.50 M

28. What is the approximate K_b for ethylamine?

   (A) 1.0×10^{-2}
   (B) 1.0×10^{-4}
   (C) 1.0×10^{-6}
   (D) 1.0×10^{-8}

GO ON TO THE NEXT PAGE.

29. Which species is present in the greatest concentration at the equivalence point?

    (A) H^+
    (B) $CH_3CH_2NH_2$
    (C) $CH_3CH_2NH_3^+$
    (D) HCl

30. If the ethylamine were replaced with an identical volume of NaOH, how would that affect the volume of HCl needed to reach equivalence as well as the pH at the equivalence point?

	Volume of HCl to reach equivalence	Equivalence pH
(A)	Lower	Lower
(B)	Lower	No Change
(C)	No Change	Higher
(D)	Higher	Lower

31. When studying the structure of a molecule, infrared radiation would be the most effective at determining which of the following?

    (A) The electron configuration of a neutral atom
    (B) The molecular geometry of a molecule
    (C) The length of the bonds within a molecule
    (D) The oxidation state of atoms and ions

H   H   H   H
| | | |
H − C − H + H − C − H ⟶ H − C − C − H + H − H
| | | |
H   H   H   H

$$\Delta H = 40 \text{ kJ/mol}_{rxn}$$

Bond	Enthalpy $kJ \cdot mol^{-1}$
C-H	410
C-C	350
H-H	?

32. What would be the approximate enthalpy value for an H-H bond?

    (A) $-430 \text{ kJ} \cdot mol^{-1}$
    (B) $190 \text{ kJ} \cdot mol^{-1}$
    (C) $430 \text{ kJ} \cdot mol^{-1}$
    (D) $780 \text{ kJ} \cdot mol^{-1}$

$$HClO(aq) \rightleftharpoons H^+(aq) + ClO^-(aq) \qquad K_a = 1.1 \times 10^{-2}$$
$$HNO_2(aq) \rightleftharpoons H^+(aq) + NO_2^-(aq) \qquad K_a = 4.0 \times 10^{-4}$$

33. 100. mL of 0.10 M HClO is mixed with 100. mL of 0.10 M HNO_2. After the solutions are fully mixed, which of the following concentration relationships is true?

    (A) $[HClO] > [HNO_2] > [H^+]$
    (B) $[H^+] > [HNO_2] > [HClO]$
    (C) $[HNO_2] > [HClO] > [H^+]$
    (D) $[H^+] > [HClO] > [HNO_2]$

34. Which of the following graphs correctly displays the relationship between gas pressure and volume at constant temperature?

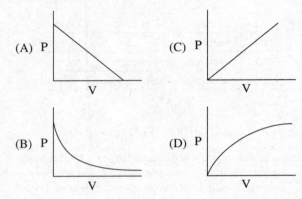

$$CH_4(g) + NH_3(g) \rightarrow HCN(g) + 3\ H_2 \qquad \Delta H = 256 \text{ kJ/mol}_{rxn}$$

35. Which of the reaction coordinates shown below correctly labels both the activation energy, E_a, and the enthalpy change, ΔH, for the above reaction?

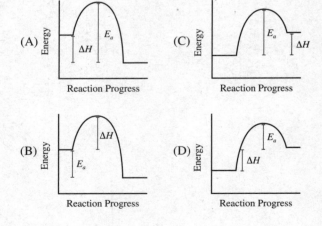

GO ON TO THE NEXT PAGE.

36. Atoms of which element, when bonded with an atom of nitrogen, would cause the nitrogen atom to have the strongest positive partial charge?

    (A) Phosphorus
    (B) Carbon
    (C) Oxygen
    (D) Arsenic

Use the following information to answer questions 37–39.

Three chambers are filled with gases, sealed, and connected using several tubes as shown below. The system is held at constant temperature throughout and the stopcocks are initially closed.

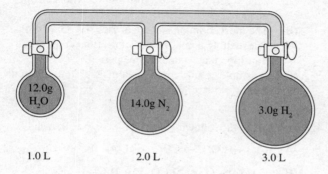

| 12.0g H_2O | 14.0g N_2 | 3.0g H_2 |
| 1.0 L | 2.0 L | 3.0 L |

37. Which gas would exert the greatest pressure afterward?

    (A) H_2
    (B) H_2O
    (C) N_2
    (D) All three gases would exert the same pressure

38. The stopcocks are then opened, and the gases are allowed to mix fully. Which gas would exert the greatest pressure afterward?

    (A) H_2
    (B) H_2O
    (C) N_2
    (D) All three gases would exert the same pressure

39. The velocity distribution for all three gases is plotted on the diagram below. Which gas belongs to which distribution?

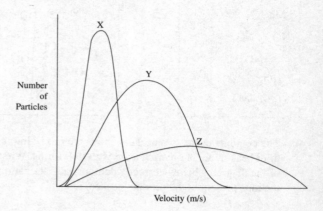

	Gas X	Gas Y	Gas Z
(A)	H_2	H_2O	N_2
(B)	H_2	N_2	H_2O
(C)	N_2	H_2O	H_2
(D)	H_2O	N_2	H_2

GO ON TO THE NEXT PAGE.

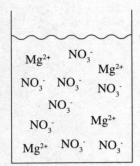

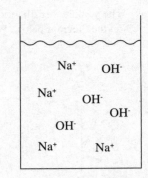

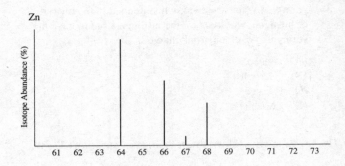

40. The contents of the two beakers above are mixed into a third beaker, and a precipitate of $Mg(OH)_2$ forms. Which of the diagrams below correctly demonstrates the ratio of the ions present in the product beaker?

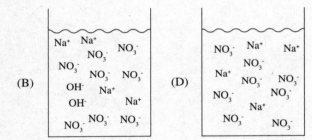

41. The mass spectrum for zinc is shown above. Based on the spectrum, which of the following can be correctly concluded about zinc?

(A) Zinc atoms will always form an ion with a charge of +2.

(B) The most common isotope of zinc has 35 neutrons.

(C) The melting point of zinc is lower than the other transition metals in the same period.

(D) A sample of zinc contains atoms with four different atomic masses.

42. In which of the following reactions is carbon reduced?

(A) $C(s) + H_2O(l) \rightarrow CO(g) + H_2(g)$

(B) $C_2H_2(g) + O_2(g) \rightarrow CO_2(g) + H_2O(g)$

(C) $CO_2(g) + H_2O(l) \rightarrow H_2CO_3(aq)$

(D) $CO_2(g) \rightarrow CO(g) + \frac{1}{2} O_2(g)$

Substance	$S°$ ($J \cdot mol^{-1} \cdot K^{-1}$)
$NO_2(g)$	240
$H_2(g)$	130
$NH_3(g)$	190
$H_2O(l)$	70

$$2\ NO_2(g) + 7\ H_2(g) \rightarrow 2\ NH_3(g) + 4\ H_2O(l)$$

43. Using the given values, calculate the entropy change, $\Delta S°$, for the above reaction.

(A) $-730\ J \cdot mol^{-1} \cdot K^{-1}$
(B) $-520\ J \cdot mol^{-1} \cdot K^{-1}$
(C) $-110\ J \cdot mol^{-1} \cdot K^{-1}$
(D) $630\ J \cdot mol^{-1} \cdot K^{-1}$

GO ON TO THE NEXT PAGE.

44. A current of 0.15 A is run through a solution containing Cu^{2+} ions, causing solid copper to plate out on an electrode. Which of the following expressions can be used to correctly calculate the mass of copper that will plate out over 300. seconds?

    (A) $\dfrac{(0.15)(96,500)(63.55)}{(300)(2)}$

    (B) $\dfrac{(0.15)(300)(63.55)}{(96,500)(2)}$

    (C) $\dfrac{(2)(63.55)(300)}{(96,500)(0.15)}$

    (D) $\dfrac{(96,500)(63.55)(2)}{(0.15)(300)}$

45. Which molecule below would be completely nonpolar?

    (A)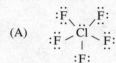

    (C) :F̈ — N̈ — F̈:
            |
            :F̈:

    (B) (Xe with four F atoms)

    (D) :F̈ — Ö — F̈:

46. There are 200 mL of 1.0 M HCN ($pK_a = 9.2$) present in a 400 mL beaker. Adding 100. mL of which of the following 1.0 M solutions would create a buffer with a pH of 9.2?

    (A) $HC_2H_3O_2$ ($K_a = 1.8 \times 10^{-5}$)
    (B) NH_3 ($K_b = 1.8 \times 10^{-5}$)
    (C) HCl
    (D) NaOH

$C(graphite) + O_2(g) \rightarrow CO_2(g) \qquad \Delta H = -394 \text{ kJ/mol}_{rxn}$
$4 Fe(s) + 3 O_2(g) \rightarrow 2 Fe_2O_3(s) \qquad \Delta H = -1,650 \text{ kJ/mol}_{rxn}$

47. What would the enthalpy change for the below reaction be?

    $2 Fe_2O_3(s) + 3 C(graphite) \rightarrow 4 Fe(s) + 3 CO_2(g)$

    (A) 468 kJ·mol^{-1}
    (B) $1,256 \text{ kJ·mol}^{-1}$
    (C) $-2,044 \text{ kJ·mol}^{-1}$
    (D) $-2,832 \text{ kJ·mol}^{-1}$

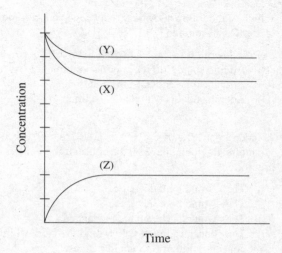

48. A reaction between substances X and Y yields a product of Z, and the concentration of all three species is monitored over time. What would the correct balanced equation for the reaction be?

    (A) $X + Y \rightleftharpoons Z$
    (B) $2X + Y \rightleftharpoons 2Z$
    (C) $X + 2Y \rightleftharpoons Z$
    (D) $2X + 2Y \rightleftharpoons Z$

Use the following information to answer questions 49–53.

A 400 mL beaker is filled with 250 mL of water, and 5.0 g of $Cd(IO_3)_2$ is added. The solution is stirred, and some solid $Cd(IO_3)_2$ settles to the bottom.

$$Cd(IO_3)_2(s) \rightleftharpoons Cd^{2+} + 2 IO_3^- \qquad K_{sp} = 4.0 \times 10^{-6}$$

49. What will the concentration of each ion be in the solution?

    (A) $[Cd^{2+}] = 0.010\ M$ $[IO_3^-] = 0.020\ M$
    (B) $[Cd^{2+}] = 0.010\ M$ $[IO_3^-] = 0.040\ M$
    (C) $[Cd^{2+}] = 0.040\ M$ $[IO_3^-] = 0.080\ M$
    (D) $[Cd^{2+}] = 0.020\ M$ $[IO_3^-] = 0.040\ M$

GO ON TO THE NEXT PAGE.

50. Which of the following would reduce the concentration of the IO_3^- in solution?

    (A) Allowing some water to evaporate
    (B) Adding some $Cd(NO_3)_2$ to the solution
    (C) Adding some $NaIO_3$ to the solution
    (D) Adding some $NaNO_3$ to the solution

51. An additional 1.0 g of $Cd(IO_3)_2$ is added to the beaker. What effect will that have on the concentration of the ions in solution?

    (A) They would both decrease.
    (B) They would both increase.
    (C) They would both remain unchanged.
    (D) The concentration of Cd^{2+} would increase, while the concentration of IO_3^- would increase.

52. 5.0 g of another salt, $Ni(IO_3)_2$ ($K_{sp} = 5.0 \times 10^{-5}$), is added to a separate beaker that contains the same amount of water, and some solid settles to the bottom. Which solution, of the two, will contain the higher concentration of IO_3^- ions?

    (A) The $Cd(IO_3)_2$ solution
    (B) The $Ni(IO_3)_2$ solution
    (C) Both solutions would have the same concentration of IO_3^-
    (D) Whichever solution has a smaller volume of water

53. Of the three ions in either beaker (Ni^{2+}, Cd^{2+}, and IO_3^-), which would form the strongest ion-dipole attractions with water molecules?

    (A) Ni^{2+}
    (B) Cd^{2+}
    (C) IO_3^-
    (D) All three ions would have identical attractive forces

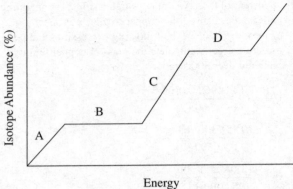

54. 1.0 mole of a pure substance is heated, and the temperature change is tracked. Given the rate of heating, which section of the graph above could be used to find the molar heat of fusion?

    (A) Section A
    (B) Section B
    (C) Section C
    (D) Section D

55. A certain chemical process is favored at 500 K, but not favored at 300 K. What must be true about the signs for ΔH and ΔS for this process?

	ΔH	ΔS
(A)	> 0	> 0
(B)	> 0	< 0
(C)	< 0	> 0
(D)	< 0	< 0

GO ON TO THE NEXT PAGE.

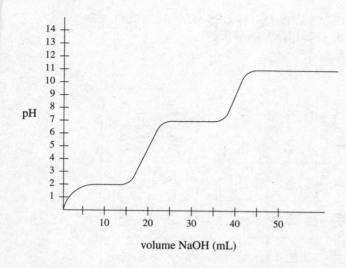

volume NaOH (mL)

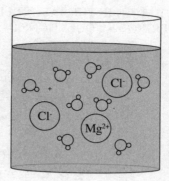

56. 20. mL of 1.0 M H_2SO_3, a diprotic acid, is titrated with 1.0 M NaOH, creating the above graph. What are the approximate pK_a values for each dissociation of H_2SO_3?

	pK_{a1}	pK_{a2}
(A)	5.0	9.0
(B)	2.0	7.0
(C)	2.0	9.0
(D)	5.0	7.0

$$H^+(aq) + Fe^{3+}(aq) + SCN^-(aq) \rightleftharpoons FeSCN^{2+}(aq) \quad \Delta H < 0$$
Yellow   Red

57. The above system is at equilibrium. Predict whether the solution would become more yellow, more red, or experience no color change when it is either heated or when it is diluted with extra water.

	Heating	Diluting
(A)	Yellow	No Change
(B)	Yellow	Yellow
(C)	Red	No Change
(D)	Red	Yellow

58. Some magnesium chloride is fully dissolved in water, as shown above. Which of the forces involved in the solution would be the strongest?

(A) The attraction between the hydrogen and the Cl^- anions

(B) The attraction between the oxygen and the Mg^{2+} ion

(C) The hydrogen bonding between water molecules

(D) The bond between the Mg^{2+} and Cl^- ions

59. If 1.0 g of $Mg(OH)_2$ ($K_{sp} = 1.8 \times 10^{-11}$) were added to 100. mL of the following solutions, in which would the largest mass of salt dissolve?

(A) Water

(B) 1.0 M NaOH

(C) 1.0 M HCl

(D) 1.0 M $Mg(NO_3)_2$

60. In which of the following situations would the entropy of the system increase?

(A) A bond forms between two atoms

(B) A pure liquid freezes

(C) The volume of a gas increases

(D) A precipitate forms when two aqueous solutions are mixed

END OF SECTION I

INFORMATION IN THE TABLE BELOW AND ON THE FOLLOWING PAGES MAY BE USEFUL IN ANSWERING THE QUESTIONS IN THIS SECTION OF THE EXAMINATION

DO NOT DETACH FROM BOOK.

PERIODIC TABLE OF THE ELEMENTS

1	2	3	4	5	6	7	8	9	10	11	12	13	14	15	16	17	18
1 H 1.008																	2 He 4.00
3 Li 6.94	4 Be 9.01											5 B 10.81	6 C 12.01	7 N 14.01	8 O 16.00	9 F 19.00	10 Ne 20.18
11 Na 22.99	12 Mg 24.30											13 Al 26.98	14 Si 28.09	15 P 30.97	16 S 32.06	17 Cl 35.45	18 Ar 39.95
19 K 39.10	20 Ca 40.08	21 Sc 44.69	22 Ti 47.87	23 V 50.94	24 Cr 52.00	25 Mn 54.94	26 Fe 55.85	27 Co 58.93	28 Ni 58.69	29 Cu 63.55	30 Zn 65.38	31 Ga 69.72	32 Ge 72.63	33 As 74.92	34 Se 78.97	35 Br 79.90	36 Kr 83.80
37 Rb 85.47	38 Sr 87.62	39 Y 88.91	40 Zr 91.22	41 Nb 92.91	42 Mo 95.95	43 Tc	44 Ru 101.07	45 Rh 102.91	46 Pd 106.42	47 Ag 107.87	48 Cd 112.41	49 In 114.82	50 Sn 118.71	51 Sb 121.76	52 Te 127.60	53 I 126.90	54 Xe 131.29
55 Cs 132.91	56 Ba 137.33	57 57-71 *	72 Hf 178.49	73 Ta 180.95	74 W 183.94	75 Re 186.21	76 Os 190.23	77 Ir 192.22	78 Pt 195.08	79 Au 196.97	80 Hg 200.59	81 Tl 204.38	82 Pb 207.2	83 Bi 208.98	84 Po	85 At	86 Rn
87 Fr	88 Ra	89-103 †	104 Rf	105 Db	106 Sg	107 Bh	108 Hs	109 Mt	110 Ds	111 Rg	112 Cn	113 Nh	114 Fl	115 Mc	116 Lv	117 Ts	118 Og

*Lanthanoids	57 La 138.91	58 Ce 140.12	59 Pr 140.91	60 Nd 144.24	61 Pm	62 Sm 150.36	63 Eu 151.97	64 Gd 157.25	65 Tb 158.93	66 Dy 162.50	67 Ho 164.93	68 Er 167.26	69 Tm 168.93	70 Yb 173.05	71 Lu 174.97
†Actinoids	89 Ac	90 Th 232.04	91 Pa 231.04	92 U 238.03	93 Np	94 Pu	95 Am	96 Cm	97 Bk	98 Cf	99 Es	100 Fm	101 Md	102 No	103 Lr

GO ON TO THE NEXT PAGE.

AP® CHEMISTRY EQUATIONS & CONSTANTS

Throughout the exam the following symbols have the definitions specified unless otherwise noted.

L, mL	=	liter(s), milliliter(s)	mm Hg =	millimeters of mercury
g	=	gram(s)	J, kJ =	joule(s), kilojoule(s)
nm	=	nanometer(s)	V =	volt(s)
atm	=	atmosphere(s)	mol =	mole(s)

ATOMIC STRUCTURE

$E = h\nu$

$c = \lambda \nu$

E = energy

ν = frequency

λ = wavelength

Planck's constant, $h = 6.626 \times 10^{-34}$ J s

Speed of light, $c = 2.998 \times 10^8$ m s^{-1}

Avogadro's number $= 6.022 \times 10^{23}$ mol^{-1}

Electron charge, $e = -1.602 \times 10^{-19}$ coulomb

EQUILIBRIUM

$K_c = \dfrac{[C]^c[D]^d}{[A]^a[B]^b}$, where $a\,A + b\,B \rightleftarrows c\,C + d\,D$

$K_p = \dfrac{(P_C)^c(P_D)^d}{(P_A)^a(P_B)^b}$

$K_a = \dfrac{[H^+][A^-]}{[HA]}$

$K_b = \dfrac{[OH^-][HB^+]}{[B]}$

$K_w = [H^+][OH^-] = 1.0 \times 10^{-14}$ at 25°C

$\quad = K_a \times K_b$

$pH = -\log[H^+]$, $pOH = -\log[OH^-]$

$14 = pH + pOH$

$pH = pK_a + \log \dfrac{[A^-]}{[HA]}$

$pK_a = -\log K_a$, $pK_b = -\log K_b$

Equilibrium Constants

K_c (molar concentrations)

K_p (gas pressures)

K_a (weak acid)

K_b (weak base)

K_w (water)

KINETICS

$[A]_t - [A]_0 = -kt$

$\ln[A]_t - \ln[A]_0 = -kt$

$\dfrac{1}{[A]_t} - \dfrac{1}{[A]_0} = kt$

$t_{1/2} = \dfrac{0.693}{k}$

k = rate constant

t = time

$t_{1/2}$ = half-life

GO ON TO THE NEXT PAGE.

GASES, LIQUIDS, AND SOLUTIONS

$$PV = nRT$$

$$P_A = P_{total} \times X_A, \text{ where } X_A = \frac{\text{moles A}}{\text{total moles}}$$

$$P_{total} = P_A + P_B + P_C + \cdots$$

$$n = \frac{m}{M}$$

$$K = °C + 273$$

$$D = \frac{m}{V}$$

$$KE_{molecule} = \frac{1}{2}mv^2$$

Molarity, M = moles of solute per liter of solution

$$A = \varepsilon bc$$

P = pressure
V = volume
T = temperature
n = number of moles
m = mass
M = molar mass
D = density
KE = kinetic energy
v = velocity
A = absorbance
ε = molar absorptivity
b = path length
c = concentration

Gas constant, $R = 8.314 \text{ J mol}^{-1} \text{K}^{-1}$

$= 0.08206 \text{ L atm mol}^{-1} \text{K}^{-1}$

$= 62.36 \text{ L torr mol}^{-1} \text{K}^{-1}$

$1 \text{ atm} = 760 \text{ mm Hg} = 760 \text{ torr}$

STP = 273.15 K and 1.0 atm

Ideal gas at STP = 22.4 L mol^{-1}

THERMOCHEMISTRY/ ELECTROCHEMISTRY

$$q = mc\Delta T$$

$$\Delta S° = \sum S° \text{ products} - \sum S° \text{ reactants}$$

$$\Delta H° = \sum \Delta H_f° \text{ products} - \sum \Delta H_f° \text{ reactants}$$

$$\Delta G° = \sum \Delta G_f° \text{ products} - \sum \Delta G_f° \text{ reactants}$$

$$\Delta G° = \Delta H° - T\Delta S°$$

$$= -RT \ln K$$

$$= -nFE°$$

$$I = \frac{q}{t}$$

$$E_{cell} = E_{cell}° - \frac{RT}{nF} \ln Q$$

q = heat
m = mass
c = specific heat capacity
T = temperature
$S°$ = standard entropy
$H°$ = standard enthalpy
$G°$ = standard Gibbs free energy
n = number of moles
$E°$ = standard reduction potential
I = current (amperes)
q = charge (coulombs)
t = time (seconds)
Q = reaction quotient

Faraday's constant, F = 96,485 coulombs per mole of electrons

$$1 \text{ volt} = \frac{1 \text{ joule}}{1 \text{ coulomb}}$$

GO ON TO THE NEXT PAGE.

CHEMISTRY

SECTION II

Time—1 hour and 45 minutes

7 Questions

Directions: Questions 1–3 are long free-response questions that require about 23 minutes each to answer and are worth 10 points each. Questions 4–7 are short free-response questions that require about 9 minutes each to answer and are worth 4 points each.

On test day, you will be asked to show your work for each part in the space provided after that part. For this practice test, you may use scrap paper. Examples and equations may be included in your responses where appropriate. For calculations, clearly show the method used and the steps involved in arriving at your answers. You must show your work to receive credit for your answer. Pay attention to significant figures.

$$H_2O_2(aq) + OCl^-(aq) \rightarrow Cl^-(aq) + H_2O(aq) + O_2(g)$$

1. The active ingredient in most commercial bleaches is sodium hypochlorite, NaClO. When it is mixed with hydrogen peroxide, H_2O_2, the above reaction occurs.

    (a) Identify the oxidation state on the oxygen atom(s) present in each species.

    H_2O_2 OCl^- H_2O O_2

    (b) How many moles of electrons are transferred when one mole of reaction goes to completion? Justify your answer.

    When some bleach is titrated into 200.0 mL of 0.085 M H_2O_2, it is observed that gas is no longer evolved after 33.92 mL of bleach is added. No other species in the bleach reacts with the H_2O_2, and the bleach solution has a density of 1.02 g/mL.

    (c) (i) Assuming the NaClO was limiting, how many moles of NaClO are present in the original bleach sample?
          (ii) What is the percent mass of NaClO in the bleach sample?

    (d) Complete the Lewis electron-dot diagram of hydrogen peroxide in the box below. Show all valence electrons.

    > H   O   O   H

    (e) Based on your Lewis diagram, what types of intermolecular forces would be present in a solution of H_2O_2?

    (f) If a solution of H_2O_2 is heated, would you expect the vapor pressure of the solution to increase, decrease, or stay the same? Justify your answer with reference to any impacts on molecular activity.

GO ON TO THE NEXT PAGE.

2. Hydrofluoric acid, HF, has a K_a value of 7.2×10^{-4}.

   (a)　(i)　Write out the equilibrium reaction that occurs when HF is added to water.

   　　　(ii)　Write out the equilibrium constant expression for the above reaction.

   (b)　A sample of HF is found to have a pH of 1.67. What was the initial concentration of the HF?

   (c)　A buffer is created by mixing solutions of HF and NaF. Circle the diagram that accurately represents the correct particulate ratio between HF and F^- when the buffer has a pH of 4.14.

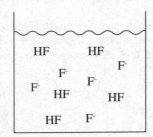 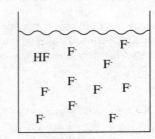 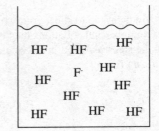

   (d)　HF is considered a weak acid, but subsequent halogen acids (HCl, HBr, and HI) are strong acids. Explain this trend in terms of atomic structure.

   (e)　A 100. mL sample of 0.10 M HF is diluted with 100. mL of water and the temperature remains constant. What effect, if any, will the dilution have on the following values? Justify your answer.

   　　　(i)　The K_a value

   　　　(ii)　The percent dissociation of the acid

   (f)　Boric acid, H_3BO_3, has a K_{a1} value of 5.8×10^{-10}.

   　　　(i)　100 mL of two samples of identical concentration of HF and H_3BO_3 are placed in two separate beakers. Which acid would have a higher pH and why?

   　　　(ii)　Boric acid is tri-protic. How would the magnitude of successive K_a values compare to the first dissociation value (K_{a1})? Justify your answer.

GO ON TO THE NEXT PAGE.

3. 2.15 grams of solid magnesium bromide, $MgBr_2$, are fully dissolved in 500. mL of water.

   (a) Which piece of volume-measuring equipment from the list below would create a solution of the greatest accuracy?

500 mL Erlenmeyer Flask

500 mL beaker

500 mL volumetric flask

   (b) Which ion, Mg^{2+} or Br^-, would have a greater Coulombic attraction to water? Provide TWO justifications for your answer.

The $MgBr_2$ solution is then mixed with excess aqueous Na_3PO_4, causing a precipitate to form.

   (c) Write out and balance the net ionic reaction that occurs.
   (d) How many grams of precipitate would be created?

The student then filters the precipitate out of solution before massing it. The precipitate is found to have a value that is larger than the value calculated in part (d). Would the following errors account for the observed results? Justify your answers.

   (e) (i)   The particles of precipitate were smaller than the holes in the filter paper.
        (ii) The precipitate was not fully dried before massing it.

   (f) Instead of filtering, another student suggests isolating the precipitate in its pure form by allowing all of the water to evaporate. Would this be effective? Why or why not?

$$NH_4^+(aq) + NO_2^-(aq) \rightarrow N_2(g) + 2\,H_2O(l)$$

4. In order to determine the rate law of the above reaction, a student runs several trials at 25°C, tracking the rate of the reaction by measuring the change in pressure of the nitrogen gas as it is produced.

Trial	$[NH_4^+]$	$[NO_2^-]$	Rate (M/s)
1	0.010 M	0.010 M	0.025
2	0.020 M	0.010 M	0.050
3	0.015 M	0.020 M	0.150

   (a) What is the order with respect to each reactant?

   (b) What is the rate constant for this reaction under the given conditions? Include units.

   (c) In order to increase the initial rate of reaction, the student suggests rerunning all of the trials at 50°C. Would this be effective? Justify your answer on a particulate level.

GO ON TO THE NEXT PAGE.

5. A solution held at 25°C contains a mixture of three different substances:

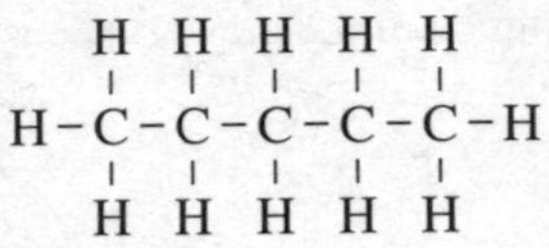

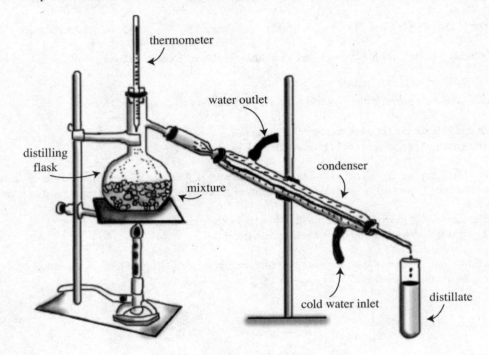

pentane 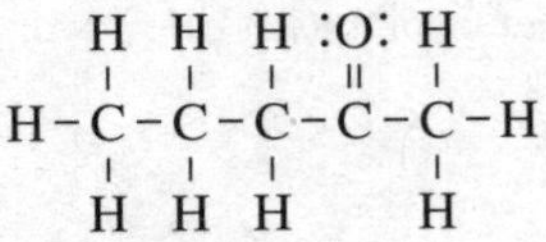   pentanone 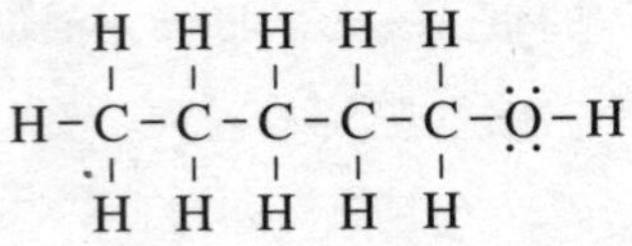   pentanol

(a)   In which, if any, of the substances would hydrogen bonding be present between molecules?

In order to separate the three substances, the student places the solution into a distillation apparatus as shown below.

thermometer

water outlet

distilling flask

condenser

mixture

cold water inlet

distillate

(b)   (i)   The temperature of the apparatus is slowly increased until the solution starts to boil.
Identify the distillate that would first be produced.
(ii)   Would the distillate be completely pure? Why or why not?

GO ON TO THE NEXT PAGE.

6.

Substance	$S°$ (J/mol·K)	$H_f°$ (kJ/mol)
$Br_2(l)$	152.2	0.0
$Br_2(g)$	245.5	30.9

The above table gives information about the thermodynamic properties of bromine.

(a) Why is the absolute entropy value greater for bromine in its gaseous state than its liquid state?

A sample of bromine originally at room temperature is heated until it boils.

$$Br_2(l) \rightarrow Br_2(g)$$

(b) What is the boiling point of bromine?

(c) As the bromine boils, it is observed that the temperature remains constant even as additional heat is added. Explain why this would be true.

$$CS_2(g) + 4 H_2(g) \rightleftharpoons CH_4(g) + 2 H_2S(g) \qquad K_c = 0.28 \text{ at } 900°C$$

7. The four gases in the reaction above are mixed together in a reaction vessel at the following concentrations:

$$[CS_2] = 0.30\ M \quad [H_2] = 0.60\ M \quad [CH_4] = 0.15\ M \quad [H_2S] = 0.25\ M$$

(a) What will happen to the $[H_2]$ as the reaction moves towards equilibrium? Justify your answer.

(b) Once equilibrium is established, the reaction vessel is expanded. Which way will the reaction shift (if at all) to return to equilibrium?

(c) What would the K_c value be for $CH_4(g) + 2 H_2S(g) \rightleftharpoons CS_2(g) + 4 H_2(g)$ at 900°C?

STOP

END OF EXAM

Practice Test 1: Diagnostic Answer Key and Explanations

PRACTICE TEST 1: DIAGNOSTIC ANSWER KEY

Let's take a look at how you did on Practice Test 1. Follow the three-step process in the diagnostic answer key below, and go read the explanations for any questions you got wrong or struggled with but got correct. Once you finish working through the answer key and the explanations, go to the next chapter to make your study plan.

STEP 1 ≫ Check your answers and mark any correct answers with a ✔ in the appropriate column.

Q #	Ans.	✔	Unit #, Section Title	Q #	Ans.	✔	Unit #, Section Title
			Section 1—Multiple Choice				
1	A		**5,** Reaction Mechanisms	23	A		**4,** Chemical Equations—Chemical Equations and Calculations
2	D		**5,** Catalysts	24	B		**7,** The Reaction Quotient, Q
3	B		**1,** Moles—Empirical and Molecular Formulas	25	D		**5,** Rate Law Using Concentration and Time—Second-Order Rate Laws
4	B		**9,** Gibbs Free Energy—Reduction Potentials	26	C		**2,** Ionic Bonds
5	D		**9,** Galvanic Cells	27	A		**8,** Titration
6	A		**9,** Galvanic Cells—Non-Standard Conditions	28	B		**8,** Titration
7	D		**9,** Gibbs Free Energy—Standard Free Energy Change and the Equilibrium Constant	29	C		**8,** Neutralization Reactions
8	B		**9,** Galvanic Cells	30	C		**8,** Titration
9	B		**2,** Metallic Bonds	31	C		**1,** Photoelectron Spectroscopy
10	C		**1,** Photoelectron Spectroscopy	32	C		**6,** Bond Energy
11	C		**1,** Photoelectron Spectroscopy—Spectra	33	C		**8,** Acid Strengths—Weak Acids
12	A		**1,** Photoelectron Spectroscopy—Spectra	34	B		**3,** The Ideal Gas Equation
13	D		**1,** Photoelectron Spectroscopy	35	C		**6,** Energy Diagrams—Exothermic and Endothermic Reactions
14	A		**2,** Molecular Covalent Bonding—Internuclear Distance	36	C		**3,** Polarity
15	D		**8,** Acid Strengths—Acid/Base Structure	37	B		**3,** The Ideal Gas Equation
16	D		**7,** The Equilibrium Constant, K_{eq}—Manipulating K_{eq}	38	A		**3,** Dalton's Law
17	D		**3,** Solution Separation—Solutes and Solvents	39	C		**3,** Maxwell-Boltzmann Diagrams
18	D		**6,** Calorimetry	40	A		**4,** Types of Reactions
19	A		**3,** Intermolecular Forces	41	D		**1,** The Periodic Table
20	C		**6,** Calorimetry	42	D		**4,** Oxidation States, Oxidation-Reduction Reactions
21	A		**6,** Thermodynamics of Phase Change—Enthalpy of Vaporization	43	A		**9,** Entropy
22	A		**1,** Photoelectron Spectroscopy—Electron Configuration	44	B		**9,** Electrolytic Cells—Electroplating

Section 1—Multiple Choice, Continued

Q #	Ans.	✔	Unit #, Section Title	Q #	Ans.	✔	Unit #, Section Title
45	B		**3,** Polarity—Molecular Polarity	53	A		**6,** Enthalpy of Solution
46	D		**8,** Buffers	54	B		**6,** Heating Curves
47	A		**6,** Hess's Law	55	A		**9,** Gibbs Free Energy—ΔG, ΔH, and ΔS
48	B		**7,** The Equilibrium Constant, K_{eq}	56	B		**8,** Titration
49	A		**7,** Solubility—The Solubility Product (K_{sp})	57	B		**7,** Le Châtelier's Principle—Temperature, Dilution
50	B		**7,** The Common Ion Effect, and Le Châtelier's Principle—Concentration	58	B		**6,** Enthalpy of Solution
51	C		**7,** The Equilibrium Constant, K_{eq}	59	C		**8,** pH
52	B		**7,** Solubility—The Solubility Product (K_{sp})	60	C		**9,** Entropy

Section 2—Free Response

Q #	Ans.	✔	Unit #, Section Title
1a	See Explanation		**4,** Oxidation States
1b	See Explanation		**4,** Chemical Equations—Chemical Equations and Calculations
1c-i	See Explanation		**4,** Chemical Equations—Chemical Equations and Calculations
1c-ii	See Explanation		**4,** Gravimetric Analysis
1d	See Explanation		**2,** Lewis Dot Structures—Drawing Lewis Dot Structures
1e	See Explanation		**3,** Intermolecular Forces—Dipole-Dipole Forces, Hydrogen Bonds, and London Dispersion Forces
1f	See Explanation		**3,** Intermolecular Forces—IMF Strength, Vapor Pressure
2a-i	See Explanation		**8,** Acid Strengths—Weak Acids
2a-ii	See Explanation		**8,** Acid Strengths—Weak Acids
2b	See Explanation		**8,** Acid Strengths—Weak Acids
2c	See Explanation		**8,** Acid Strengths—Weak Acids
2d	See Explanation		**8,** Acid Strengths—Percent Dissociation
2e-i	See Explanation		**7,** Changes in the Equilibrium Constant
2e-ii	See Explanation		**8,** Acid Strengths—Percent Dissociation
2f-i	See Explanation		**8,** Acid Strengths—Weak Acids
2f-ii	See Explanation		**8,** Acid Strengths—Polyprotic Acids
3a	See Explanation		**Laboratory Overview,** Laboratory Equipment
3b	See Explanation		**6,** Enthalpy of Solution
3c	See Explanation		**4,** Types of Reactions
3d	See Explanation		**4,** Chemical Equations and Calculations
3e-i	See Explanation		**Laboratory Overview,** Experimental Design
3e-ii	See Explanation		**Laboratory Overview,** Experimental Design
3f	See Explanation		**Laboratory Overview,** Experimental Design

			Section 2—Free Response, Continued
Q #	Ans.	✔	Unit #, Section Title
4a	See Explanation		**5,** Rate Law Using Initial Concentrations
4b	See Explanation		**5,** Rate Law Using Initial Concentrations
4c	See Explanation		**5,** Collision Theory
5a	See Explanation		**3,** Intermolecular Forces—Hydrogen Bonds
5b-i	See Explanation		**3,** Solution Separation—Distillation
5b-ii	See Explanation		**3,** Solution Separation—Distillation
6a	See Explanation		**9,** Entropy
6b	See Explanation		**9,** Gibbs Free Energy—ΔGo and Phase Changes
6c	See Explanation		**9,** Thermodynamics of Phase Change—Enthalpy of Vaporization
7a	See Explanation		**7,** The Reaction Quotient, Q
7b	See Explanation		**7,** Le Châtelier's Principle—Pressure
7c	See Explanation		**7,** The Equilibrium Constant, K_{eq}—Manipulating K_{eq}

 Tally your correct answers from Step 1 by unit. For each unit, write the number of correct answers in the appropriate box. Then, divide your correct answers by the number of total questions (which we've provided) to get your percent correct.

UNIT 1 TEST SELF-EVALUATION

CORRECT ANSWERS

$$\frac{}{8} = \boxed{} \%$$

TOTAL QUESTIONS   PERCENT CORRECT

UNIT 2 TEST SELF-EVALUATION

CORRECT ANSWERS

$$\frac{}{4} = \boxed{} \%$$

TOTAL QUESTIONS   PERCENT CORRECT

UNIT 3 TEST SELF-EVALUATION

CORRECT ANSWERS

$$\frac{}{13} = \boxed{} \%$$

TOTAL QUESTIONS   PERCENT CORRECT

UNIT 4 TEST SELF-EVALUATION

CORRECT ANSWERS

$$\frac{}{9} = \boxed{} \%$$

TOTAL QUESTIONS   PERCENT CORRECT

UNIT 5 TEST SELF-EVALUATION

CORRECT ANSWERS

$$\frac{}{6} = \boxed{} \%$$

TOTAL QUESTIONS   PERCENT CORRECT

UNIT 6 TEST SELF-EVALUATION

CORRECT ANSWERS

$$\frac{}{10} = \boxed{} \%$$

TOTAL QUESTIONS   PERCENT CORRECT

UNIT 7 TEST SELF-EVALUATION

CORRECT ANSWERS

$$\frac{}{12} = \boxed{} \%$$

TOTAL QUESTIONS   PERCENT CORRECT

UNIT 8 TEST SELF-EVALUATION

CORRECT ANSWERS

$$\frac{}{17} = \boxed{} \%$$

TOTAL QUESTIONS   PERCENT CORRECT

UNIT 9 TEST SELF-EVALUATION

CORRECT ANSWERS

$$\frac{}{12} = \boxed{} \%$$

TOTAL QUESTIONS   PERCENT CORRECT

LAB OVERVIEW TEST SELF-EVALUATION

CORRECT ANSWERS

$$\frac{}{4} = \boxed{} \%$$

TOTAL QUESTIONS   PERCENT CORRECT

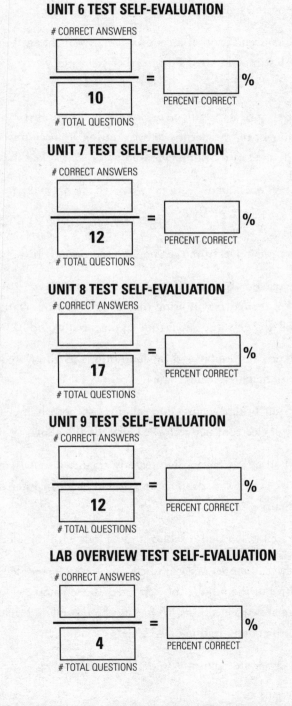

STEP 3 ≫ Use the results above to customize your study plan. You may want to start with, or give more attention to, the units with the lowest percents correct.

PRACTICE TEST 1: ANSWERS AND EXPLANATIONS

Section I—Multiple-Choice Answers and Explanations

1. **A**   The coefficients on the reactants in a rate-determining step become the order of those reactants in the overall rate law.

2. **D**   The concentration of a catalyst remains unchanged during a chemical reaction. While some of the catalyst may be used up in earlier elementary steps, it will then be regenerated in later elementary steps.

3. **B**   The empirical formula is the lowest whole number ratio of the atoms in the molecule. The molecular formula can be determined by counting atoms and would be $C_8H_{10}N_4O_2$. All of those subscripts can be divided by a common denominator (2) to yield the empirical formula.

4. **B**   Always do cathode – anode when calculating cell potential.

$$0.80 \text{ V} - (-0.44 \text{ V}) = 1.24 \text{ V}$$

5. **D**   Electrons travel from the anode to the cathode through the wire connecting them.

6. **A**   The reaction quotient for this reaction is $Q = [Fe^{2+}]/[Ag^+]^2$. Any change in the reaction quotient that makes it smaller will bring the cell further from equilibrium and thus increase the voltage, and the smaller Q gets, the larger the cell potential would be.

7. **D**   Voltaic cells are favored by definition, and all favored reactions have negative values for ΔG and K values that are greater than 1.

8. **B**   The salt bridge ensures the solutions in both half-cells stay electrically neutral. This facilitates the flow of electrons between the half-cells and allows the cell to operate.

9. **B**   Carbon atoms have a significantly smaller radius than chromium or iron atoms, which are similar in size. Thus, the carbon atoms would fit in the interstices between the other atoms as shown in this diagram.

10. **C**   Infrared radiation is used to study bonds within molecules.

11. **C**   Each peak represents a subshell, and lowest energy levels always fill first. The heights of the peak are relative to the number of electrons. The leftmost peak must represent the 1s subshell, and we know there are two electrons in that subshell. Applying that knowledge to the rest of the PES, we get an electron configuration of $1s^2 2s^2 2p^6 3s^1$- sodium.

12. **A**   The closer an electron is to the nucleus, the greater the binding energy will be.

13. **D** The energy absorbed by the atoms here goes into two things. First, it has to break their attraction with the nuclei, and the amount of energy needed to do that is equal to the binding energy of that electron. Any energy that does not go into breaking that attraction is left over and goes into the kinetic energy of the ejected electron. The lower the binding energy, the greater the kinetic energy (and thus velocity) of the electrons will be.

14. **A** The nuclei are both positively charged and would thus have a repulsive force between them. That value would be greatest when they are closest together.

15. **D** Hydrogen atoms at the end of hydroxyl (–OH) groups are the most likely to dissociate.

16. **D** To get the target reaction, we have to both flip the original reaction and multiply it by two. Flipping the reaction means we take the reciprocal of the original K value, and if we then multiply the reaction by two, we have to square the K value.

$$\frac{1.0 \times 10^0}{2.0 \times 10^{-3}} = (0.50 \times 10^3)^2 = 0.25 \times 10^6 = 2.5 \times 10^5$$

17. **D** In terms of intermolecular forces, like dissolves like. C_6H_{14} is completely nonpolar, and would only be fully miscible in another nonpolar substance such as CCl_4.

18. **D** Energy can be calculated via $q = mc\Delta T$. Using the values for the vanadium, we get $q = (10.0)(0.50)(77)$, which simplifies to $q = (5.0)(77)$, or 385 J.

19. **A** When the temperature of a substance is changed, the kinetic energy of the particles that make up the substance also changes. If the particles of the substance interact very strongly, it will be difficult to change their kinetic energy, and that substance will have a higher specific heat.

20. **C** As aluminum has a greater specific heat capacity, the temperature of the aluminum will not change as much as the temperature of the vanadium before thermal equilibrium is established. This means the final temperature of the water itself will be higher.

21. **A** The water is changing phase from liquid to gas, and in order for this to occur the intermolecular forces between the water molecules must be neutralized. While this is occurring, the energy being added goes into breaking those IMFs instead of adding kinetic energy to the water.

22. **A** Valence electrons are those in the outermost energy level, and core electrons are all the others. The mass number (top number) is irrelevant when counting electrons; all that matters is the atomic number (bottom number). A neutral nitrogen atom has five valence electrons and only two core electrons.

23. **A** We start with two moles of both reactants, and as iron is present in a higher stoichiometric ratio in the balanced equation, it must be limiting. We next need to figure out how much O_2 reacts with the Fe.

$$2 \text{ mol Fe} \times \frac{3 \text{ mol O}_2}{4 \text{ mol Fe}} \times \frac{32 \text{ g O}_2}{1 \text{ mol O}_2} = 48 \text{ g O}_2 \text{ reacts. } 64 \text{ g} - 48 \text{ g} = 16 \text{ g O}_2 \text{ in excess}$$

24. **B** To determine where the reaction is with respect to equilibrium, the reaction quotient needs to be calculated.

$$Q = \frac{(P_{NO_2})^4(P_{O_2})}{(P_{N_2O_5})^2} = \frac{(1.0 \times 10^{-1})^4(2.0 \times 10^{-1})}{(2.0 \times 10^{-1})^2} = \frac{2.0 \times 10^{-5}}{4.0 \times 10^{-2}} = 5.0 \times 10^{-4}$$

Given that $Q < K$, the reaction will need to shift right to achieve equilibrium, which will cause the concentration of the reactant to decrease and the concentration of both products to increase.

25. **D** For a second order process, $1/[A]$ vs. t produces a straight line.

26. **C** The melting point of an ionic substance is based on its lattice energy, which is a type of Coulombic energy. Given that all of the charges involved are identical, the only other factor that affects Coulombic energy is radius, with a smaller radius leading to a greater energy (and thus higher melting point).

27. **A** At the equivalence point, moles of acid = moles of base. We can use $M_aV_a = M_bV_b$ here because it is a 1:1 ratio in this reaction. $M_a(30.0 \text{ mL}) = (1.0 \text{ M})(20.0 \text{ mL})$. $M_a = 0.67 \text{ } M$.

28. **B** Via Henderson-Hasselbalch, we know that at half-equivalence, $pOH = pK_b$. Half-equivalence is 15.0 mL in this titration, and at the point the pH is 10. Since $pOH + pH = 14$, that makes the $pOH = 4$. So, $pK_b = 4$, meaning $-\log(K_b) = 4$ and thus $K_b = 1.0 \times 10^{-4}$.

29. **C** The reaction taking place here is $CH_3CH_2NH_2(aq) + H^+(aq) \rightarrow CH_3CH_2NH_3^+(aq)$. At equivalence, both reactants have been added in equal amounts and completely reacted, so are no longer present. The conjugate acid created does in turn dissociate to some extent to produce more H^+ ions, but most of the acid itself will remain intact.

30. **C** Whether the base is strong or weak has no effect on the number of moles present, and thus no effect on the amount of acid needed to bring it to equivalence. However, replacing the weak base with a strong base changes the reaction taking place to $OH^-(aq) + H^+(aq) \rightarrow H_2O(l)$, meaning that at equivalence all that is present is water, making the pH 7, a value that is greater than the pH shown for the original titration.

31. **C** Infrared radiation is used to study bonds within molecules.

32. **C** Broken reactant bonds are assigned positive values, and bonds formed in the product are assigned negative values. In this reaction, there are 8 C-H bonds in the reactants and 6 C-H bonds in the products, so 6 of those cancel out. We then have 2 C-H bonds in the reactants to worry about, and one C-C and one H-H bond in the products. Doing the math yields: $40 = 2(410) - (350) - (H-H)$. H-H = 430 kJ/mol.

33. **C** Both acids are weak acids, meaning most of the molecules do not dissociate. So, H^+ will be of the lowest concentration. Between the two acids, the one with the greater K_a value will dissociate more, so there will be fewer HClO molecules left in solution than HNO_2 molecules.

34. **B** As pressure increases, volume will decrease. The relevant equation is $P_1V_1 = P_2V_2$, which describes an inverse relationship that creates a curve with a decreasing slope.

35. **C** ΔH is the difference in enthalpy between the products and reactants, and in this reaction it is positive, meaning the products are at a higher energy level than the reactants. The activation energy is the difference between the reactants and the activated state.

36. **C** To create a positive partial charge on a nitrogen atom, it must be bonded to another atom that is more electronegative than it. Of the options listed, only oxygen has a smaller radius (and thus greater electronegativity) value than nitrogen.

37. **B** When temperature is constant, the ratio of moles:volume determines the pressure. The H_2 is 1.5 moles/3.0 L (0.50), the H_2O is 0.67 moles/1.0 L (0.67), and the N_2 is 0.50 moles/2.0 L (0.25). The gas with the greatest ratio (and thus greatest pressure) would be the H_2O.

38. **A** When all the gases are mixed, they take up identical volumes. At this point, the gas that has the most moles present will exert the greatest pressure.

39. **C** At the same temperature, all gases have the same kinetic energy. As $KE = \frac{1}{2}mv^2$, the gas with the lowest molar mass (H_2) will have the highest average velocity, and vice versa.

40. **A** We start with the same amount of Mg^{2+} and OH^-, but the OH^- reacts twice as often, meaning it will be limiting and not appear in the final solution. There will, however, be excess Mg^{2+} present in the final solution.

41. **D** The mass spectrum shows four different isotopes of zinc, meaning that in any given zinc sample you will find atoms with four different atomic masses.

42. **D** The carbon goes from an oxidation state of +4 to +2, which means it gains electrons and is reduced.

43. **A** When given absolute entropy values, we calculate the ΔS by subtracting the entropy of the reactants from that of the products, making sure to consider the coefficients as well.

$\Delta S = 2(190) + 4(70) - 2(240) - 7(130)$ $\Delta S = 380 + 280 - 480 - 910 = -730 \text{ J·mol}^{-1}\text{·K}^{-1}$

44. **B** For any electroplating problem, the key is getting the units to come out properly.

$$300. \text{ s} \times \frac{0.15 \text{ C}}{\text{s}} \times \frac{1 \text{ mol e}^-}{96{,}500 \text{ C}} \times \frac{1 \text{ mol Cu}}{2 \text{ mol e}^-} \times \frac{63.55 \text{ g Cu}}{1 \text{ mol Cu}} = \text{grams of Cu}$$

45. **B** XeF_4 has a molecular geometry of square planar, which is a nonpolar molecule.

46. **D** When NaOH is added to the solution, the OH^- will react with the HCN, turning exactly half of it into CN^- via $HCN(aq) + OH^-(aq) \rightarrow CN^-(aq) + H_2O(l)$. When the concentration of the acid and its conjugate base are equal, the pH of the solution is equal to the pK_a of the acid via Henderson-Hasselbalch.

47. **A** To get the target reaction, we have to first multiply the top reaction by three to get the correct coefficients on the C and CO_2, which also multiplies its enthalpy value by 3 as well. Then, we have to flip the bottom reaction to get the Fe_2O_3 and Fe on the correct sides, which flips the sign on that enthalpy value. We then combine both new enthalpies: $3(-394) + 1,650 = 468$ kJ/mol.

48. **B** The change in the concentration of each reactant before equilibrium is established will be directly proportional to the coefficients in the balanced equation. Both X and Z change twice as much as Y, meaning that the coefficients on X and Z will be the same, and twice that of Y.

49. **A** $K_{sp} = [Cd^{2+}][IO_3^-]^2$. At equilibrium, $[IO_3^-] = 2[Cd^{2+}]$, so $K_{sp} = (x)(2x)^2$. $4.0 \times 10^{-6} = 4x^3$, $x^3 = 1.0 \times 10^{-6}$ and finally $x = 1.0 \times 10^{-2}$ M, so $2x = 2.0 \times 10^{-2}$ M.

50. **B** Adding $Cd(NO_3)_2$ to the solution would cause the increase in the concentration of Cd^{2+} ions, as nitrates are fully soluble. Via the common ion effect, increasing the concentration of Cd^{2+} would cause a shift to the left and a reduction in the concentration of the IO_3^-.

51. **C** The solution is already saturated, so adding more of the solid would have no effect on the concentration of the ions.

52. **B** The ratio of the IO_3^- is identical in both salts (2:1), so the salt with the higher K_{sp} would also have the greater concentration of IO_3^- ions in solution.

53. **A** The Coulombic attraction between the ions and the water molecules depends on both their charge and their size. Ni^{2+} has the same charge as Cd^{2+}, but as Ni^{2+} is smaller it will form a stronger attraction to the water molecules.

54. **B** The heat of fusion is the amount of energy needed to change a substance from a solid to a liquid. During that time, the temperature will remain constant. (Line D is the liquid-to-gas phase change, aka the heat of vaporization).

55. **A** Favorability can be determined using $\Delta G = \Delta H - T\Delta S$. When ΔG is negative, the process is favored. If the process is favored only at higher temperatures, that means that as temperature increases, ΔG is becoming more negative. That is only true when ΔS is positive (making the $-T\Delta S$ term negative). For the reaction to be unfavored at lower temperatures, ΔH must be positive and greater in magnitude than the negative $-T\Delta S$ term.

56. **B** Via Henderson-Hasselbalch, pK_a = pH at the halfway point of each H^+ being fully reacted. For the first H^+, it fully reacts at 20 mL, meaning the pK_a for the first dissociation would be equal to the pH at 10 mL– so about 2. The second H^+ is fully reacted at 40 mL, meaning the pK_a for the second dissociation will be equal to the pH at 30 mL; that makes it about 7.

57. **B** ΔH being negative means the reaction is exothermic, and heating an exothermic reaction drives it to the left, in this case making it more yellow. Dilution of an aqueous equilibrium drives the reaction to the side with more aqueous ions, which is also the left, so the color shift would be the same.

58. **B** The Mg^{2+} and Cl^- ions are no longer bonded, and hydrogen bonding is weaker than ion-dipole attractions. Mg^{2+} has a greater charge and a smaller radius than Cl^-, creating stronger Coulombic attractions with the surrounding water molecules.

59. **C** Hydroxide salts are going to be more soluble in acidic solutions, as the H^+ in solution will react with the OH^- from the salt, causing more salt to dissolve.

60. **C** An increase in entropy means the matter becomes more dispersed. This will occur when the volume of a gas is increased, meaning there will be more room (on average) between the gas particles.

Section II—Free-Response Answers and Explanations

1. (a) H_2O_2: O = −1 in peroxides

OCl$^-$: O = −2 when not in a peroxide

H_2O: 2(+1) + x = 0, x = −2 (also not a peroxide like OCl$^-$)

O_2: O = 0 in the elemental form

(b) Chlorine starts with an oxidation state of +1 and ends at −1, which is the loss of two moles of electrons in the reduction half-reaction. On the other end of things, the oxygens in peroxide start at −1 and they both end up at 0, meaning they lose one mole of electron each, for a total of two moles of electrons in the oxidation half-reaction. So, either way you look at it, the answer is two moles.

(c) (i) $0.085\ M\ H_2O_2 \times 0.200\ L \times \dfrac{1\ mol\ ClO^-}{1\ mol\ H_2O_2} \times \dfrac{1\ mol\ NaClO}{1\ mol\ ClO^-} = 0.017\ mol\ NaClO$

(ii) $0.017\ mol\ NaClO \times \dfrac{74.44\ g\ NaClO}{1\ mol\ NaClO} = 1.27\ g\ NaClO$

$\dfrac{1.02\ g}{mL} \times 33.92\ mL = 34.60\ g\ bleach$ $\qquad$ $\dfrac{1.27\ g\ NaClO}{34.60\ g\ bleach} \times 100\% = 3.7\%\ NaClO$

(d) $H - \ddot{\underset{..}{O}} - \ddot{\underset{..}{O}} - H$

(e) London dispersion forces and hydrogen bonding. Dipole-dipole (aka permanent dipole) is also an acceptable answer here, although not required.

(f) The vapor pressure would increase. As the temperature rises, the molecules are moving faster and are thus more likely to be able to break the IMFs holding them in solution, allowing them to evaporate. The more molecules that have evaporated, the greater the vapor pressure will be.

2. (a) (i) $HF(aq) + H_2O(l) \rightleftharpoons F^-(aq) + H_3O^+(aq)$

(ii) $K_c = \dfrac{[F^-][H_3O^+]}{[HF]}$

(b) First we figure out the concentration of H^+, using $pH = -\log[H^+]$: $10^{-1.67} = 0.0214$ M $= [H^+]$. Then we go to an ICE chart to figure out the variables at equilibrium.

	HF	H⁺	F⁻
I	[HF]	~0	~0
C	$-x$	$+x$	$+x$
E	[HF] $- x$	0.0214	0.0214

We can take the values from the ICE chart and plug them into the equilibrium constant expression to solve.

$7.2 \times 10 - 4 = \dfrac{(0.0214 \text{ M})(0.0214 \text{ M})}{([HF] - 0.0214 \text{ M})}$ $\qquad$ $[HF] = \dfrac{(0.0214 \text{ M})^2}{7.2 \times 10^{-4}} + 0.0214 \text{ M} = 0.657$ M

(c) Via Henderson-Hasselbalch, we know that when $[HF] = [F^-]$, the pH of the buffer will be equal to the pK_a of HF, which is $-\log(7.2 \times 10^{-4}) = 3.14$. For the pH of the buffer to be above the pK_a of HF, there must be more conjugate base in solution than acid. Thus, the middle beaker is correct.

(d) Of the four halogens, fluorine has the smallest atoms and will create the shortest bond with HF. Shorter bonds are stronger bonds, which means the H^+ is less likely to dissociate. That makes HF a weaker acid than the other three halogen acids, which have larger halogen atoms that will not be able to hold onto the H^+ ions as effectively.

(e) (i) The K_a value describes the ratio of the products to the reactants at equilibrium, and that will not change if the acid is diluted. The only way for the K_a value to change would be through a temperature change.

(ii) The percent dissociation of the acid would increase. When diluting an aqueous equilibrium, that causes a shift to the side with more aqueous species. In this case, that would mean a shift to the right, meaning more HF would dissociate.

(f) (i) The boric acid sample would have the higher pH. As the K_a value for the boric acid is lower than that of the HF, fewer boric acid molecules would dissociate, meaning the $[H^+]$ in solution will be lower, leading to a higher pH.

(ii) It is more difficult to remove an H^+ ion from a negatively charged ion than from a neutral molecule, and more difficult to remove it from an ion with a -2 charge than from an ion with a -1 charge. Therefore the K_a decreases with each successive removal.

3. (a) The volumetric flask is considered to be infinitely accurate to the volume that it is calibrated for and is the best option here.

    (b) The Mg^{2+} ion will experience a greater attraction because it has both a greater charge and a smaller radius than the Br^- ion.

    (c) $3\,Mg^{2+}(aq) + 2\,PO_4^{3-}(aq) \rightarrow Mg_3(PO_4)_2(s)$

    (d)

$$2.15g\ MgBr_2 \times \frac{1\ mol\ MgBr_2}{184\ g\ MgBr_2} \times \frac{1\ mol\ Mg^{2+}}{1\ mol\ MgBr_2} \times \frac{1\ mol\ Mg_3(PO_4)_2}{3\ mol\ Mg^{2+}} \times \frac{263\ g\ Mg_3(PO_4)_2}{1\ mol\ Mg_3(PO_4)_2} = 1.02\ g\ Mg_3(PO_4)_2$$

    (e) (i) No. If some of the particles of the precipitate fit through the holes in the filter paper, less precipitate would be collected, not more.

         (ii) Yes. If the precipitate is still wet, the mass of the water will add to the mass of the precipitate, making it seem artificially high.

    (f) This will not be effective due to the fact that the spectator ions originally in solution (Br^- and Na^+) would become part of the precipitate mass if all of the water they are dissolved in evaporates.

4. (a) In trials 1 and 2, the concentration of the NO_2^- is held constant while the concentration of the NH_4^+ doubles. The rate also doubles, which means the reaction is first order with respect to NH_4^+. To determine the order with respect to NO_2^-, we have to do some more involved calculations. We will use trial 1 and trial 3 here. Note that in the calculations below, we already know the order with respect to NH_4^+ is 1, and that the rate constant k will cancel out.

$$\frac{Rate = k[NH_4^+][NO_2^-]^n}{Rate = k[NH_4^+][NO_2^-]^n} \quad \frac{0.025 = (0.010)(0.010)^n}{0.150 = (0.015)(0.020)^n} \qquad 0.17 = (0.67)(0.50)^n \qquad 0.25 = (0.50)^n \qquad n = 2$$

The reaction is second order with respect to NO_2^-.

    (b) The rate law is rate = $k[NH_4^+][NO_2^-]^2$. We can use data from any trial to determine the rate constant. Using Trial 2: $0.050\ Ms^{-1} = k(0.020\ M)(0.010\ M)^2$ $k = 2.5 \times 10^4\ M^{-2}s^{-1}$

    (c) Yes, this would be effective. When running this reaction at a higher temperature, the average velocity of all of the reactant ions would increase. This would lead to both more collisions and enough collisions where the collision energy exceeds the activation energy of the reaction. Both factors contribute to an increased initial rate of reaction.

5. (a) Hydrogen bonding would be present in the pentanol only. In order for hydrogen bonding to occur between molecules, a hydrogen must be directly bonded to a highly electronegative element, in this case oxygen. That will create a very strong positive partial charge due to the (effectively) unshielded hydrogen proton, which is necessary for hydrogen bonding to occur between molecules.

(b) (i) The pentane would have the lowest boiling point, and thus would be the primary component of the distillate. The pentane has the lowest boiling point because it is completely nonpolar and would only experience London dispersion forces. Both pentanone and pentanol would have permanent dipoles and would have stronger IMFs than pentane.

(ii) No, it would not be. Even though neither the pentanone nor the pentanol will be boiling, some of each substance will still be evaporating due to normal molecular motion in a liquid. Thus, some of both of those substances would also be present in the distillate.

6. (a) Entropy measures dispersion of matter, and in the liquid phase the bromine molecules are going to be less dispersed than they would be in a gaseous phase. As there is a greater dispersion in the gaseous phase, the entropy value of the bromine molecules would be greater.

(b) During phase changes, the ΔG value = 0. We can then use $\Delta G = \Delta H - T\Delta S$ to determine the temperature value at which that would be true. To calculate ΔH and ΔS, we simply need to use the values from the table and do products − reactants.

$\Delta H = 30.9$ kJ/mol − 0 = 30.9 kJ/mol

$\Delta S = 245.5$ J/mol K − 152.2 J/mol K = 93.3 J/mol K = 0.0933 kJ/mol K

$0 = (30.9) - T(0.0933)$ T = 331 K

(c) When a substance is undergoing a phase change from liquid to gas, the energy added goes into breaking the IMFs between the molecules and does NOT increase the kinetic energy of the substance at all. Thus, the temperature will remain constant.

7. (a) In order to determine which way the reaction must proceed to reach equilibrium, the reaction quotient must be calculated.

$$Q = \frac{[CH_4][H_2S]^2}{[CS_2][H_2]^4} \qquad Q = \frac{(0.15)(0.25)^2}{(0.30)(0.60)^4} \qquad Q = 0.24$$

As the reaction quotient value is less than the equilibrium constant, the reaction must shift right to get to equilibrium, decreasing the concentration of H_2.

(b) This will decrease the pressure of all gases equally, and when that occurs the equilibrium will shift to the side with more gaseous molecules. In this case, that causes a shift to the left.

(c) This reaction flips the products and reactants from the original reaction. The new equilibrium constant will thus be a reciprocal of the equilibrium constant of the original reaction. So, 1/0.28 = 3.6.

Part III
About the
AP Chemistry
Exam

- The Structure of the AP Chemistry Exam
- Overview of Content Topics
- How AP Exams Are Used
- Other Resources
- Designing Your Study Plan

THE STRUCTURE OF THE AP CHEMISTRY EXAM

The AP Chemistry Exam is a three-hour-long, two-section test that attempts to cover the material you would learn in a college first-year chemistry course. The first section is a 90-minute, 60-question multiple-choice section. The second section is 105 minutes and consists of three long-form free-response questions and four short-form free-response questions.

The multiple-choice section is scored by a computer, and the free-response questions are scored by a committee of high school and college teachers. The free-response questions are graded according to a standard set at the beginning of the grading period by the chief faculty consultants. Inevitably, the grading of Section II is never as consistent or accurate as the grading of Section I.

Take a look at the following data on how students scored on the 2022 test:

Score	Percentage 2022	Credit Recommendation	College Grade Equivalent
5	12.5%	Extremely Well Qualified	A
4	17.0%	Well qualified	A−, B+, B
3	24.5%	Qualified	B−, C+, C
2	23.6%	Possibly Qualified	−
1	22.5%	No Recommendation	−

Scores from the May 2022 test administration. Data taken from the College Board website.

> ### Stay Up to Date!
> For late-breaking information about test dates, exam formats, and any other changes pertaining to AP Chemistry, make sure to check the College Board's website at apstudents.collegeboard.org/courses/ap-chemistry/assessment

OVERVIEW OF CONTENT TOPICS

The College Board's Course and Exam Description (CED) organizes the AP Chemistry course content into nine units. These are shown below, along with an estimation of what percentage of the exam is allocated to each unit.

This book is organized to reflect this unit structure.

Unit Number	Unit Title	Exam Percentage
1	Atomic Structure and Properties	7–9%
2	Molecular and Ionic Compound Structure and Properties	7–9%
3	Intermolecular Forces and Properties	18–22%
4	Chemical Reactions	7–9%
5	Kinetics	7–9%
6	Thermodynamics	7–9%
7	Equilibrium	7–9%
8	Acids and Bases	11–15%
9	Applications of Thermodynamics	7–9%

HOW AP EXAMS ARE USED

Different colleges use AP Exam scores in different ways, so it is important that you go to a particular college's website to determine how it uses AP Exam scores. The three items below represent the main ways in which AP Exam scores can be used.

- **College Credit**. Some colleges will give you college credit if you score well on an AP Exam. These credits count toward your graduation requirements, meaning that you can take fewer courses while in college. Given the cost of college, this could be quite a benefit, indeed.

- **Satisfy Requirements.** Some colleges will allow you to "place out" of certain requirements if you do well on an AP Exam, even if they do not give you actual college credits. For example, you might not need to take an introductory-level course, or perhaps you might not need to take a class in a certain discipline at all.

- **Admissions Plus**. Even if your AP Exam will not result in college credit or even allow you to place out of certain courses, most colleges will respect your decision to push yourself by taking an AP course or even an AP Exam outside of a course. A high score on an AP Exam shows proficiency with more difficult content than is taught in many high school courses, and colleges may take that into account during the admissions process.

More Great Books
For more information on colleges, you might want to check out some of our guide books, which include *The Best 389 Colleges, The Complete Guide to College Essays, Paying for College*, and many more!

OTHER RESOURCES

There are many resources available to help you improve your score on the AP Chemistry Exam, not the least of which are your teachers. If you are taking an AP class, you may be able to get extra attention from your teacher, such as obtaining feedback on your free-response questions. If you are not in an AP course, reach out to a teacher who teaches chemistry, and ask if the teacher will review your free-response questions or otherwise help you with content.

Another wonderful resource is **AP Students,** the official site of the AP Exams. The scope of the information at this site is quite broad and includes:

- the course description, which provides details on what content is covered as well as sample questions

- free-response question prompts from previous AP Chemistry Exams along with scoring guidelines

- access to AP Classroom if you are enrolled in a course (teacher assistance required)

- exam practice tips

The AP Students web page is apstudents.collegeboard.org.

Finally, The Princeton Review offers tutoring and small group instruction. Our expert instructors can help you refine your strategic approach and add to your content knowledge. For more information, call 1-800-2REVIEW.

Looking to Guarantee a 4 or 5?
We now offer one-on-one tutoring for a guaranteed 5 or an online course for a guaranteed 4 on the AP Chemistry Exam. For information on rates and availability, visit PrincetonReview. com/college/ap-test-prep

DESIGNING YOUR STUDY PLAN

As part of the Introduction, you identified some areas of potential improvement. Let's now delve further into your performance on Test 1, with the goal of developing a study plan appropriate to your needs and time commitment.

Read the answers and explanations associated with the multiple-choice questions (starting at page 35). After you have done so, respond to the following:

- Review your results from the calculations in the Practice Test 1 Diagnostic Answer Key, and make a list of the content units. You might even want to include the subtopics from Step 1. Next to each topic, indicate your rank of the topic as follows: "1" means "I need a lot of work on this," "2" means "I need to beef up my knowledge," and "3" means "I know this topic well."

- How many days/weeks/months away is your exam?

- What time of day is your best, most focused study time?

- How much time per day/week/month will you devote to preparing for your exam?

- When will you do this preparation? (Be as specific as possible: Mondays and Wednesdays from 3:00 P.M.–4:00 P.M., for example.)

- Based on the answers above, will you focus on strategy (Part IV) or content (Part V) or both?

- What are your overall goals in using this book?

Part IV
Test-Taking Strategies for the AP Chemistry Exam

Chapter 1
How to Approach Multiple-Choice Questions

THE BASICS

Section I of the test is composed of 60 multiple-choice questions for which you are allotted 90 minutes. This part is worth 50% of your total score.

For this section, you will be given a periodic table of the elements along with a sheet that lists common chemistry formulas, and (new as of 2023), you MAY use a calculator.

On the multiple-choice section, you receive 1 point for a correct answer. There is no penalty for leaving a question blank or getting a question wrong.

PACING

According to the College Board, the multiple-choice section of the AP Chemistry Exam covers more material than any individual student is expected to know. Nobody is expected to get a perfect or even near perfect score. What does that mean to you?

Use the Two-Pass System

Go through the multiple-choice section twice. The first time, do all the questions that you can get answers to immediately. That is, do the questions with little or no math and questions on chemistry topics in which you are well versed. Skip questions on topics that make you uncomfortable. Circle the questions that you skip in your test booklet so you can find them easily during the second pass. Once you've done all the questions that come easily to you, go back and pick out the tough ones that you have the best shot at.

Proven Techniques

In general, the questions near the end of the section are tougher than the questions near the beginning. You should keep that in mind, but be aware that each person's experience will be different. If you can do acid-base questions in your sleep, but you'd rather have your teeth drilled than draw a Lewis diagram, you may find questions near the end of the section easier than questions near the beginning.

That's why the Two-Pass System is so handy. By using it, you make sure you get to see all the questions you can get right, instead of running out of time because you got bogged down on questions you couldn't do earlier in the test.

Don't Turn a Question into a Crusade!

Most people don't run out of time on standardized tests because they work too slowly. Instead, they run out of time because they spend half the test wrestling with two or three particular questions.

You should never spend more than a minute or two on any question. If a question doesn't involve calculation, then either you know the answer, you can make an educated guess, or you don't know the answer. Figure out where you stand on a question, make a decision, and move on.

Any question that requires more than two minutes worth of calculations probably isn't worth doing. Remember, skipping a question early in the section is a good thing if it means that you'll have time to get two correct answers later on.

GUESSING

You get one point for every correct answer on the multiple-choice section. Guessing randomly neither helps you nor hurts you. Educated guessing, however, will help you.

Use Process of Elimination (POE) to Find Wrong Answers

There is a fundamental weakness to a multiple-choice test. The test-makers must show you the right answer, along with three wrong answers. Sometimes seeing the right answer is all you need. Other times you may not know the right answer, but you may be able to identify one or two of the answers that are clearly wrong. Here is where you should use Process of Elimination (POE) to take an educated guess.

Look at this hypothetical question.

1. Which of the following compounds will produce a purple solution when added to water?

   (A) Brobogdium rabelide
   (B) Diblythium perjuvenide
   (C) Sodium chloride
   (D) Carbon dioxide

You should have no idea what the correct answer is because two of these compounds are made up, but you do know something about the obviously wrong answers. You know that sodium chloride, (C), and carbon dioxide, (D), do not turn water purple. So, using POE, you have a 50 percent chance at guessing the correct answer. Now the odds are in your favor. Now you should guess.

Guess and Move On

Remember that you're guessing. Pondering the possible differences between brobogdium rabelide and diblythium perjuvenide is a waste of time. Once you've taken POE as far as it will go, pick your favorite letter and move on.

The multiple-choice section is the exact opposite of the free-response section. It's scored by a machine. There's no partial credit. The computer doesn't know, or care if you know, why an answer is correct. All the computer cares about is whether you blackened in the correct oval on your score sheet. You get the same number of points for picking (B) because you know that (A) is wrong and that (B) is a nicer letter than (C) or (D) as you would get for picking (B) because you fully understood the subtleties of an electrochemical process.

ABOUT CALCULATORS

The College Board recommends that you use a scientific or graphing calculator; a four-function calculator is allowed but not recommended. Be aware that some models have unapproved features and capabilities and are not permitted. For more information, including a list of approved calculators, check the AP Students website at apstudents.collegeboard.org/exam-policies-guidelines/calculator-policies.

If you're still unsure, you can bring two calculators so that if one is rejected (or not functioning), you do not have to rely on a school-provided backup that you may be unfamiliar with.

REFLECT

Respond to the following questions:

- How long will you work on multiple-choice questions?

- How will you change your approach to multiple-choice questions?

- What is your multiple-choice guessing strategy?

Chapter 2
How to Approach
Free-Response
Questions

OVERVIEW OF THE FREE-RESPONSE SECTION

Section II is composed of seven free-response questions. You will be given 105 minutes to complete this section, which is worth 50 percent of your total score.

The first three free-response questions are longer, and are worth 10 points each. The College Board recommends you take about 22 minutes per question on these. The last four questions are much shorter, and only worth 4 points each. For these, the College Board recommends you take about 9 minutes each.

The suggested times are just that—suggestions. Your results may vary based on how comfortable you are with the topic being tested within each question. They do give you a good guideline, though. You probably shouldn't be spending more than 20 minutes on any of the first three questions. Don't be afraid to cut yourself off and come back to questions later if time allows.

APPROACHING THE MATH PROBLEMS

You want to show the graders that you understand the math behind these chemistry problems, so here are some suggestions.

Show Every Step of Your Calculations on Paper

This section is the opposite of multiple-choice. You don't just get full credit for writing the correct answer. You get most of your points on this section for showing the process that got you to the answer. The graders give you partial credit when you show them that you know what you're doing. So even if you can do a calculation in your head, you should set it up and show it on the page.

By showing every step, or explaining what you're doing in words, you ensure that you'll get all the partial credit possible, even if you screw up a calculation.

> **Room to Write**
> On the actual test, you will be given space along with the bubble sheet to record your answers for each free-response question. You should use scrap paper for the free-responses on the practice tests in this book. After you've gotten a hang of the timing, be aware of how much space each response is taking up, in case you need to write in smaller print or use fewer words on the test.

Include Units in All Your Calculations

Scientists like units in calculations. Units make scientists feel secure. You'll get points for including them and you may lose points for leaving them out.

Remember Significant Figures

You can lose 1 point per question if your answer is off by more than one significant figure. Without getting too bent out of shape about it, try to remember that a calculation is only as accurate as the least accurate number in it.

(For a detailed explanation of significant figures, see the Laboratory Overview, pages 322–324.)

The Graders Will Follow Your Reasoning, Even If You've Made a Mistake

Often, you are asked to use the result of a previous part of a problem in a later part. If you got the wrong answer in part (a) and used it in part (c), you can still get full credit for part (c), as long as your work is correct based on the number that you used. That's important, because it means that botching the first part of a question doesn't necessarily sink the whole question.

Remember the Mean!

So let's say that you could complete only parts (a) and (b) on the required equilibrium problem. That's 4 or 5 points out of 10, tops. Are you doomed? Of course not. You're above average. If this test is hard on you, it's probably just as hard on everybody else. Remember, you don't need anywhere near a perfect score to get a 5, and you can leave half the test blank and still get a 3!

STRATEGIES FOR THE FREE-RESPONSE SECTION

This section is here to test whether you can translate chemistry into English. Most of the questions can be answered in two or three simple sentences, or with a simple diagram or two. Here are some tips for answering the seven free-response questions on Part II.

Show That You Understand the Terms Used in the Question

If you are asked why sodium and potassium have differing first ionization energies, the first thing you should do is tell them what ionization energy is. That's probably worth the first point of partial credit. Then you should tell them how the differing structures of the atoms make for differing ionization energies. That leads to the next tip.

Take a Step-by-Step Approach

Grading these tests is hard work. Breaking a question into parts in this way makes it easier on the grader, who must match your response to a set of guidelines they have been given that describe how to assign partial and full credit.

Each grader scores each test based on these rough guidelines that are established at the beginning of the grading period. For instance, if a grader has 3 points for the question about ionization energies, the points might be distributed the following way:

*   one point for understanding ionization energy
*   one point for explaining the structural difference between sodium and potassium
*   one point for showing how this difference affects the ionization energy

You can get all three points for this question if the grader thinks that all three concepts are addressed implicitly in your answer, but by taking a step-by-step approach, you improve your chances of explicitly addressing the things that a grader has been instructed to look for. Once again, grading these tests is hard work; graders won't know for sure if you understand something unless you tell them.

Write Neatly

This simple concept cannot be stressed enough: write neatly, even if that means working at half-speed. You can't get points for answers if the graders can't understand them. Of course, this applies to the rest of the free-response section as well.

REFLECT

Respond to the following questions:

- How much time will you spend on the short free-response questions? What about the long free-response questions?

- What will you do before you begin writing your free-response answers?

- Will you seek further help, outside of this book (such as a teacher, tutor, or AP Students), on how to approach the questions that you will see on the AP Chemistry Exam?

Part V
Content Review for the AP Chemistry Exam

Chapter 3
Unit 1: Atomic Structure and Properties

THE PERIODIC TABLE

The most important tool you will use on this test is the periodic table of the elements.

PERIODIC TABLE OF THE ELEMENTS

1																	18
1 **H** 1.008	2											13	14	15	16	17	2 **He** 4.00
3 **Li** 6.94	4 **Be** 9.01											5 **B** 10.81	6 **C** 12.01	7 **N** 14.01	8 **O** 16.00	9 **F** 19.00	10 **Ne** 20.18
11 **Na** 22.99	12 **Mg** 24.30	3	4	5	6	7	8	9	10	11	12	13 **Al** 26.98	14 **Si** 28.09	15 **P** 30.97	16 **S** 32.06	17 **Cl** 35.45	18 **Ar** 39.95
19 **K** 39.10	20 **Ca** 40.08	21 **Sc** 44.69	22 **Ti** 47.87	23 **V** 50.94	24 **Cr** 52.00	25 **Mn** 54.94	26 **Fe** 55.85	27 **Co** 58.93	28 **Ni** 58.69	29 **Cu** 63.55	30 **Zn** 65.38	31 **Ga** 69.72	32 **Ge** 72.63	33 **As** 74.92	34 **Se** 78.97	35 **Br** 79.90	36 **Kr** 83.80
37 **Rb** 85.47	38 **Sr** 87.62	39 **Y** 88.91	40 **Zr** 91.22	41 **Nb** 92.91	42 **Mo** 95.95	43 **Tc**	44 **Ru** 101.07	45 **Rh** 102.91	46 **Pd** 106.42	47 **Ag** 107.87	48 **Cd** 112.41	49 **In** 114.82	50 **Sn** 118.71	51 **Sb** 121.76	52 **Te** 127.60	53 **I** 126.90	54 **Xe** 131.29
55 **Cs** 132.91	56 **Ba** 137.33	57-71 *	72 **Hf** 178.49	73 **Ta** 180.95	74 **W** 183.94	75 **Re** 186.21	76 **Os** 190.23	77 **Ir** 192.22	78 **Pt** 195.08	79 **Au** 196.97	80 **Hg** 200.59	81 **Tl** 204.38	82 **Pb** 207.2	83 **Bi** 208.98	84 **Po**	85 **At**	86 **Rn**
87 **Fr**	88 **Ra**	89-103 †	104 **Rf**	105 **Db**	106 **Sg**	107 **Bh**	108 **Hs**	109 **Mt**	110 **Ds**	111 **Rg**	112 **Cn**	113 **Nh**	114 **Fl**	115 **Mc**	116 **Lv**	117 **Ts**	118 **Og**

	57 **La** 138.91	58 **Ce** 140.12	59 **Pr** 140.91	60 **Nd** 144.24	61 **Pm**	62 **Sm** 150.36	63 **Eu** 151.97	64 **Gd** 157.25	65 **Tb** 158.93	66 **Dy** 162.50	67 **Ho** 164.93	68 **Er** 167.26	69 **Tm** 168.93	70 **Yb** 173.05	71 **Lu** 174.97
*Lanthanoids															
†Actinoids	89 **Ac**	90 **Th** 232.04	91 **Pa** 231.04	92 **U** 238.03	93 **Np**	94 **Pu**	95 **Am**	96 **Cm**	97 **Bk**	98 **Cf**	99 **Es**	100 **Fm**	101 **Md**	102 **No**	103 **Lr**

Bad Joke Alert
To prepare for the AP Chemistry Exam, familiarize yourself with the periodic table of elements by looking at it…periodically.

The periodic table gives you very basic but very important information about each element.

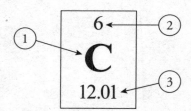

1. This is the **symbol** for the element: carbon, in this case. On the test, the symbol for an element is used interchangeably with the name of the element.
2. This is the **atomic number** of the element. The atomic number is the same as the number of protons in the nucleus of an element; it is also the same as the number of electrons surrounding the nucleus of an element when it is neutrally charged.
3. This number represents the average atomic mass of a single atom of carbon, measured in atomic mass units (amus). It also represents the average mass for a mole (see page 68) of carbon atoms, measured in grams. Thus, one mole of carbon atoms has a mass of 12.01 g. This is called the **molar mass** of the element.

The horizontal rows of the periodic table are called **periods.**

The vertical columns of the periodic table are called **groups.**

Groups can be numbered in two ways. The old system used Roman numerals to indicate groups. The new system simply numbers the groups from 1 to 18. While it is not important to know the specific group numbers, it is important to know the names of some groups.

Group IA/1—Alkali Metals
Group IIA/2—Alkaline Earth Metals
Group B/3–12—Transition Metals
Group VIIA/17—Halogens
Group VIIIA/18—Noble Gases

In addition, the two rows offset beneath the table are alternatively called the lanthanides and actinides, the rare Earth elements, or the inner transition metals.

The identity of an atom is determined by the number of protons contained in its nucleus. The nucleus of an atom also contains neutrons. The **mass number** of an atom is the sum of its neutrons and protons. Electrons have significantly less mass than protons or neutrons and do not contribute to an element's mass.

Atoms of an element with different numbers of neutrons are called **isotopes**; for instance, carbon-12, which contains 6 protons and 6 neutrons, and carbon-14, which contains 6 protons and 8 neutrons, are isotopes of carbon. The molar mass given on the periodic table is the average of the mass numbers of all known isotopes weighted by their percent abundance.

The mass of various isotopes of an element can be determined by a technique called mass spectrometry. A mass spectrum of selenium looks like the following:

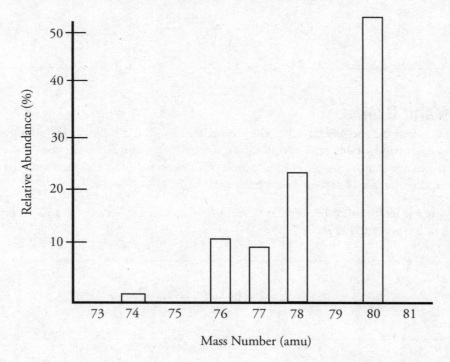

As you can see, the most abundant isotope of selenium has a mass of 80, but there are four other naturally occurring isotopes of selenium. The average atomic mass is the weighted average of all five isotopes of selenium shown on this spectrum.

The molar mass of an element will give you a pretty good idea of the most common isotope of that element. For instance, the molar mass of carbon is 12.01 and about 99 percent of the carbon in existence is carbon-12.

MOLES

The mole is the most important concept in chemistry, serving as a bridge that connects all the different quantities that you'll come across in chemical calculations. The coefficients in chemical reactions tell you about the reactants and products in terms of moles, so most of the stoichiometry questions you'll see on the test will be exercises in converting between moles and grams, liters, molarities, and other units.

Moles and Molecules

Avogadro's number describes the number of atoms that are in a single mole of any given element. Much like a dozen is always 12, Avogadro's number is always 6.022×10^{23}. While it technically can be used to count anything, due to its extremely large value, it is usually only used to count extraordinarily small things. Within the confines of this book, it will be used to count atoms, molecules, electrons, or ions, depending on the problem.

$$1 \text{ mole} = 6.022 \times 10^{23} \text{ particles}$$

$$\text{Moles} = \frac{\text{particles}}{(6.022 \times 10^{23})}$$

Moles and Grams

Moles and grams can be related using the atomic masses given in the periodic table. Atomic masses on the periodic table are given in terms of atomic mass units (amu); however, they also signify how many grams are present in one mole of an element. So, if 1 carbon atom has a mass of 12 amu, then 1 mole of carbon atoms has a mass of 12 grams.

You can use the relationship between amu and g/mol to convert between grams and moles by using the following equation:

$$\text{Moles} = \frac{\text{grams}}{\text{molar mass}}$$

Moles and Gases

We'll talk more about the ideal gas equation in Unit 3, but for now, you should know that you can use it to calculate the number of moles of a gas if you know some of the gas's physical properties. All you need to remember at this point is that in the equation $PV = nRT$, n stands for moles of gas.

$$\text{Moles} = \frac{PV}{RT}$$

P = pressure (atm)
V = volume (L)
T = temperature (K)
R = the gas constant,
0.0821 $\text{L·atm·mol}^{-1}\cdot\text{K}^{-1}$

The equation above gives the general rule for finding the number of moles of a gas. Many gas problems will take place at STP, or standard temperature and pressure, where P = 1 atmosphere and T = 273 K. At STP, the situation is much simpler, and you can convert directly between the volume of a gas and the number of moles. That's because at STP, one mole of gas always occupies 22.4 liters.

$$\text{Moles} = \frac{\text{liters}}{\left(22.4\,\text{L/mol}\right)}$$

Molarity

Molarity (M) expresses the concentration of a solution in terms of volume. It is the most widely used unit of concentration, turning up in calculations involving equilibrium, acids and bases, and electrochemistry, among others. When you see a chemical symbol in brackets on the test, that means they are talking about molarity. For instance, "$[Na^+]$" is the same as "the molar concentration (molarity) of sodium ions."

$$\text{Molarity}\ (M) = \frac{\text{moles of solute}}{\text{liters of solution}}$$

Percent Composition

Percent composition is the percent by mass of each element that makes up a compound. It is calculated by dividing the mass of each element or component in a compound by the total molar mass for the substance.

Calculate the percent composition of each element in calcium nitrate, $Ca(NO_3)_2$.

To do this, you need to first separate each element and count the number of atoms present. Subscripts outside of parentheses apply to all atoms inside of those parentheses.

Calcium: 1
Nitrogen: 2
Oxygen: 6

Then, multiply the number of atoms by the atomic mass of each element.

Ca: 40.08 g/mol × 1 = 40.08 g/mol

N: 14.01 g/mol × 2 = 28.02 g/mol

O: 16.00 g/mol × 6 = 96.00 g/mol

Add up the masses of the individual elements to get the atomic mass of that compound. Divide each individual mass by the total molar mass to get your percent composition.

40.08 g/mol + 28.02 g/mol + 96.00 g/mol = 164.10 g/mol

Ca: 40.08/164.10 × 100% = 24.42%

N: 28.02/164.10 × 100% = 17.07%

O: 96.00/164.10 × 100% = 58.50%

You can check your work at the end by making sure your percents add up to 100% (taking rounding into consideration).

24.42% + 17.07% + 58.50% = 99.99%. Close enough!

Empirical and Molecular Formulas

You will also need to know how to determine the empirical and molecular formulas of a compound given masses or mass percents of the components of that compound. Remember that the empirical formula represents the simplest ratio of one element to another in a compound (e.g., CH_2O), while the molecular formula represents the actual formula for the substance (e.g., $C_6H_{12}O_6$).

Let's take a look at the following example:

A compound is found to contain 56.5% carbon, 7.11% hydrogen, and 36.4% phosphorus.

a) Determine the empirical formula for the compound.

We start by assuming a 100 gram sample; this allows us to convert those percentages to grams. After we have that done, each element needs to be converted to moles.

$$C: 56.5 \text{ g C} \times \frac{1 \text{ mol C}}{12.01 \text{ g C}} = 4.71 \text{ mol C}$$

$$H: 7.11 \text{ g H} \times \frac{1 \text{ mol H}}{1.01 \text{ g H}} = 7.04 \text{ mol H}$$

$$P: 36.4 \text{ g P} \times \frac{1 \text{ mol P}}{30.97 \text{ g P}} = 1.18 \text{ mol P}$$

We then divide each mole value by the lowest of the values. In this example, that would be the phosphorus. It is acceptable to round your answers if they are close (within 0.1) to a whole number.

$$C: \frac{4.71 \text{ mol}}{1.18 \text{ mol}} = 4$$

$$H: \frac{7.04 \text{ mol}}{1.18 \text{ mol}} = 6$$

$$P: \frac{1.18 \text{ mol}}{1.18 \text{ mol}} = 1$$

Those values become subscripts, so the empirical formula for the compound is C_4H_6P.

b) If the compound has a molar mass of 170.14 g/mol, what is its molecular formula?

First, we determine the molar mass of the empirical formula.

$$(12.01 \text{ g/mol} \times 4) + (1.01 \text{ g/mol} \times 6) + 30.97 \text{ g/mol} = 85.07 \text{ g/mol}$$

Then, we divide that mass into the molar mass.

$$\frac{170.14}{85.07} = 2$$

Finally, multiply all subscripts in the empirical formula by that value. So, the molecular formula is $C_8H_{12}P_2$.

Electron Configurations and the Periodic Table

The positively charged nucleus is always pulling at the negatively charged electrons around it, and the electrons have potential energy that increases with their distance from the nucleus. It works the same way that the gravitational potential energy of a brick on the third floor of a building is greater than the gravitational potential energy of a brick nearer to ground level.

The energy of electrons, however, is **quantized.** That's important. It means that electrons can exist only at specific energy levels, separated by specific intervals. It's similar to a situation in which the brick in the building could be placed only on the first, second, or third floor of the building, but not in between.

COULOMB'S LAW

The attraction between opposite charges is known as an electrostatic force, and it depends on the charges and distances involved. This can be calculated using Coulomb's law.

$$F_{Coulombic} \propto \frac{q_1 q_2}{r^2}$$

F = electrostatic force between the nucleus and an electron

q_1 = magnitude of the positive charge (nucleus)

q_2 = magnitude of the negative charge (electron)

r = distance between the charges

While on the exam you will not be required to mathematically calculate the force between these particles, you should be able to qualitatively apply Coulomb's law. Essentially, the greater the charge of the nucleus, the greater the force is between it and an electron. As a result, electrons feeling the influence of a more positive charge feel a stronger pull from the nucleus and have lower potential energy. Regardless of the charge magnitudes, if two opposite charges are closer together, meaning the value of r is lower, the force between them is higher. Electrons closer to the nucleus feel a stronger pull from it, and thus have lower potential energy than electrons further away. In order to remove an electron from an atom, this electrostatic attraction must be overcome by adding energy—the energy needed is called the binding energy of the electron and is always a positive value.

The Bohr Model

Neils Bohr took the quantum theory and used it to predict that electrons orbit the nucleus at specific, fixed radii, like planets orbiting the Sun.

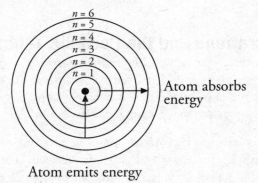

Each energy level is represented by a row on the periodic table. There are currently seven known energy levels, which correspond with $n = 1$ to $n = 7$. The closer an energy level is to an atomic nucleus, the less energy electrons on that level have. While the Bohr model is not a perfect model of the atom, it serves as an excellent basis to understand atomic structure.

When atoms absorb energy in the form of **electromagnetic radiation,** electrons jump to higher energy levels. When electrons drop from higher to lower energy levels, atoms give off energy in the form of electromagnetic radiation.

PHOTOELECTRON SPECTROSCOPY

If an atom is exposed to electromagnetic radiation at an energy level that exceeds the various binding energies of the electrons of that atom, the electrons can be ejected. The amount of energy necessary to do that is called the **ionization energy** for that electron. For the purposes of this exam, ionization energy and binding energy can be considered to be synonymous terms. When examining the spectrum for electrons from a single atom or a small number of atoms, this energy is usually measured in electronvolts, eV (1 eV = 1.60×10^{-19} Joules). If moles of atoms are studied, the unit for binding energy is usually either kJ/mol or MJ/mol.

All energy of the incoming radiation must be conserved and any of that energy that does not go into breaking the electron free from the nucleus will be converted into **kinetic energy** (the energy of motion) for the ejected electron. So:

Incoming Radiation Energy =
Binding Energy + Kinetic Energy (of the ejected electron)

The faster an ejected electron is going, the more kinetic energy it has. Electrons that were originally further away from the nucleus require less energy to eject, and thus will be moving faster. So, by examining the speed of the ejected electrons, we can determine how far they were from the nucleus of the atom in the first place. Usually, it takes electromagnetic radiation in either the visible or ultraviolet range to cause electron emission, while radiation in the infrared range is often used to study chemical bonds. Radiation in the microwave region is used to study the shape of molecules.

Spectra

If the amount of ionization energy for all electrons ejected from a nucleus is charted, you get what is called a **photoelectron spectrum** (PES) that looks like the following:

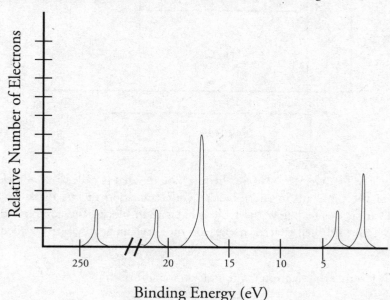

The *y*-axis describes the relative number of electrons that are ejected from a given energy level, and the *x*-axis shows the binding energy of those electrons. Unlike most graphs, binding energy (ionization energy) decreases going from left to right in a PES. The spectrum above is for sulfur.

Each section of peaks in the PES represents a different energy level. Because there are multiple peaks in each section, we can tell that not all electrons in the same energy level are located at the same distance from the nucleus. With each energy level, there are **subshells**, which describe the shape of the space the electron can be found in (remember, we are in three-dimensional space here). The Bohr model is limited to two dimensions and does not represent the true positions of electrons due to that reason. Electrons do not orbit the nucleus as planets orbit the Sun. Instead, they are found moving about in a certain area of space (the subshell) a given distance (the energy level) away from the nucleus.

In all energy levels, the first subshell is called the *s*-subshell and can hold a maximum of two electrons. The second subshell is called the *p*-subshell and can hold a maximum of six electrons. In the spectrum on the previous page, we can see the peak for the *p*-subshell in energy level 2 is three times higher than that of the *s*-subshell. The relative height of the peaks helps determine the number of electrons in that subshell.

In the area for the third energy level, the *p*-subshell peak is only twice as tall as the *s*-subshell. This indicates there are only four electrons in the *p*-subshell of this particular atom.

Electron Configuration

Studying the PES of elements allows scientists to understand more about the structure of the atom. In addition to the *s* and *p* subshells, two others exist: *d* (10 electrons max) and *f* (14 electrons max). The periodic table is designed so that each area is exactly the length of one particular subshell.

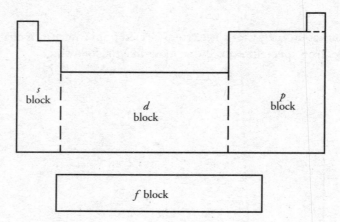

As you can see, the first two groups (plus helium) are in what is called the *s*-block. The groups on the right of the table are in the *p*-block, while transition metals make up the *d*-block. The inner transition metals below the table inhabit the *f*-block. The complete description of the energy level and subshell that each electron on an element inhabits is called its **electron configuration**.

Example 1: Determine the electron configuration for sulfur.

Sulfur has 16 electrons. The first two go into energy level 1 subshell *s*; this is represented by $1s^2$. The next two go into energy level 2 subshell *s* ($2s^2$). Six more fill energy level *p* ($2p^6$), then two more go into 3*s* ($3s^2$), and the final four enter into 3*p* ($3p^4$). So the final configuration is $1s^2 2s^2 2p^6 3s^2 3p^4$.

Using the periodic table as a reference can allow for the determination of any electron configuration. One thing to watch out for is that the energy level drops by 1 when entering the *d*-block (transition metals) and drops by 2 when entering the *f*-block (inner transition metals). The why behind that isn't important right now, but you should still be able to apply that rule.

Example 2: Determine the electron configuration for nickel.

$$1s^2 2s^2 2p^6 3s^2 3p^6 4s^2 3d^8$$

Electron configurations can also be written "shorthand" by replacing parts of them with the symbol for the noble gas at the end of the highest energy level that has been filled.

For example, the shorthand notation for Si is $[Ne]3s^2 3p^2$ and the shorthand for nickel would be $[Ar]4s^2 3d^8$.

CONFIGURATION RULES

The Aufbau Principle

The **Aufbau Principle** states that when building up the electron configuration of an atom, electrons are placed in orbitals, subshells, and shells in order of increasing energy.

The Pauli Exclusion Principle

The **Pauli Exclusion Principle** states that the two electrons which share an orbital cannot have the same spin. One electron must spin clockwise, and the other must spin counterclockwise.

Hund's Rule

Hund's Rule says that when an electron is added to a subshell, it will always occupy an empty orbital if one is available. Electrons always occupy orbitals singly if possible and pair up only if no empty orbitals are available.

Watch how the 2*p* subshell fills as we go from boron to neon.

	$1s$	$2s$	$2p$		
Boron	⇅	⇅	↑		
Carbon	⇅	⇅	↑	↑	
Nitrogen	⇅	⇅	↑	↑	↑
Oxygen	⇅	⇅	⇅	↑	↑
Fluorine	⇅	⇅	⇅	⇅	↑
Neon	⇅	⇅	⇅	⇅	⇅

PREDICTING IONIC CHARGES

One of the great rules of chemistry is that the most stable configurations from an energy standpoint are those in which the outermost energy level is full. For anything in the s or p blocks, that means achieving a state in which there are eight electrons in the outermost shell (2 in the s subshell and 6 in the p subshell). The electrons in the outermost s and p subshells are called the **valence electrons** of that atom.

Elements that are close to a full energy level, such as the halogens or those in the oxygen group, tend to gain electrons to achieve a stable configuration. An **ion** is an atom that has either gained or lost electrons, while the number of protons and neutrons remains constant. Halogens need only gain one electron to achieve a stable configuration, and as such, typically form ions with a charge of negative one: F^-, Cl^-, etc. Any particle with more electrons than protons is called an **anion** (negatively charged ion). Those elements in the oxygen group need two electrons for stability; thus, they have a charge of negative 2.

> It's tempting to say that fluorine atoms "want" one electron or are "happy" with a full valence shell, but in reality, atoms don't have feelings and don't want anything! Instead, say fluorine atoms need only attract one electron to have a stable energy level. Assigning feelings and desires to inanimate particles won't do you any favors on the exam!

On the other end of the table, the alkali metals can most easily achieve a full valence shell by losing a single electron, rather than by gaining seven. So, they will form positively charged ions (**cations**), which have more protons than electrons. Alkali metals typically have a charge of +1, alkaline Earth metals of +2, and so forth. The table below gives a good overview of the common ionic charges for various groups:

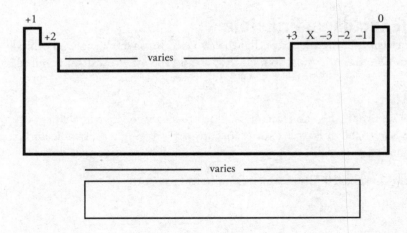

In general, transition metals can form ions of varying charges. All transition metals lose electrons to form cations, but how many electrons they lose will vary depending on the compound they are in. For example, in $CuBr_2$, we know that the charge on the bromide is -1, and with two bromides the total negative charge present in the compound is -2. The total compound charge must be zero for the compound to be stable, so it stands to reason that the charge on the single copper ion is +2. That compound would be called copper (II) bromide. However, in CuBr, the charge on the copper would only need to be +1 to balance the single bromide, so that compound would be called copper (I) bromide.

There are a few transition metals that only form ions with one possible charge. Two important examples are zinc, which always forms ions with a charge of +2, and silver, which always forms ions with a charge of +1.

There's also a slightly different rule as to how transition metals lose electrons when forming those cations. Transition metals, when losing electrons, will lose their higher-level s electrons before losing any of the lower-level d electrons. Thus, while an iron atom has a configuration of $[Ar]4s^23d^6$, an iron ion that has lost two electrons (Fe^{2+}) would have a configuration of $[Ar]3d^6$. If it were to lose further electrons, those would then come from the d-orbital.

PERIODIC TRENDS

You can make predictions about certain behavior patterns of an atom and its electrons based on the position of the atom in the periodic table. All the periodic trends can be understood in terms of three basic rules.

1. Electrons are attracted to the protons in the nucleus of an atom.
    a. The closer an electron is to the nucleus, the more strongly it is attracted.
    b. The more protons in a nucleus, the more strongly an electron is attracted.
2. Electrons are repelled by other electrons in an atom. So, if other electrons are between a valence electron and the nucleus, the valence electron will be less attracted to the nucleus. That's called shielding.
3. Completed shells (and to a lesser extent, completed subshells) are very stable. Atoms will add or subtract valence electrons to create complete shells if possible.

The atoms in the left-hand side of the periodic table are called metals. Metals give up electrons when forming bonds. Most of the elements in the table are metals. The elements in the upper right-hand portion of the table are called nonmetals. Nonmetals generally gain electrons when forming bonds. The metallic character of the elements decreases as you move from left to right across the periodic table. The elements in the borderline between metal and nonmetal, such as silicon and arsenic, are called metalloids.

Atomic Radius

The atomic radius is the approximate distance from the nucleus of an atom to its valence electrons.

Moving from Left to Right Across a Period (Li to Ne, for Instance), Atomic Radius Decreases

Moving from left to right across a period, protons are added to the nucleus, so the valence electrons are more strongly attracted to the nucleus; this decreases the atomic radius. Electrons are also being added, but they are all in the same shell at about the same distance from the nucleus, so there is not much of a shielding effect.

Moving Down a Group (Li to Cs, for Instance), Atomic Radius Increases

Moving down a group, shells of electrons are added to the nucleus. Each shell shields the more distant shells from the nucleus and the valence electrons get farther away from the nucleus. Protons are also being added, but the shielding effect of the negatively charged electron shells cancels out the added positive charge.

Cations (Positively Charged Ions) Are Smaller than Atoms

Generally, when electrons are removed from an atom to form a cation, the outer shell is lost, making the cation smaller than the atom. Also, when electrons are removed, electron–electron repulsions are reduced, allowing all of the remaining valence electrons to move closer to the nucleus.

Anions (Negatively Charged Ions) Are Larger than Atoms

When an electron is added to an atom, forming an anion, electron–electron repulsions increase, causing the valence electrons to move farther apart, which increases the radius.

Ionization Energy

Electrons are attracted to the nucleus of an atom, so it takes energy to remove an electron. The energy required to remove an electron from an atom is called the first ionization energy. Once an electron has been removed, the atom becomes a positively charged ion. The energy required to remove the next electron from the ion is called the second ionization energy, and so on.

A reaction showing the first ionization energy for a phosphorus atom would look like this:

$$P(g) \rightarrow P^+(g) + e^-$$

Moving from Left to Right Across a Period, Ionization Energy Increases

Moving from left to right across a period, protons are added to the nucleus, which increases its positive charge. The electrons being added are all in the same energy level at about the same distance from the nucleus, so they do not effectively shield these added protons. For this reason, the negatively charged valence electrons are more strongly attracted to the nucleus, which increases the energy required to remove them.

Moving Down a Group, Ionization Energy Decreases

Moving down a group, shells of electrons are added to the nucleus. Each inner shell shields the more distant shells from the nucleus, reducing the pull of the nucleus on the valence electrons and making them easier to remove. Protons are also being added, but the shielding effect of the negatively charged electron shells cancels out the added positive charge.

The Second Ionization Energy Is Greater than the First Ionization Energy

When an electron has been removed from an atom, electron–electron repulsion decreases, and the remaining valence electrons move closer to the nucleus. This increases the attractive force between the electrons and the nucleus, increasing the ionization energy.

As Electrons Are Removed, Ionization Energy Increases Gradually Until a Shell Is Empty, Then Makes a Big Jump

- For each element, when the valence shell is empty, the next electron must come from a shell that is much closer to the nucleus, making the ionization energy for that electron much larger than for the previous ones.
- For Na, the second ionization energy is much larger than the first.
- For Mg, the first and second ionization energies are comparable, but the third is much larger than the second.
- For Al, the first three ionization energies are comparable, but the fourth is much larger than the third.

Electronegativity

Electronegativity refers to how strongly the nucleus of an atom attracts the electrons of other atoms in a bond. Electronegativity is affected by two factors. The smaller an atom is, the more effectively its nuclear charge will be felt past its outermost energy level and the higher its electronegativity will be. Second, the closer an element is to having a full energy level, the more likely it is to attract the necessary electrons to complete that level. In general:

- Moving from left to right across a period, electronegativity increases.
- Moving down a group, electronegativity decreases.

The various periodic trends are summarized in the diagram below. The primary exception to these trends is the electronegativity values for the three smallest noble gases. As helium, neon, and argon do not form bonds, they have zero electronegativity. The larger noble gases, however, can form bonds under certain conditions and do follow the general trends as outlined in this section. This will be discussed in more detail in the next unit.

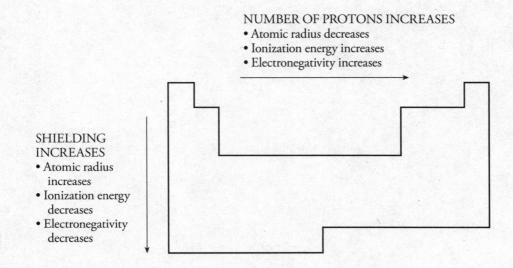

NUMBER OF PROTONS INCREASES
- Atomic radius decreases
- Ionization energy increases
- Electronegativity increases

SHIELDING INCREASES
- Atomic radius increases
- Ionization energy decreases
- Electronegativity decreases

Electron Affinity

Electron affinity describes the energy change that occurs when an electron is added to an atom in its gaseous state. Electron affinity is usually an exothermic process during which energy is released. The reaction showing the first electron affinity of a phosphorus atom would look like this:

$$P(g) + e^- \rightarrow P^-(g)$$

Note that unlike the other topics in this heading, the trends regarding electron affinity are very muddled indeed. There are a great number of variables that go into predicting electron affinity, many of which are beyond the scope of this exam. It is unlikely that the exam will ask for any information regarding electron affinity trends, but if it does, follow the same logic as you would for electronegativity trends.

UNIT 1 QUESTIONS

Multiple-Choice Questions

Use the PES spectrum below to answer questions 1–4.

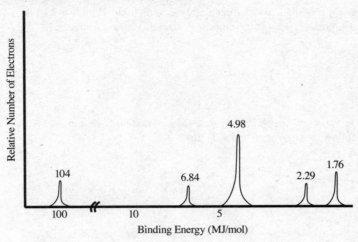

1. What element does this spectrum represent?

   (A) Boron
   (B) Nitrogen
   (C) Aluminum
   (D) Phosphorus

2. Which peak represents the 2*s* subshell?

   (A) The peak at 104 MJ/mol
   (B) The peak at 6.84 MJ/mol
   (C) The peak at 2.29 MJ/mol
   (D) The peak at 1.76 MJ/mol

3. An electron from which peak would have the greatest velocity after ejection?

   (A) The peak at 104 MJ/mol
   (B) The peak at 6.84 MJ/mol
   (C) The peak at 4.98 MJ/mol
   (D) The peak at 1.76 MJ/mol

4. How many valence electrons does this atom have?

   (A) 2
   (B) 3
   (C) 4
   (D) 5

5. Why does an ion of phosphorus, P^{3-}, have a larger radius than a neutral atom of phosphorus?

   (A) There is a greater Coulombic attraction between the nucleus and the electrons in P^{3-}.
   (B) The core electrons in P^{3-} exert a weaker shielding force than those of a neutral atom.
   (C) The nuclear charge is weaker in P^{3-} than it is in P.
   (D) The electrons in P^{3-} have a greater Coulombic repulsion than those in the neutral atom.

6. A compound is entirely made up of silicon and oxygen atoms. If there are 14.0 g of silicon and 32.0 g of oxygen present, what is the empirical formula of the compound?

   (A) SiO_2
   (B) SiO_4
   (C) Si_2O
   (D) Si_2O_3

7. The diagram below shows the relative atomic sizes of three different elements from the same period. Which of the following statements must be true?

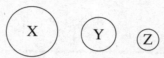

   (A) The effective nuclear charge will be the greatest in element X.
   (B) The first ionization energy will be greatest in element X.
   (C) The electron shielding effect will be greatest in element Z.
   (D) The electronegativity value will be greatest in element Z.

8. The first ionization energy for a neutral atom of chlorine is 1.25 MJ/mol and the first ionization energy for a neutral atom of argon is 1.52 MJ/mol. How would the first ionization energy value for a neutral atom of potassium compare to those values?

   (A) It would be greater than both because potassium carries a greater nuclear charge than either chlorine or argon.
   (B) It would be greater than both because the size of a potassium atom is smaller than an atom of either chlorine or argon.
   (C) It would be less than both because there are more electrons in potassium, meaning they repel each other more effectively and less energy is needed to remove one.
   (D) It would be less than both because a valence electron of potassium is farther from the nucleus than one of either chlorine or argon.

9. Neutral atoms of chlorine are bombarded by high-energy photons, causing the ejection of electrons from the various filled subshells. Electrons originally from which subshell would have the highest velocity after being ejected?

   (A) $1s$
   (B) $2p$
   (C) $3p$
   (D) $3d$

10. The average mass, in grams, of one mole of carbon atoms is equal to

   (A) the average mass of a single carbon atom, measured in amus
   (B) the ratio of the number of carbon atoms to the mass of a single carbon atom
   (C) the number of carbon atoms in one amu of carbon
   (D) the mass, in grams, of the most abundant isotope of carbon

11. Which of the following statements is true regarding sodium and chlorine?

    (A) Sodium has greater electronegativity and a larger first ionization energy.
    (B) Sodium has a larger first ionization energy and a larger atomic radius.
    (C) Chlorine has a larger atomic radius and greater electronegativity.
    (D) Chlorine has greater electronegativity and a larger first ionization energy.

12. Approximately how many neutrons are present in a 10 g sample of argon atoms which have a mass number of 40?

    (A) 3.3×10^{23}
    (B) 4.5×10^{23}
    (C) 3.3×10^{24}
    (D) 6.0×10^{24}

13. A photoelectron spectrum for which of the following atoms would show peaks at exactly three different binding energies?

(A)

4p
5n

(C)

11p
12n

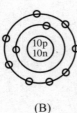

(B)

10p
10n

(D)

13p
14n

14. Which of the following can be inferred from examining mass spectroscopy data?

    (A) The common oxidation states of elements
    (B) Atomic size trends within the periodic table
    (C) Ionization energy trends within the periodic table
    (D) The existence of isotopes

15. In general, do metals or nonmetals from the same period have higher ionization energies? Why?

    (A) Metals have higher ionization energies because they usually have more protons than nonmetals.
    (B) Nonmetals have higher ionization energies because they are larger than metals and harder to ionize.
    (C) Metals have higher ionization energies because there is less electron shielding than there is in nonmetals.
    (D) Nonmetals have higher ionization energies because they are closer to having filled a complete energy level.

16. The ionization energies for an element are listed in the table below.

First	Second	Third	Fourth	Fifth
8 eV	15 eV	80 eV	109 eV	141 eV

Based on the ionization energy table, the element is most likely to be

(A) sodium
(B) magnesium
(C) aluminum
(D) silicon

17. Nitrogen's electronegativity value is between those of phosphorus and oxygen. Which of the following correctly describes the relationship between the three values?

(A) The value for nitrogen is less than that of phosphorus because nitrogen is larger, but greater than that of oxygen because nitrogen has a greater effective nuclear charge.
(B) The value for nitrogen is less than that of phosphorus because nitrogen has fewer protons, but greater than that of oxygen because nitrogen has fewer valence electrons.
(C) The value for nitrogen is greater than that of phosphorus because nitrogen has fewer electrons, but less than that of oxygen because nitrogen is smaller.
(D) The value for nitrogen is greater than that of phosphorus because nitrogen is smaller, but less than that of oxygen because nitrogen has a smaller effective nuclear charge.

18. Which of the following ions would have the most unpaired electrons?

(A) Mn^{2+}
(B) Ni^{3+}
(C) Ti^{2+}
(D) Cr^{6+}

19. Most transition metals share a common oxidation state of +2. Which of the following best explains why?

(A) Transition metals all have a minimum of two unpaired electrons.
(B) Transition metals have unstable configurations and are very reactive.
(C) Transition metals tend to gain electrons when reacting with other elements.
(D) Transition metals will lose their outermost s-block electrons when forming bonds.

Free-Response Questions

1. Explain each of the following in terms of atomic and molecular structures and/or forces.

    (a) The first ionization energy for magnesium is greater than the first ionization energy for calcium.

    (b) The first and second ionization energies for calcium are comparable, but the third ionization energy is much greater.

    (c) There are three peaks of equal height in the PES of carbon, but on the PES of oxygen the last peak has a height twice as high as all the others.

    (d) The first ionization energy for aluminum is lower than the first ionization energy for magnesium.

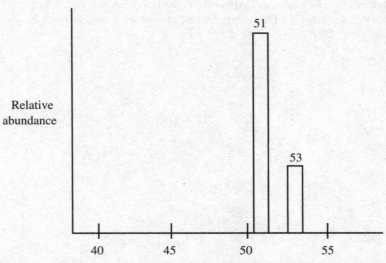

2. The above mass spectrum is for the hypochlorite ion, ClO⁻. Oxygen has only one stable isotope, which has a mass of 16 amu.

    (a) How many neutrons does the most common isotope of chlorine have?

    (b) Using the spectrum, calculate the average mass of a hypochlorite ion.

    (c) Does the negative charge on the ion affect the spectrum? Justify your answer.

    (d) The negative charge in the ion is located around the oxygen atom. Speculate as to why.

3. The table below gives data on four different elements, in no particular order:

Carbon, Oxygen, Phosphorus, and Chlorine

	Atomic radius (pm)	First Ionization Energy (kJ/mol^{-1})
Element 1	170	1,086.5
Element 2	180	1,011.8
Element 3	175	1,251.2
Element 4	152	1,313.9

(a) Which element is number 3? Justify your answer using both properties.

(b) What is the outermost energy level that has electrons in element 2? How many valence electrons does element 2 have?

(c) Which element would you expect to have the highest electronegativity? Why?

(d) How many peaks would the PES for element 4 have and what would the relative heights of those peaks be to each other?

4. The photoelectron spectrum of an element is given below:

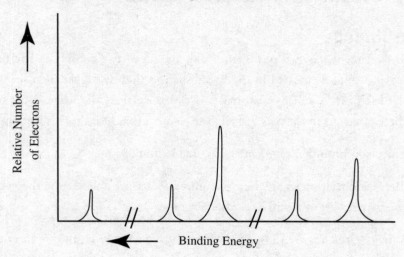

(a) Identify the element this spectrum most likely belongs to and write out its full electron configuration.

(b) Using your knowledge of atomic structure, explain the following:
    (i) The reason for the three discrete areas of ionization energies
    (ii) The justification for there being a total of five peaks
    (iii) The relative heights of the peaks when compared to one another

UNIT 1 ANSWERS AND EXPLANATIONS

Multiple-Choice

1. **D** This element has five peaks, meaning a total of five subshells. The final peak, which would be located in the $3p$ subshell, is slightly higher than the $3s$ peak to the left of it. A full $3s$ peak has two electrons; therefore, there must be at least three electrons in the $3p$ subshell. The element that best fits this is phosphorus.

2. **B** The peaks, in order, represent $1s$, $2s$, $2p$, $3s$, and $3p$.

3. **D** The less ionization energy that is required to remove an electron, the more kinetic energy that electron will have after ejection.

4. **D** Valence electrons are those in the outermost energy level. In this case, that is the third level, which has five valence electrons in it (two in $3s$ and three in $3p$).

5. **D** The ion has three more electrons than the neutral atom, meaning the overall repulsion will be greater. The electrons will "push" one another away more effectively, creating a bigger radius.

6. **B** 14.0 g of Si is 0.50 mol, and 32.0 g of oxygen is 2.0 mol. Converting that to a whole number ratio gives you 1 mol of Si for every 4 moles of O.

7. **D** Moving across a period, atomic size decreases. Therefore, element Z will be farthest to the right (have the most protons), and thus will have the highest electronegativity value.

8. **D** Potassium's first valence electron is in the fourth energy level, but both chlorine and argon's first valence electron is in the third energy level.

9. **C** The farther away an electron is from the nucleus, the less ionization energy that is required to eject it, and as a result, the electron will have more kinetic energy after it is ejected. The $3p$ subshell is the farthest one that a neutral chlorine atom would have electrons in. Beware of (D); chlorine does not have a $3d$ subshell.

10. **A** This is a straightforward concept to understand. The average atomic mass of an element on the periodic table measures the average mass of a mole of atoms of that element in grams, as well as being the average mass of a single atom of that element in amus.

11. **D** As you move from left to right across the periodic table within a single period (from sodium to chlorine), you add protons to the nuclei, which progressively increases the pull of each nucleus on its electrons. So chlorine will have a larger first ionization energy, greater electronegativity, and a smaller atomic radius.

12. **C** The 10 g sample contains (10 g)/(40 g/mol) = 0.25 mol of argon atoms. Argon's atomic number is 18, so each atom with a mass number of 40 has 22 neutrons. Multiplying, we get $(0.25 \text{ mol})(6.022 \times 10^{23} \text{ atoms/mol})(22 \text{ neutrons/atom}) \approx 33 \times 10^{23}$ neutrons = 3.3×10^{24} neutrons.

13. **B** Diagram B represents an atom of neon, which has three subshells: 1s, 2s, and 2p. Electrons from each subshell would have a different binding energy, yielding three peaks on a PES.

14. **D** Mass spectrometry is used to determine the masses for individual atoms of an element. Through mass spectrometry, it is proven that each element has more than one possible mass.

15. **D** Nonmetals appear on the right side of the periodic table, and so tend to be smaller than the other elements in their period. For this reason, it is easier for them to attract additional electrons, which means they have higher ionization energy values.

16. **B** The ionization energy will show a large jump when an electron is removed from a full shell. In this case, the jump occurs between the second and third electrons removed, so the element is stable after two electrons are removed. Magnesium (Mg) is the only element on the list with exactly two valence electrons.

17. **D** Nitrogen only has two shells of electrons, while phosphorus has three, making nitrogen smaller and more able to attract additional electrons, meaning a higher electronegativity. Nitrogen and oxygen both have two shells, but oxygen has more protons and an effective nuclear charge of +6 versus nitrogen's effective nuclear charge of +5. Thus, oxygen has a higher electronegativity.

18. **A** Remember, transition metals lose their s-electrons first when forming an ion. A manganese atom is initially [Ar]$4s^2 3d^5$, but the ion becomes [Ar]$3d^5$, which has a total of five unpaired electrons (as Hund's Rule states, the electrons will remain unpaired as long as there are empty orbitals for them to enter).

19. **D** The outermost s-block electrons in a transition metal tend to be lost before the d-block electrons. Additionally, the other options do not accurately describe the properties of transition metals.

Free-Response

1. (a) Ionization energy is the energy required to remove an electron from an atom. The outermost electron in Ca is at the $4s$ energy level. The outermost electron in Mg is at the $3s$ level. The outermost electron in Ca is at a higher energy level and is more shielded from the nucleus, making it easier to remove.

    (b) Calcium has two electrons in its outer shell. The second ionization energy will be larger than the first but still comparable because both electrons are being removed from the same energy level. The third electron is much more difficult to remove because it is being removed from a lower energy level, so it will have a much higher ionization energy than the other two.

    (c) The height of the peaks on a PES represents the relative number of electrons in each subshell. In carbon, all three subshells hold two electrons ($1s^2 2s^2 2p^2$), and thus all peaks are the same height. In oxygen, the $2p$ subshell has four electrons, meaning its peak will be twice as high as the other two.

    (d) The valence electron to be removed from magnesium is located in the completed $3s$ subshell, while the electron to be removed from aluminum is the lone electron in the $3p$ subshell. It is easier to remove the electron from the higher-energy $3p$ subshell than from the lower energy (completed) $3s$ subshell, so the first ionization energy is lower for aluminum.

2. (a) The most common mass of a ClO^- ion is 51 amu. 51 amu – 16 amu = 35 amu, which must be the mass of the most common isotope of chlorine. As mass number is equal to protons + neutrons, and chlorine has 17 protons (its atomic number), 35 – 17 = 18 neutrons.

    (b) 51(0.75) + 53(0.25) = 51.5 amu

    (c) No. The only subatomic particles that contribute to the mass of any atom are neutrons and protons. Changing the number of electrons does not change the mass significantly.

    (d) An oxygen atom is smaller than a chlorine atom, and as such is more electronegative. The electrons in the bond are thus more attracted to oxygen than chlorine.

3. (a) Element 3 is chlorine. Chlorine and phosphorus would have the largest atomic radii as they both have three energy levels with electrons present. However, chlorine would be smaller than phosphorus because it has more protons (a higher effective nuclear charge). Additionally, chlorine would have a higher ionization energy than phosphorus due to its smaller size and greater number of protons.

    (b) Element 2 is phosphorus, and therefore the outermost energy level would be $n = 3$. Phosphorus has two electrons in $3s$ and three electrons in $3p$, for a total of five valence electrons.

    (c) Electronegativity increases as atomic radius decreases, so it is expected that element 4 (oxygen) would have the highest electronegativity value. Alternatively, electronegativity increases as an energy level comes close to being full, so it is possible that element 3 (chlorine) may have the highest electronegativity, as it is only one electron away from filling its outermost energy level. (Either answer is acceptable with the proper justification.)

    (d) Element 4 is oxygen, so it would be expected to have three peaks in a PES, one for each subshell. The first two peaks would be the same height because there are two electrons each in the $1s$ and $2s$ subshells. The final peak would be twice the height of the others, as there are four electrons in oxygen's $2p$ subshell.

4. (a) This PES belongs to sulfur: $1s^2 2s^2 2p^6 3s^2 3p^4$.

    (b) (i) Each discrete area of ionization energy represents a different energy level of the electrons. The closer the electrons are to the nucleus, the more ionization energy will be required to remove the electrons. Sulfur has electrons present at three different energy levels; thus, there are three different areas for the peaks.

        (ii) Within each energy level (except for the first), there are subshells which are not the exact same distance from the nucleus. Both energy levels 2 and 3 have s and p subshells, and while the electrons in those shells will have similar ionization energy values, they will not be identical. Thus, the five peaks represent $1s$, $2s$, $2p$, $3s$, and $3p$.

        (iii) The heights of the peaks represent the ratio of electrons present in each of them. All three s peaks are exactly one-third the height of the $2p$ peak, meaning there are three times more electrons in $2p$ than in any of the s subshells. The $3p$ peak is only twice as high as the s peaks; therefore, there are twice as many electrons in $3p$ than in any of the s subshells.

Chapter 4
Unit 2: Molecular and Ionic Compound Structure and Properties

BONDS OVERVIEW

Atoms engage in chemical reactions in order to reach a more stable, lower-energy state. This requires the transfer or sharing of electrons, a process that is called **bonding**. Atoms of elements are usually at their most stable when they have eight electrons in their valence shells. As a result, atoms with too many or too few electrons in their valence shells will find one another and pass the electrons around until all the atoms in the molecule have stable outer shells. Sometimes an atom will give up electrons completely to another atom, forming an ionic bond. Sometimes atoms share electrons, forming covalent bonds.

IONIC BONDS

An ionic solid is held together by the electrostatic attractions between ions that are next to one another in a lattice structure. They often occur between metals and nonmetals. In an ionic bond, electrons are not shared. Instead, the cation gives up an electron (or electrons) to the anion.

The two ions in an ionic bond are held together by electrostatic forces. In the diagram below, a sodium atom has given up its single valence electron to a chlorine atom, which has seven valence electrons and uses the electron to complete its outer shell (with eight). The two atoms are then held together by the positive and negative charges on the ions.

$$\left[\mathrm{Na}\right]^{+}\left[:\ddot{\mathrm{Cl}}:\right]^{-}$$

The electrostatic attractions that hold together the ions in the NaCl lattice are very strong, and any substance held together by ionic bonds will usually be a solid at room temperature and have very high melting and boiling points.

Two factors affect the melting points of ionic substances. The primary factor is the charge on the ions. According to Coulomb's Law, a greater charge leads to a greater bond energy (often called **lattice energy** in ionic bonds), so a compound composed of ions with charges of +2 and −2 (such as MgO) will have a higher melting point than a compound composed of ions with charges of +1 and −1 (such as NaCl). If both compounds are made up of ions with equal charges, then the size of the ions must be considered. Smaller ions will have greater Coulombic attraction (remember, size is inversely proportional to bond energy), so a substance like LiF would have a greater melting point than KBr.

In an ionic solid, each electron is localized around a particular atom, so electrons do not move around the lattice; this makes ionic solids poor conductors of electricity. Ionic liquids, however, do conduct electricity because the ions themselves are free to move about in the liquid phase, although the electrons are still localized around particular atoms. Salts are held together by ionic bonds.

METALLIC BONDS

When examining metals, the sea of electrons model can be used. The positively charged core of a metal, consisting of its nucleus and core electrons, is generally stationary, while the valence electrons on each atom do not belong to a specific atom and are very mobile. These mobile electrons explain why metals are such good conductors of electricity. The delocalized structure of a metal also explains why metals are both malleable and ductile, as deforming the metal does not change the environment immediately surrounding the metal cores.

Metals can also bond with each other to form alloys. This typically occurs when two metals are melted into their liquid phases, and are then poured together before cooling and creating the alloy. In an **interstitial alloy,** metal atoms with two vastly different radii combine. Steel is one such example—the much smaller carbon atoms occupy the interstices of the iron atoms. A **substitutional alloy** forms between atoms of similar radii. Brass is a good example; atoms of zinc are substituted for some copper atoms to create the alloy.

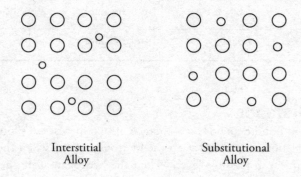

Interstitial
Alloy

Substitutional
Alloy

MOLECULAR COVALENT BONDING

In a covalent bond, two atoms share electrons. Each atom counts the shared electrons as part of its valence shell. In this way, both atoms achieve complete outer shells. Covalent bonds occur between nonmetal atoms.

In the diagram below, two fluorine atoms, each of which has seven valence electrons and needs one electron to complete its valence shell, form a covalent bond. Each atom donates an electron to the bond, which is considered to be part of the valence shell of both atoms.

$$\ddot{\ddot{\text{:}}}\text{F}\cdot \ + \ \cdot\ddot{\ddot{\text{F}}}\text{:} \ \Rightarrow \ \ddot{\ddot{\text{:}}}\text{F:}\ddot{\ddot{\text{F}}}\text{:}$$

When two or more atoms bond together by sharing valence electrons, they create a molecule. Molecules can be as small as just two atoms (as in F_2, above), and there is no upper limit on how many atoms may be in a molecule. One example of a larger molecule is glucose, which has the formula of $C_6H_{12}O_6$. That's 24 atoms in a single molecule, and even that is small compared to some of the very large molecules studied in organic chemistry. All molecules have a definite composition. For instance, every water molecule has exactly two hydrogen atoms and one oxygen atom. A given sample of water may have a LOT of molecules, but each molecule is self-contained when it comes to sharing bonding electrons.

> Single bonds have one sigma (σ) bond and a bond order of one. The single bond has the longest bond length and the least bond energy.

The first covalent bond formed between two atoms is called a sigma (σ) bond. All single bonds are sigma bonds. If additional bonds between the two atoms are formed, they are called pi (π) bonds. The second bond in a double bond is a pi bond and the second and third bonds in a triple bond are also pi bonds. Double and triple bonds are stronger and shorter than single bonds, but they are not twice or triple the strength.

Summary of Multiple Bonds			
Bond type:	Single	Double	Triple
Bond designation:	One sigma (σ)	One sigma (σ) and one pi (π)	One sigma (σ) and two pi (π)
Bond order:	One	Two	Three
Bond length:	Longest	Intermediate	Shortest
Bond energy:	Least	Intermediate	Greatest

Internuclear Distance

The length of a covalent bond depends on a balance of two forces. Let's look at a bond between two hydrogen atoms. A bond forms between the hydrogen atoms when the potential energy of that bond is at the minimum possible level. When the hydrogen atoms are too close together, the potential energy value is very high, because the nuclei repel each other. When the atoms are too far apart, the potential energy value is close to zero, because the protons in the nucleus of one hydrogen atom are unable to attract the electrons around the other (and vice-versa).

The minimum potential energy will occur when the repulsive and attractive forces are balanced. A graph of potential energy as a function of the distance between two hydrogen atoms is shown below.

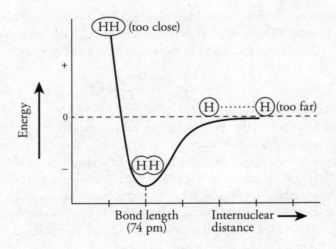

Note that the dashed line marks zero potential energy. However, in this case zero potential energy is not the lowest value. Whenever we are dealing with potential energy, the more negative that value is, the lower the energy is. This is the same concept as looking at the potential energy of electrons at varying distances from the nucleus, as we did on page 72.

In this case, the bond between the hydrogen atoms will form at the point that is lowest on the curve. That represents a distance of 74 picometers. The bond length of any covalent bond will always occur at the point of minimum potential energy.

Network Covalent Bonds

In a network solid, atoms are held together in a lattice of covalent bonds. You can visualize a network solid as one big molecule. Network solids are very hard and have very high melting and boiling points.

The electrons in a network solid are localized in covalent bonds between particular atoms, so they are not free to move about the lattice. This makes network solids poor conductors of electricity.

The most commonly seen network solids contain either carbon (such as diamond or graphite) and silicon (SiO_2—quartz). This is because both carbon and silicon have four valence electrons, meaning they are able to form a large number of covalent bonds.

CONDUCTIVITY

One way to differentiate the type of bonding in various compounds is by looking at whether or not a substance is a good conductor of electricity. The chart below sums up the conductivity of various substances in their different phases.

	Solid	Aqueous	Liquid	Gas
Ionic	No	Yes	Yes	No
Molecular Covalent	No	No	No	No
Network Covalent	No	N/A	No	No
Metallic	Yes	N/A	Yes	No

As you can see, covalent substances never conduct electricity. This includes water! Many students think that water is a good conductor, however, pure water is actually NOT a good conductor at all. That being said, most water that comes from the tap is anything but pure; there are a lot of dissolved ions floating around in the water, which is what gives the water its conductive properties.

When it comes to ionic substances, when they are in their solid phase, the ions are locked in place in the lattice, and no electrons can move around. However, when those ions are freed to move around in the liquid phase, electrons can transfer between them and thus the substance can conduct. Those ions are also free to move around when an ionic substance is dissolved in water (aqueous phase). How well aqueous ionic solutions conduct electricity is dependent on two factors: their concentration, and the number of ions they dissociate into.

When it comes to concentration, a 1.0 M solution of NaCl would be a better conductor than a 0.10 M solution of NaCl. There are simply more Na^+ and Cl^- ions present per unit of solution in the 1.0 M solution, and thus it would be a better conductor. If, on the other hand, you were comparing that 1.0 M NaCl to 1.0 M $CaCl_2$, you would have to consider what it looks like when both substances dissociate.

$$NaCl(s) \rightarrow Na^+(aq) + Cl^-(aq)$$
$$CaCl_2(s) \rightarrow Ca^{2+}(aq) + 2\ Cl^-(aq)$$

As you can see, the $CaCl_2$ dissociates into three ions per unit, whereas the NaCl only dissociates into two. Thus, a 1.0 M solution of $CaCl_2$ would be a better conductor than a 1.0 M solution of NaCl.

Note that while molecular covalent substances can also exist in the aqueous phase (sugar dissolved in water is an example), an aqueous solution of a molecular covalent compound would still not conduct electricity. Neither network covalent nor metallic substances would dissolve in water, thus the aqueous phase does not exist for those substances.

LEWIS DOT STRUCTURES

Drawing Lewis Dot Structures

At some point on the test, you'll be asked to draw the Lewis structure for a molecule or polyatomic ion. Here's how to do it.

1. Count the valence electrons in the molecule or polyatomic ion; refer to page 66 for the periodic table.
2. If a polyatomic ion has a negative charge, add electrons equal to the charge of the total in (1). If a polyatomic ion has a positive charge, subtract electrons equal to the charge of the electrons from the total in (1).
3. Draw the skeletal structure of the molecule and place two electrons (or a single bond) between each pair of bonded atoms. If the molecule contains three or more atoms, the least electronegative atom will usually occupy the central position.
4. Add electrons to the surrounding atoms until each has a complete outer shell.
5. Add the remaining electrons to the central atom.
6. Look at the central atom.
    (a) If the central atom has fewer than eight electrons, remove an electron pair from an outer atom and add another bond between that outer atom and the central atom. Do this until the central atom has a complete octet.
    (b) If the central atom has a complete octet, you are finished.
    (c) Some central atoms can have more than eight, but not more than twelve, electrons. We'll talk more about those on page 100.

Let's find the Lewis dot structure for the CO_3^{2-} ion.

1. Carbon has 4 valence electrons; oxygen has 6.
    4 + 6 + 6 + 6 = 22
2. The ion has a charge of –2, so add 2 electrons.
    22 + 2 = 24
3. Carbon is the central atom.

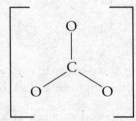

4. Add electrons to the oxygen atoms.

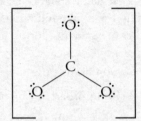

5. We've added all 24 electrons, so there's nothing left to put on the carbon atom.
6. (a) We need to give carbon a complete octet, so we take an electron pair away from one of the oxygens and make a double bond instead. Place a bracket around the model and add a charge of negative two.

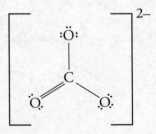

Resonance Forms

When we put a double bond into the CO_3^{2-} ion, we place it on any one of the oxygen atoms, as shown below.

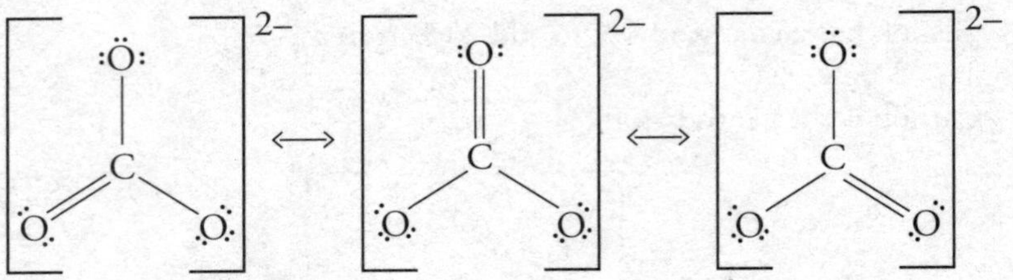

Don't Be Fooled by Appearances
Despite the conventional way the Lewis diagrams have been drawn, remember that the bonds shown here are actually identical. The only reason it looks like a mix of single and double bonds is because any attempt to illustrate a bond somewhere between these two strengths and lengths would be awkward.

The strength and length of all three bonds in the carbonate ion are the same: somewhere between the strength and length of a single bond and a double bond.

To determine the relative length and strength of a bond in a resonance structure, a bond order calculation can be used. A single bond has a bond order of 1, and a double bond has an order of 2. When resonance occurs, pick one of the bonds in the resonance structure and add up the total bond order across the resonance forms; then divide that sum by the number of resonance forms.

For example, in the carbonate ion above, the top C–O bond would have a bond order of $\frac{1+2+1}{3}$, or 1.33. Bond order can be used to compare the length and strength of any bonds, whether they exhibit resonance or not.

Incomplete Octets

Some atoms are stable with less than eight electrons in their outer shell. Hydrogen only requires two electrons, as does helium (although helium never forms bonds). Boron is considered to be stable with six electrons, as in the BF_3 diagram below. All other atoms involved in covalent bonding require a minimum of eight electrons to be considered stable.

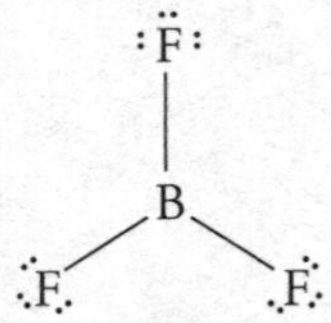

Expanded Octets

In molecules that have *d* subshells available, the central atom can have more than eight valence electrons, but never more than twelve. This means any atom of an element from $n = 3$ or greater can have expanded octets (especially common for Si, P, S, and Cl), but NEVER elements in $n = 2$ (C, N, O, etc.). Expanded octets also explains why some noble gases can actually form bonds; the extra electrons go into the empty *d*-orbital.

Here are some examples.

PCl₅

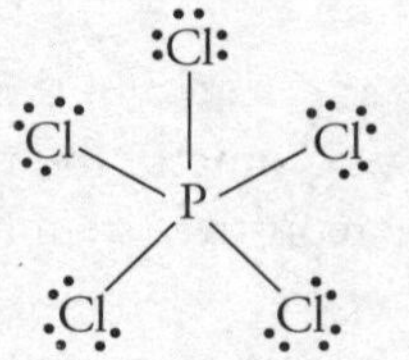

SF₄

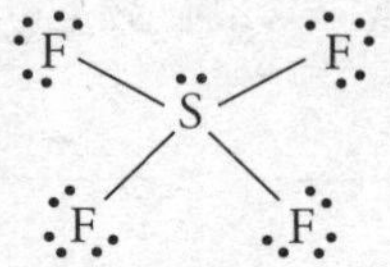

XeF₄

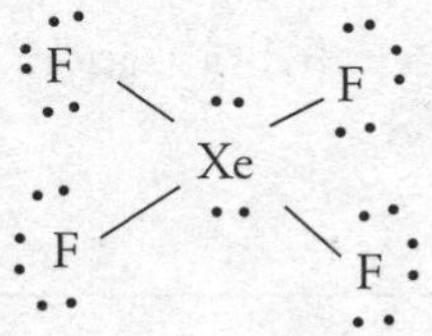

Formal Charge

Sometimes, there is more than one valid Lewis structure for a molecule. Take CO_2; it has two valid structures as shown below. To determine the more likely structure, a formal charge is used. To calculate the formal charge on atoms in a molecule, take the number of valence electrons for that atom and subtract the number of assigned electrons in the Lewis structure. When counting assigned electrons, lone pairs count as two and bonds count as one.

6	4	6	valence e^-	6	4	6
− 6	4	6	assigned e^-	− 7	4	5
0	0	0	formal charge	−1	0	+1

The total formal charge for a neutral molecule should be zero, which it is on both diagrams. Additionally, the fewer number of atoms there are with an actual formal charge, the more likely the structure will be—so the left structure is the more likely one for CO_2. For polyatomic ions, the sum of the formal charges on each atom should equal the overall charge on the ion.

Molecular Geometry

Electrons repel one another, so when atoms come together to form a molecule, the molecule will assume the shape that keeps its different electron pairs as far apart as possible. When we predict the geometries of molecules using this idea, we are using the valence shell electron pair repulsion (VSEPR) model.

In a molecule with more than two atoms, the shape of the molecule is determined by the number of electron pairs on the central atom. The central atom forms hybrid orbitals, each of which has a standard shape. Variations on the standard shape occur depending on the number of bonding pairs and lone pairs of electrons on the central atom.

Here are some things you should remember when dealing with the VSEPR model.

- Double and triple bonds are treated in the same way as single bonds in terms of predicting overall geometry for a molecule; however, multiple bonds have slightly more repulsive strength and will therefore occupy a little more space than single bonds.
- Lone electron pairs have a little more repulsive strength than bonding pairs, so molecules with lone pairs will have slightly reduced bond angles between terminal atoms.

The following pages show the different hybridizations and geometries that you might see on the test.

If the central atom has 2 electron pairs, then it has sp hybridization and its basic shape is **linear**.

Number of lone pairs	Geometry	Examples
0	B–A–B	$BeCl_2$
	linear	CO_2

If the central atom has 3 electron pairs, then it has sp^2 hybridization and its basic shape is **trigonal planar**; its bond angles are about 120°.

The angle between the terminal atoms in the bent shape is slightly less than 120° because of the extra lone pair repulsion.

Number of lone pairs	Geometry	Examples
0	B | A / \\ B B trigonal planar	BF_3 SO_3 NO_3^- CO_3^{2-}
1	¨ | A / \\ B B bent	SO_2

If the central atom has 4 electron pairs, then it has sp^3 hybridization and its basic shape is **tetrahedral**; its bond angles are about 109.5°.

Number of lone pairs	Geometry	Examples
0	tetrahedral	CH_4 NH_4^+ ClO_4^- SO_4^{2-} PO_4^{3-}
1	trigonal pyramidal	NH_3 PCl_3 AsH_3 SO_3^{2-}
2	bent	H_2O OF_2 NH_2^-

The angle between the terminal atoms in the trigonal pyramidal and bent shapes is slightly less than 109.5° because of the extra lone pair repulsion.

If the central atom has 5 electron pairs, its basic shape is **trigonal bipyramidal**.

Number of lone pairs	Geometry	Examples
0		PCl_5
		PF_5
	trigonal bipyramidal	
1		SF_4
		IF_4^+
	folded square, seesaw, distorted tetrahedron	
2		ClF_3
		ICl_3
	T-shaped	
3		XeF_2
		I_3^-
	linear	

If the central atom has 6 electron pairs, its basic shape is **octahedral**.

Number of lone pairs	Geometry	Examples
0	octahedral	SF_6
1	square pyramidal	BrF_5 IF_5
2	square planar	XeF_4 ICl_4^-

UNIT 2 QUESTIONS

Multiple-Choice Questions

1. Why can a molecule with the structure of NBr_5 not exist?

   (A) Nitrogen only has two energy levels and is thus unable to expand its octet.
   (B) Bromine is much larger than nitrogen and cannot be a terminal atom in this molecule.
   (C) It is impossible to complete the octets for all six atoms using only valence electrons.
   (D) Nitrogen does not have a low enough electronegativity to be the central atom of this molecule.

2. Which of the following compounds would have the highest lattice energy?

   (A) LiF
   (B) $MgCl_2$
   (C) $CaBr_2$
   (D) C_2H_6

3. Lewis diagrams for the nitrate and nitrite ions are shown below. Choose the statement that correctly describes the relationship between the two ions in terms of bond length and bond energy.

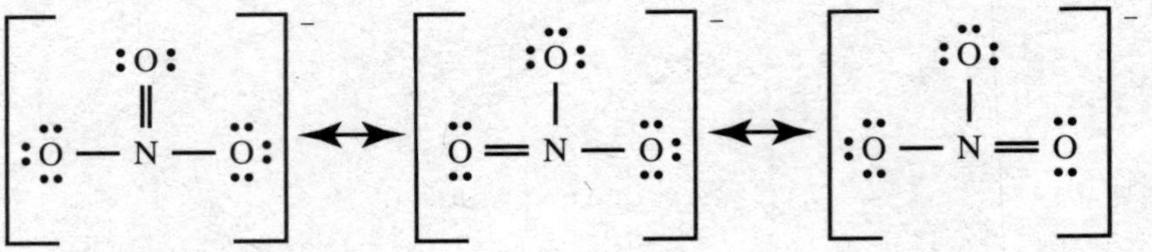

   Nitrate

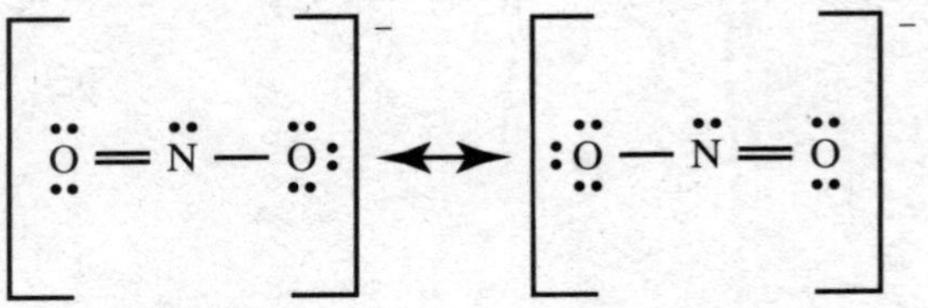

   Nitrite

   (A) Nitrite has longer and stronger bonds than nitrate.
   (B) Nitrite has longer and weaker bonds than nitrate.
   (C) Nitrite has shorter and stronger bonds than nitrate.
   (D) Nitrite has shorter and weaker bonds than nitrate.

4. The graph below shows the amount of potential energy between two hydrogen atoms as the distance between them changes. At which point in the graph would a molecule of H_2 be the most stable?

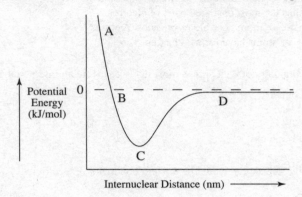

(A) Point A
(B) Point B
(C) Point C
(D) Point D

Use the following information to answer questions 5–8.

There are several different potential Lewis diagrams for the sulfate ion, two of which are below.

Structure A   Structure B

5. What is the molecular geometry in Structure A?

(A) Tetrahedral
(B) Trigonal Planar
(C) Trigonal Pyramidal
(D) Octahedral

6. What is the S–O bond order in Structure B?

(A) 1.0
(B) 1.33
(C) 1.5
(D) 1.67

7. Which of the following statements regarding Structure B is true?

(A) The double bonds must be located opposite each other due to additional electron repulsion.
(B) There are fewer sigma bonds compared to Structure A.
(C) The bonds in the molecule are weaker than those in Structure A.
(D) All bonds in the molecule are identical to each other.

8. Which structure is more likely to correspond with the actual Lewis diagram for the sulfate ion?

    (A) Structure A; single bonds are more stable than double bonds
    (B) Structure A; it has the most unshared pairs of electrons
    (C) Structure B; there are more possible resonance structures
    (D) Structure B; fewer atoms have formal charges

9. Which of the following pairs of elements is most likely to create an interstitial alloy?

    (A) Titanium and copper
    (B) Aluminum and lead
    (C) Silver and tin
    (D) Magnesium and calcium

10. Which of the following substances would be the best conductor of electricity?

    (A) 1.0 M KF
    (B) 1.0 M CCl_4
    (C) 1.0 M BaS
    (D) 1.0 M $SrBr_2$

11. An unknown substance is found to have a high melting point. In addition, it is a poor conductor of electricity and does not dissolve in water. The substance most likely contains

    (A) ionic bonding
    (B) nonpolar covalent bonding
    (C) covalent network bonding
    (D) metallic bonding

12.

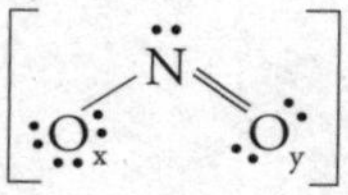

    One of the resonance structures for the nitrite ion is shown above. What is the formal charge on each atom?

	O_x	N	O_y
(A)	−1	+1	−1
(B)	+1	−1	0
(C)	0	0	−1
(D)	−1	0	0

Use the following information to answer questions 13–15.

    Consider the Lewis structures for the following molecules:

    CO_2, CO_3^{2-}, NO_2^-, and NO_3^-

13. Which molecule would have the shortest bonds?

    (A) CO_2
    (B) CO_3^{2-}
    (C) NO_2^-
    (D) NO_3^-

14. Which molecule or molecules exhibit sp^2 hybridization around the central atom?

    (A) CO_2 and $CO_3{}^{2-}$
    (B) $NO_2{}^-$ and $NO_3{}^-$
    (C) $CO_3{}^{2-}$ and $NO_3{}^-$
    (D) $CO_3{}^{2-}$, $NO_2{}^-$, and $NO_3{}^-$

15. Which molecule would have the smallest bond angle between terminal atoms?

    (A) CO_2
    (B) CO_3^{2-}
    (C) NO_2^-
    (D) NO_3^-

Free-Response Questions

1. The ion below, known as an enolate, has a formula of $C_2H_3O^-$ and two possible Lewis electron-dot diagram representations:

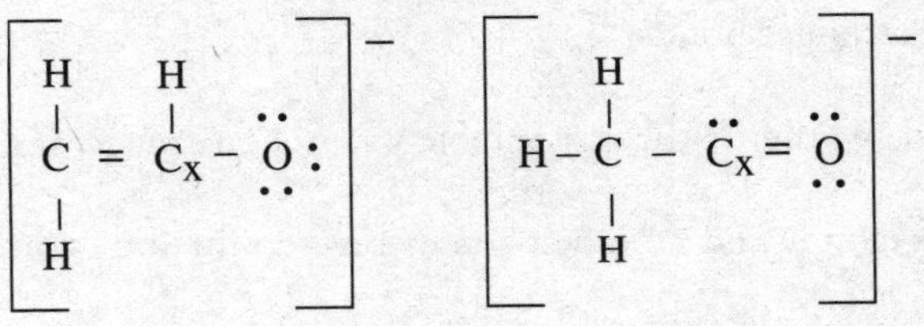

    (a) Using formal charge, determine which structure is the most likely correct structure.
    (b) For carbon atom "x" in the structure you chose:
         (i)   What is the hybridization around the atom?
         (ii) How many sigma and pi bonds has the atom formed?
    (c) A hydrogen ion attaches itself to the enolate ion, creating C_2H_4O. Draw the Lewis diagram of the new molecule.

UNIT 2 ANSWERS AND EXPLANATIONS

Multiple-Choice

1. **A**   Only atoms with at least three energy levels ($n = 3$ and above) have empty *d*-orbitals that additional electrons can fit into, thus expanding their octet.

2. **B**   First, C_2H_6 is not an ionic substance and thus has no lattice energy. Next, LiF is composed of ions with charges +1 and −1, and will not be as strong as the two compounds that have ions with charges of +2 and −1. Finally, $MgCl_2$ is smaller than $CaBr_2$, meaning it will have a higher lattice energy, as (according to Coulomb's Law) atomic radius is inversely proportional to bond energy.

3. **C**   The N-O bonds in nitrate have a bond order of $\dfrac{(2 + 1 + 1)}{3}$ = 1.33. The N-O bonds in nitrite have a bond order of $\dfrac{(2 + 1)}{2}$ = 1.5. A higher bond order means shorter and stronger bonds.

4. **C**   The molecule would be the most stable when it has the least potential energy.

5. **A**   Four charge clouds and no lone pairs means tetrahedral geometry.

6. **C**   Six total bonds divided by four locations gives a bond order of 1.5.

7. **D**   Each bond can be written as either a single or double an equal number of times; they all have the same bond order and are indistinguishable.

8. **D**   The formal charge tables for each diagram are below (note: for Structure B, the double-bonded oxygens are the first two, and the single bonded are the last two).

Structure A						Structure B				
S	O	O	O	O		S	O	O	O	O
6	6	6	6	6	**Valence e^-**	6	6	6	6	6
4	7	7	7	7	**Assigned e^-**	6	6	6	7	7
+2	−1	−1	−1	−1	**Formal charge**	0	0	0	−1	−1

    The total formal charge on each potential structure is −2, which is correct, as that is the charge on a sulfate ion. However, the right-hand structure has fewer atoms with formal charges, making it the more likely structure.

9. **B**   Interstitial alloys form when atoms of greatly different sizes combine. The aluminum atoms would have a chance to fit between the comparatively larger lead atoms.

10. **D** $SrBr_2$ will dissociate into three ions: one Sr^{2+} and two Br^-. Both the KF and BaS only dissociate into two ions, and the CCl_4 is a covalent substance (two nonmetals!) and thus would not conduct at all.

11. **C** An ionic substance would dissolve in water, and a nonpolar covalent substance would have a low melting point. A metallic substance would be a good conductor. The only type of bonding that meets all the criteria is covalent network bonding.

12. **D** O_x has 6 valence electrons and 7 assigned electrons: $6 - 7 = -1$. Both O_y and the N atoms have the same number of valence and assigned electrons, making their formal charges zero.

13. **A** The following Lewis structures are necessary to answer questions 13–15:

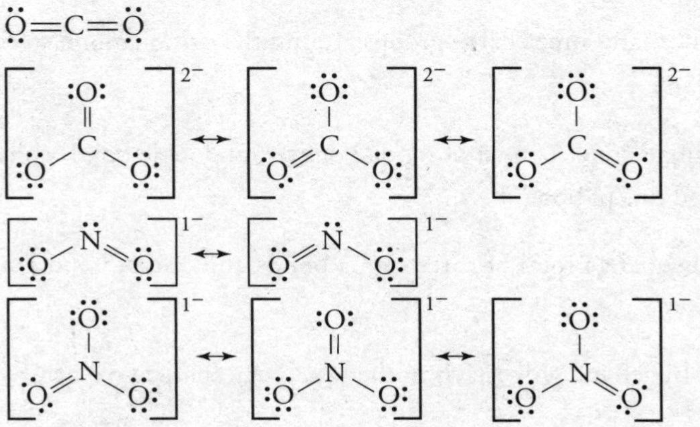

   CO_2 has a bond order of 2, which exceeds the order of the other structures. Remember that a higher bond order corresponds with shorter and stronger bonds.

14. **D** All three of those structures have three electron domains, and thus sp^2 hybridization.

15. **C** The bond angle of NO_2^- would be less than that of NO_3^- or CO_3^{2-} because the unbonded pair of electrons on the nitrogen atom reduces the overall bond angle.

Free-Response

1. (a) For this formal charge calculation, the H atoms are left out as they are identically bonded/drawn in both structures:

	C	C_x	O		C	C_x	O
Valence	4	4	6		4	4	6
Assigned	−4	4	7		−4	5	6
Formal Charge	0	0	−1		0	−1	0

As oxygen is more electronegative than carbon, an oxygen atom is more likely to have the negative formal charge than a carbon atom. The left-hand structure is most likely correct.

(b) (i) There are three charge groups around the carbon atom, so the hybridization is sp^2.

(ii) Single bonds consist of sigma bonds, and double bonds consist of one sigma and one pi bond.

There are a total of three sigma bonds and one pi bond around the carbon atom.

(c) The hydrogen ion will attach to the negatively charged oxygen.

$$
\begin{array}{ccc}
\text{H} & \text{H} & \\
| & | & \\
\text{C} = \text{C} - \ddot{\text{O}} - \text{H} & & \\
| & & \\
\text{H} & &
\end{array}
$$

Chapter 5
Unit 3: Intermolecular Forces and Properties

POLARITY

When a covalent bond forms between two atoms, electrons are shared. However, just because electrons are shared does not mean they are shared equally. For instance, when an atom of chlorine and an atom of carbon bond, the chlorine atom has a stronger pull on the shared electrons due to its higher electronegativity. That means, in the bond between the carbon and the chlorine, the shared electrons (which are constantly moving) will spend more time around the chlorine atom than the carbon one. This type of covalent bond, in which electrons are unequally shared, is called a **polar covalent** bond. This unequal sharing of electrons results in a partial electric charge on each atom.

This causes something called dipoles. A **dipole** is a pair of opposite electric charges separated by some distance, like the partial charges on atoms in a polar covalent bond. In the diagram below, the carbon has a positive partial charge, and the chlorine has a negative partial charge. Note that partial charges are indicated by the lowercase version of the Greek letter delta.

$$\overset{\delta^+}{C} - \overset{\delta^-}{Cl}$$

Note that if two identical atoms bond (such as in the Cl_2 molecule), the electrons can be equally shared, creating a **nonpolar covalent** bond in which no dipole is present.

Molecular Polarity

In addition to individual bonds being polar, entire molecules can be polar as well. A molecule's polarity depends on the polarity of its bonds, but even more importantly, on its overall molecular geometry. Take the carbon tetrafluoride molecule, CF_4, below.

$$\ddot{\underset{..}{F}} - C - \ddot{\underset{..}{F}}$$

Even though the C-F bond is polar, there are four of them and the molecule is shaped in such a way that the direction of each dipole cancels. Think of it like a giant electron tug-of-war: if each of the four fluorine atoms is pulling on the electrons surrounding carbon with equal strength, all four forces would cancel out directionally, leaving the electrons in the same place. Thus, CF_4 is a nonpolar molecule.

But what if one of those fluorine atoms simply wasn't there? Take NF_3, for example. There are still three fluorine atoms pulling on nitrogen's electrons; however, the fourth possible bond location remains unfilled, instead housing a lone pair that belongs entirely to the nitrogen atom. In this case, because fluorine is more electronegative than nitrogen, each fluorine atom will gain a negative partial charge, and the nitrogen atom will gain a positive partial charge, as shown below.

$$\overset{\delta^-}{\underset{..}{\ddot{F}}} - \overset{\delta^+}{N} - \overset{\delta^-}{\underset{..}{\ddot{F}}}$$

Whether or not a molecule is polar is highly dependent on its molecular geometry and makeup. As a general rule of thumb, if the central atom has a lone pair and is different from the surrounding atoms, the resulting molecule is asymmetric and will usually be polar. If there are no lone pairs on the central atom, and the atoms around it are all the same, the molecule is symmetric and the resulting molecule will usually be nonpolar.

A few common exceptions to "lone pairs on the central atom make things polar" are trigonal bipyramidal with three lone pairs or octahedral molecules with two lone pairs, known as linear or square planar, respectively. XeF_2 is linear, and as such the fluorine atoms pull on the electrons equally, but in opposite directions. $BrCl_4^-$ is square planar, and all the chlorine atoms are in the same plane, so those dipoles cancel out completely.

Also, because Lewis diagrams are specifically drawn with the LEAST electronegative atom in the center (that is, the atom least likely to attract electrons), in polar molecules the central atom generally has a positive partial charge, while the terminal atoms have negative partial charges. An exception to this is molecules that have hydrogen as a terminal atom. Hydrogen's electronegativity value is low enough that it will usually have a positive partial charge in a polar molecule, such as in the water molecules below.

The term "exception" sure did get used a lot in this section, but when it comes down to it, understanding polarity is about combining the concepts of electronegativity and molecular geometry and figuring out the end result. While the above section gives some good quick-and-dirty tips on figuring out molecular polarity, the best way of learning this concept is to fully understand how those two concepts interact and then applying that knowledge.

INTERMOLECULAR FORCES

Intermolecular forces (IMFs) are the forces that exist between molecules in a covalently bonded substance. These forces are what need to be broken apart in order for covalent substances to change phases. Note that when ionic substances change phase, bonds between the individual ions are actually broken. When covalent substances change phase, the bonds between the individual atoms remain in place; it is just the forces that hold the molecules to other molecules that break apart.

Dipole–Dipole Forces

Dipole–dipole forces occur between polar molecules: the positive end of one polar molecule is attracted to the negative end of another polar molecule.

Molecules with greater polarity will have greater dipole–dipole attraction, so molecules with larger **dipole moments**—which refers to the measurement of the strength of electrical dipoles—tend to have higher melting and boiling points. Dipole–dipole attractions are relatively weak, however, and these substances melt and boil at very low temperatures. Most substances held together by dipole–dipole attraction are gases or liquids at room temperature.

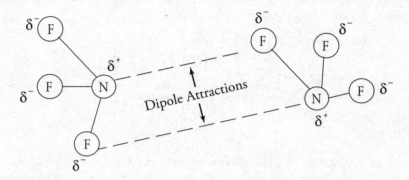

Hydrogen Bonds

Hydrogen bonds are a special type of dipole–dipole attraction. In a hydrogen bond, the positively charged hydrogen end of a molecule is attracted to the negatively charged end of another molecule containing an extremely electronegative element (fluorine, oxygen, or nitrogen—F, O, N).

Hydrogen bonds are much stronger than normal dipole–dipole forces because when a hydrogen atom gives up its lone electron to a bond, its positively charged nucleus is left virtually unshielded. Substances that have hydrogen bonds, such as water and ammonia, have higher melting and boiling points than substances that are held together only by other types of intermolecular forces.

Water is less dense as a solid than as a liquid because its hydrogen bonds force the molecules in ice to form a crystal structure, which keeps them farther apart than they are in the liquid form.

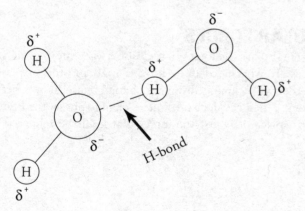

London Dispersion Forces

London dispersion forces occur between all molecules. These very weak attractions occur because of the random motions of electrons on atoms within molecules. At a given moment, a nonpolar molecule might have more electrons on one side than on the other, giving it an instantaneous polarity. For that fleeting instant, the molecule will act as a very weak dipole.

Since London dispersion forces depend on the random motions of electrons, molecules with more electrons will experience greater London dispersion forces. So among substances that experience only London dispersion forces, the one with more electrons will generally have higher melting and boiling points. London dispersion forces are even weaker than dipole–dipole forces, so substances that experience only London dispersion forces melt and boil at extremely low temperatures and tend to be gases at room temperature.

As molecules gain more electrons, the London dispersion forces between them start to become much more significant. Comparing the boiling point of a nonpolar substance with a large number of electrons versus a polar substance with fewer electrons is difficult, and there is no simple rule to follow. For instance, water has hydrogen bonds and a boiling point of 100°C. Butane (C_4H_{10}) and octane (C_8H_{18}) are both completely nonpolar molecules, and while butane's boiling point is 34°C, octane's is 125°C. Even though octane has no permanent dipoles, it has so many electrons that its London dispersion forces are significant enough that they create greater intermolecular attractions than even the hydrogen bonds in water.

The role of London dispersion forces is often determined by comparing the molar mass of molecules. However, it is not the mass itself which affects the strength of the IMFs. Rather, it is simply that as mass (based on protons and neutrons) increases, so too do the number of electrons, as the molecule must remain electrically neutral.

IMF Strength

Ionic substances are generally solids at room temperature, and turning them into liquids (melting them), requires the bonds holding the lattice together to be broken. The amount of energy needed for that is based on the Coulombic attraction between the molecules.

Covalent substances, which are liquid at room temperature, will boil when the intermolecular forces between them are broken. For molecules with similar sizes, the following IMF ranking (from strongest to weakest) can help you determine the relative strength of the IMFs within the molecules.

    a. Hydrogen bonds
    b. Non-hydrogen bond permanent dipoles
    c. London dispersion forces (temporary dipoles)
       i. Larger molecules are more polarizable and have stronger London
         dispersion forces because they have more electrons.

The melting and boiling points of covalent substances are almost always lower than the melting and boiling points of ionic ones.

Metallic bonding, which often only involves one type of atom, tends to be very strong and thus metals (particularly the transition metals) tend to have high melting points. Network covalent bonding is the strongest type of bonding there is, and it is very difficult to cause substances that exhibit network covalent bonding to melt.

Bonding and Phases

The phase of a substance is directly related to the strength of its intermolecular forces. Solids have highly ordered structures where the atoms are packed tightly together, while gases have atoms spread so far apart that most of the volume is free space.

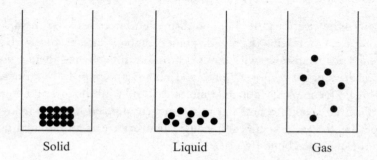

Solid   Liquid   Gas

In other words, substances that exhibit weak intermolecular forces (such as London dispersion forces) tend to be gases at room temperature. Nitrogen (N_2) is an example of this. Substances that exhibit strong intermolecular forces (such as hydrogen bonds) tend to be liquids at room temperature. A good example is water.

Ionic substances do not experience intermolecular forces. Instead, their phase is determined by the ionic bond holding the ions together in the lattice. Because ionic bonds are generally significantly stronger than intermolecular forces in covalent molecules, ionic substances are usually solid at room temperature.

VAPOR PRESSURE

Beyond helping to determine the melting point and boiling point of covalent substances, the relative strength of the intermolecular forces in a substance can also predict several other properties of that substance. The most important of these is vapor pressure. Vapor pressure arises from the fact that the molecules inside a liquid are in constant motion. If those molecules hit the surface of the liquid with enough kinetic energy, they can escape the intermolecular forces holding them to the other molecules and transition into the gas phase.

This process is called vaporization. It is not to be confused with a liquid boiling. When a liquid boils, energy (in the form of heat) is added, increasing the kinetic energy of all of the molecules in the liquid until all of the intermolecular forces are broken. For vaporization to occur, no outside energy needs to be added. Note that there is a direct relationship between temperature and vapor pressure. The higher the temperature of a liquid, the faster the molecules are moving and the more likely they are to break free of the other molecules. So, temperature and vapor pressure are directly proportional.

If two liquids are at the same temperature, the vapor pressure is dependent primarily on the strength of the intermolecular forces within that liquid. The stronger those intermolecular forces are, the less likely it is that molecules will be able to escape the liquid, and the lower the vapor pressure for that liquid will be.

SOLUTION SEPARATION

You can use intermolecular forces and the various Coulombic attractions that occur between ions and polar molecules in order to help separate various substances out from each other. There are several ways to do this.

Solutes and Solvents

There is a basic rule for remembering which solutes will dissolve in which solvents.

> Like dissolves like.

That means that polar or ionic solutes (such as salt) will dissolve in polar solvents (such as water). That also means that nonpolar solutes (such as oils) are best dissolved in nonpolar solvents. When an ionic substance dissolves, it breaks up into ions. That's dissociation. Free ions in a solution are called electrolytes because they can conduct electricity.

Paper Chromatography

Chromatography is the separation of a mixture by passing it in solution through a medium in which the components of the solution move at different rates. There are several major types of chromatography. The first is paper chromatography, in which paper is the medium through which the solution passes.

Many chemical solutions, such as the ink found in most pens, are a mixture of a number of covalent substances. Each of these substances has its own polarity value, and thus has a different affinity depending on the solvent. One of the most common paper chromatography experiments involves the separation of pigments in black ink. Black ink is usually made up of substances of several different colors, which when combined create black.

In paper chromatography, a piece of filter paper is suspended above a solvent so that the very bottom of the paper is touching the solvent. The ink in question is dotted onto a line at the bottom of the filter paper that starts out just above the solvent level. As the solvent climbs the paper, the various substances inside the ink will be attracted to the polar water molecules. The more polar the substance is, the more it will be attracted to the water molecules, and the further it will travel. You might end up with something that looks like this:

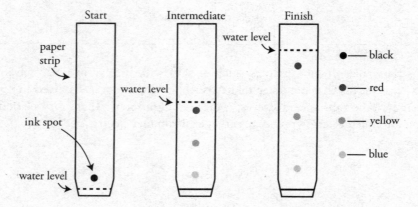

Looking at that strip, you can conclude that the ink was made of three different substances. The one that traveled the farthest with the water (the red pigment) experienced the strongest attractions and was the most polar, whereas the one that didn't travel very far from the original starting line (the blue pigment) was the least polar. Paper chromatography is the most useful with colored substances, which is why ink is used in the above example. If there were components to the ink that had no visible color, you would not be able to see them on the filter paper, and that is one major limitation of paper chromatography.

The distance the ink travels along the paper is measured via the retention (or retardation) factor, also known as the R_f value. The R_f value is calculated as such:

$$R_f = \frac{\text{Distance traveled by solute}}{\text{Distance traveled by solvent front}}$$

The stronger the attraction between the solute and the solvent front is, the larger the R_f value will be. In the diagram above, the red pigment would have the highest R_f value.

Water is not the only solvent that can be used in polar chromatography. There are many nonpolar solvents (such as cyclohexane) that can be used instead. In the case of a nonpolar solvent, the position of the various ink components in the above diagram would have been reversed—the most nonpolar substance would travel the furthest, and the most polar substance would travel the least.

Column Chromatography

Another type of chromatography is column chromatography. In this process, a column is packed with a stationary substance. Then, the solution to be separated (the analyte) is injected into the column, where it adheres to the stationary phase. After that, another solution (called the eluent) is injected into the column. As the eluent passes through the stationary phase, the analyte molecules will be attracted to it with varying degrees of strength depending on their polarity. The more attracted certain analyte molecules are to the eluent, the faster they will travel through ("elute") and leave the column.

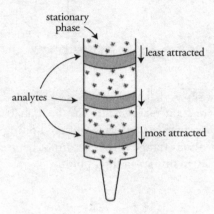

The speed at which the substances move through the column can be monitored, and if there is a sufficient polarity difference between the components, they will leave the column at different times, allowing them to be separated. Generally, after collecting the eluted mixture, it can be analyzed for compositional analysis via a variety of methods.

In column chromatography, either liquids or gases can be used as the eluent, depending on the situation.

Distillation

A third method for separating solutions is distillation. Distillation takes advantage of the different boiling points of substances in order to separate them. For instance, if you have a mixture of water (BP: 100°C) and methanol (BP: 65°C) and then heat that mixture to 70°C, the methanol will boil but the water will not.

A condenser is a piece of glassware that consists of a smaller tube running through a larger tube. The larger tube has hose connections on it, allowing for water to be run through it. This effectively cools the inner tube. At that point, the vapor, when run through the inner tube, will cool and condense back into a liquid form, which can be collected on the other side of the condenser.

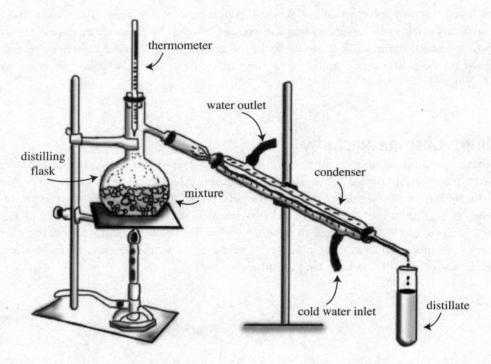

A major advantage to distillation is that the solutions need not be colored at all to separate them. Keeping the flask at a constant temperature can be a challenge, which is why the temperature must be monitored closely to ensure that you are boiling only one component of the mixture at a time. The biggest disadvantage is that it cannot be used to separate a mixture that contains substances with extremely similar boiling points.

KINETIC MOLECULAR THEORY

For ideal gases, the following assumptions can be made:

- The kinetic energy (*KE*) of an ideal gas is directly proportional to its absolute temperature: the greater the temperature, the greater the average kinetic energy of the gas molecules.

The Kinetic Energy of a Single Gas Molecule

$$KE = \frac{1}{2}mv^2$$

m = mass of the molecule (kg)
v = speed of the molecule (m/s)
KE is measured in J

- If several different gases are present in a sample at a given temperature, all the gases will have the same average kinetic energy. That is, the average kinetic energy of a gas depends only on the absolute temperature, not on the identity of the gas.

- The volume of an ideal gas particle is insignificant when compared with the volume in which the gas is contained.

- There are no forces of attraction between the gas molecules in an ideal gas.

- Gas molecules are in constant motion, colliding with one another and with the walls of their container without losing any energy.

MAXWELL-BOLTZMANN DIAGRAMS

A Maxwell-Boltzmann diagram shows the range of velocities for molecules of a gas. Molecules at a given temperature are not all moving at the same velocity. When determining the temperature, we take the average velocity of all the molecules and use that in the relevant equation to calculate temperature. You do not need to know that equation (unless you are taking AP Physics!). All you need to know here is that temperature is directly proportional to kinetic energy.

The first type of Maxwell-Boltzmann diagram involves plotting the velocity distributions for the molecules of one particular gas at multiple temperatures. In the diagram below, there are three curves representing a sample of nitrogen gas at 100 K, 300 K, and 500 K.

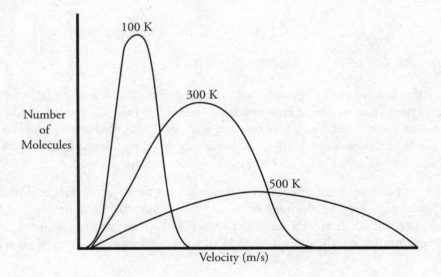

More Great Books
If you wish to dig deeper into many of these topics, check out these other books from The Princeton Review:
AP Physics 1 Prep
AP Physics 2 Prep
AP Physics C Prep

As you can see, the higher the temperature of the gas, the larger the range is for the velocities of the individual molecules. Gases at higher temperatures have greater kinetic energy (*KE*), and as all the molecules in this example have the same mass, the increased *KE* is due to the increased velocity of the gas molecules.

Maxwell-Boltzmann diagrams are also used to show a number of different gases at the same temperature. The diagram below shows helium, argon, and xenon gas, all at 300 K:

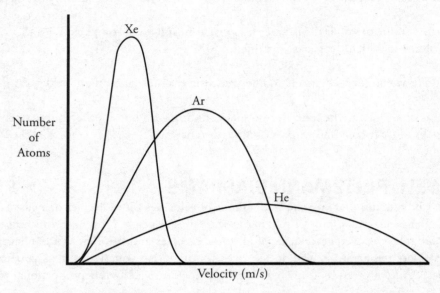

In this case, all of the gases have the same amount of total kinetic energy because they have identical temperatures. However, not all of the atoms have the same mass. If the atoms have smaller masses, they must have greater velocities in order to have a kinetic energy identical to that of atoms with greater mass. Because helium atoms have the least mass, they have the highest average velocity. Xenon atoms, which have a much greater mass, have a correspondingly lower velocity.

EFFUSION

Effusion is the rate at which a gas will escape from a container through microscopic holes in the surface of the container, from an area of high pressure to one of low pressure. For instance, even though the rubber or latex that makes up a balloon may seem solid, after the balloon is filled with a gas, it will gradually shrink over time. This is due to the fact that there are tiny holes in the surface of the balloon, through which the even tinier gas molecules can escape.

The rate at which a gas effuses from a container is dependent on the speed of the gas particles. The faster the particles are moving, the more often they hit the sides of the container, and the more likely they are to hit a hole and escape. The rate of effusion thus increases with temperature, but also, if examining gases at the same temperature, the gas with the lower molar mass will effuse first.

It is likely you have experienced this; a balloon filled with helium will deflate more rapidly than one filled with air (which is composed primarily of nitrogen and oxygen) or one filled with carbon dioxide. There is a formula that quantifies the rate at which a gas will effuse, but that is beyond the scope of the exam. As long as you understand the basic principles behind effusion, you should be able to answer any questions on this topic that may come up.

THE IDEAL GAS EQUATION

You can use the ideal gas equation to calculate any of the four variables relating to the gas, provided that you already know the other three.

The Ideal Gas Equation

$$PV = nRT$$

P = the pressure of the gas (atm)

V = the volume of the gas (L)

n = the number of moles of gas

R = the gas constant, 0.08206 L·atm·mol^{-1}·K^{-1}

T = the absolute temperature of the gas (K)

When using the Ideal Gas Law, units are important. Temperature must be measured in Kelvins, which you can get to by adding 273 to any temperature in degrees Celsius. So, 25°C + 273 = 298 K. Volume must be in liters, and pressure must be in atmospheres. The only other units you may see for pressure on the AP exam are either millimeters of mercury (mmHg) or Torricellis (torr). Those are synonymous, that is, 1 mmHg = 1 torr. The relationship between those and atmospheres is that 1 atm = 760 mmHg (or torr). So, a pressure of 745 torr = 745/760 = 0.98 atm.

You can also manipulate the ideal gas equation to figure out how changes in each of its variables affect the other variables. The following equation, often called the Combined Gas Law, can be used only when the number of moles is held constant.

Combined Gas Law

$$\frac{P_1V_1}{T_1} = \frac{P_2V_2}{T_2}$$

P = the pressure of the gas (atm)

V = the volume of the gas (L)

T = the absolute temperature of the gas (K)

You should be comfortable with the following simple relationships:

- If the volume is constant: As pressure increases, temperature increases; as temperature increases, pressure increases.
- If the temperature is constant: As pressure increases, volume decreases; as volume increases, pressure decreases. That's Boyle's Law.
- If the pressure is constant: As temperature increases, volume increases; as volume increases, temperature increases. That's Charles's Law.

DALTON'S LAW

Dalton's Law states that the total pressure of a mixture of gases is just the sum of all the partial pressures of the individual gases in the mixture.

> **Dalton's Law**
>
> $$P_{total} = P_a + P_b + P_c + \ldots$$

You should also note that the partial pressure of a gas is directly proportional to the number of moles of that gas present in the mixture. So if 25 percent of the gas in a mixture is helium, then the partial pressure due to helium will be 25 percent of the total pressure. This concept, often represented by the variable X, is called the mole fraction.

> **Partial Pressure**
>
> $$P_a = (P_{total})(X_a)$$
>
> $$X_a = \frac{\text{moles of gas A}}{\text{total moles of gas}}$$

DEVIATIONS FROM IDEAL BEHAVIOR

At low temperature and/or high pressure, gases behave in a less-than-ideal manner. That's because the assumptions made in kinetic molecular theory become invalid under conditions where gas molecules are packed too tightly together.

Two things happen when gas molecules are packed too tightly.

- *The volume of the gas molecules becomes significant.*
  The ideal gas equation does not take the volume of the gas molecules into account, and that volume can be significant compared to the amount of space in the container under high pressures. This means that under nonideal conditions, there is less free space for the molecules to move around in than predicted by the ideal gas equation.

- *Gas molecules attract one another and stick together.*
  The ideal gas equation assumes that gas molecules never stick together. When a gas is packed tightly together, intermolecular forces become significant, causing some gas molecules to stick together. When gas molecules stick together, there are fewer particles bouncing around and creating pressure, so the real pressure in a nonideal situation will be smaller than the pressure predicted by the ideal gas equation.

Even under normal conditions, gases will still show some deviation from ideal behavior due to the IMFs between the various gas molecules. These interactions are generally minimal, but when considering the likelihood that a gas will deviate from ideal behavior, stronger IMFs will lead to more deviations. H_2O, with hydrogen bonding, would be significantly more likely to deviate from ideal behavior than CH_4, which only has London dispersion forces.

If you are looking at gases that have similar IMF types, remember that the more electrons a gas has, the more polarizable it is and the more likely it is to deviate from ideal behavior. For instance, within the noble gases, which have only LDFs, argon is more likely to deviate from ideal behavior than helium, but less likely to deviate than xenon.

DENSITY

You may be asked about the density of a gas. The density of a gas is measured in the same way as the density of a liquid or solid: in mass per unit of volume.

Density of a Gas

$$D = \frac{m}{V}$$

D = density
m = mass of gas, usually in grams
V = volume occupied by a gas, usually in liters

The density of any gas sample can also be determined by combining the density equation with the Ideal Gas Law.

If $D = \frac{m}{V}$, then $V = \frac{m}{D}$.

Substituting that into the Ideal Gas Law,

$$\frac{Pm}{D} = nRT$$

A little rearrangement yields

$$D = \frac{Pm}{nRT}$$

The term (m/n) describes mass per mole, which is how molar mass (MM) is measured. Thus,

$$D = \frac{P(\text{MM})}{RT}$$

If you are given the density of a gas and need to find the molar mass, this can also be rewritten as:

$$\text{MM} = \frac{DRT}{P}$$

ELECTROMAGNETIC SPECTRUM

The relationship between the change in energy level of an electron and the electromagnetic radiation absorbed or emitted is given below.

Energy and Electromagnetic Radiation

$$E = hv$$

E = energy change (in Joules)
h = Planck's constant (6.626×10^{-34} J·s)
v = frequency (in s^{-1})

The frequency and wavelength of electromagnetic radiation are inversely proportional. Combined with the energy and electromagnetic radiation equation, we can see that higher frequencies and shorter wavelengths lead to more energy.

Frequency and Wavelength

$$c = \lambda v$$

c = speed of light (2.998×10^8 m·s^{-1})
v = frequency (in s^{-1})
λ = wavelength (in m)

While frequency is always given in s^{-1}, wavelength is often given in nanometers, and needs to be converted to meters before being used in the above equation. An example is below.

$$350 \text{ nm} \times \frac{1 \times 10^{-9} \text{ m}}{1 \text{ nm}} = 3.50 \times 10^{-7} \text{ m}$$

Let's take a look at an example using the above equations.

How much energy does light with a wavelength of 695 nm have?

First, we have to convert wavelength to frequency. Remember, wavelength has to be in meters!

$$c = \lambda v$$

$$2.998 \times 10^8 \text{ m·s}^{-1} = (6.95 \times 10^{-7} \text{ m})v$$

$$v = 4.31 \times 10^{14} \text{ s}^{-1}$$

Then, convert frequency to energy.

$$E = h\nu$$

$$E = (6.626 \times 10^{-34} \text{ J·s})(4.31 \times 10^{14} \text{ s}^{-1})$$

$$E = 2.86 \times 10^{-19} \text{ J}$$

You can also combine what you've learned about the math underlying energy to answer quantitative questions about ionization energy, such as the following:

Chlorine has a first ionization energy of 1,251 kJ/mol. What wavelength of light would be necessary to ionize a single atom of chlorine?

The ionization energy is given in kJ/mol, and that needs to be converted to Joules before it can be used in the equations that relate energy and wavelength.

$$\frac{1{,}251 \text{ kJ}}{\text{mol}} \times \frac{1 \text{ mol}}{6.02 \times 10^{23} \text{ atoms}} \times \frac{1{,}000 \text{ J}}{1 \text{ kJ}} = 2.08 \times 10^{-18} \frac{\text{J}}{\text{atom}}$$

Then, energy can be converted to frequency.

$$E = h\nu$$

$$2.08 \times 10^{-18} \text{ J} = (6.626 \times 10^{-34} \text{ J·s})\nu$$

$$\nu = 3.14 \times 10^{15} \text{ s}^{-1}$$

And finally, frequency can be converted to wavelength.

$$c = \lambda\nu$$

$$2.998 \times 10^8 \text{ m·s}^{-1} = \lambda(3.14 \times 10^{15} \text{ s}^{-1})$$

$$\lambda = 9.55 \times 10^{-8} \text{ m}$$

BEER'S LAW

To measure the concentration of a solution over time, a device called a spectrophotometer can be used in some situations. A spectrophotometer measures the amount of light at a given wavelength that is absorbed by a solution. If a solution changes color as the reaction progresses, the amount of light that is absorbed will change. Absorbance can be calculated using Beer's Law:

Please Note
It is illegal for you to use Beer's Law until you turn 21.
Only kidding! You can use Beer's Law anytime you like. It pairs well with solutions that visibly change color. (We can't stop with the jokes.)

Beer's Law

$$A = abc$$

A = absorbance
a = molar absorptivity, a constant that depends on the solution
b = path length, the distance the light is traveling through the solution
c = concentration of the solution

As molar absorptivity and path length are constants when using a spectrophotometer, Beer's Law is often interpreted as a direct relationship between absorbance and the concentration of the solution. Beer's Law is most effective with solutions that visibly change color over the course of a reaction, but if a spectrophotometer that emits light in the ultraviolet region is used, Beer's Law can be used to determine the concentrations of reactants in solutions that are invisible to the human eye.

You may also run into a device called a colorimeter while studying Beer's Law. A colorimeter is simply a spectrophotometer that can only emit light at specific frequencies, whereas a spectrophotometer can emit light at any frequency within a set range.

UNIT 3 QUESTIONS

Multiple-Choice Questions

1. A sealed, rigid container contains three gases: 28.0 g of nitrogen, 40.0 g of argon, and 36.0 g of water vapor. If the total pressure exerted by the gases is 2.0 atm, what is the partial pressure of the nitrogen?

   (A) 0.33 atm
   (B) 0.40 atm
   (C) 0.50 atm
   (D) 2.0 atm

2. The wavelength range for infrared radiation is 10^{-5} m, while that of ultraviolet radiation is 10^{-8} m. Which type of radiation has more energy, and why?

   (A) Ultraviolet has more energy because it has a higher frequency.
   (B) Ultraviolet has more energy because it has a longer wavelength.
   (C) Infrared has more energy because it has a lower frequency.
   (D) Infrared has more energy because it has a shorter wavelength.

Use the following information to answer questions 3–5.

An evacuated rigid container is filled with exactly 2.00 g of hydrogen and 10.00 g of neon. The temperature of the gases is held at 0°C and the pressure inside the container is a constant 1.0 atm.

3. What is the mole fraction of neon in the container?

   (A) 0.17
   (B) 0.33
   (C) 0.67
   (D) 0.83

4. What is the volume of the container?

   (A) 11.2 L
   (B) 22.4 L
   (C) 33.5 L
   (D) 48.8 L

5. Which gas particles have a higher average velocity and why?

   (A) Hydrogen, because it has a lower molar mass
   (B) Neon, because it has a higher molar mass
   (C) Hydrogen, because it has a larger atomic radius
   (D) Neon, because it has a smaller atomic radius

6. A sample of liquid NH_3 is brought to its boiling point. Which of the following occurs during the boiling process?

   (A) The N–H bonds within the NH_3 molecules break apart.
   (B) The overall temperature of the solution rises as the NH_3 molecules speed up.
   (C) The amount of energy within the system remains constant.
   (D) The hydrogen bonds holding separate NH_3 molecules together break apart.

7. The following diagrams show the Lewis structures of four different molecules. Which molecule would travel the farthest in a paper chromatography experiment using a polar solvent?

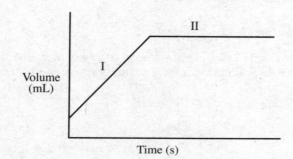

Methanol

Pentane

Acetone

Dimethyl Ether

(A) Methanol
(B) Pentane
(C) Acetone
(D) Dimethyl Ether

8.

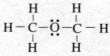

The volume of a gas is charted over time, giving the above results. Which of the following options provides a possible explanation of what was happening to the gas during each phase of the graph?

(A) During phase I, the temperature decreased while the pressure increased. During phase II, the temperature was held constant as the pressure decreased.

(B) During phase I, the temperature increased while the pressure was held constant. During phase II, the temperature and pressure both decreased.

(C) During phase I, the temperature was held constant while the pressure increased. During phase II, the temperature and pressure both decreased.

(D) During phase I, the temperature and pressure both increased. During phase II, the temperature was held constant while the pressure decreased.

Use the following information to answer questions 9–11.

$$2 SO_2(g) + O_2(g) \rightarrow 2 SO_3(g)$$

4.0 mol of gaseous SO_2 and 6.0 mol of O_2 gas are allowed to react in a sealed container.

9. Which particulate drawing best represents the contents of the flask after the reaction goes to completion?

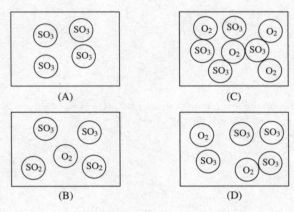

10. If the temperature remains constant, what percentage of the original pressure will the final pressure in the container be equal to?

    (A) 67%
    (B) 80%
    (C) 100%
    (D) 133%

11. Under which of the following conditions would the gases in the container most deviate from ideal conditions and why?

    (A) Low pressures, because the gas molecules would be spread far apart
    (B) High pressures, because the volume occupied by the molecules becomes more significant at low pressure
    (C) Low temperatures, because the intermolecular forces between the gas molecules would increase
    (D) High temperatures, because the gas molecules are moving too fast to interact with one another

12. $$2 CO(g) + O_2(g) \rightarrow 2 CO_2(g)$$

    2.0 mol of $CO(g)$ and 2.0 mol of $O_2(g)$ are pumped into a rigid, evacuated 4.0 L container, where they react to form $CO_2(g)$, as shown above. Which of the following values does NOT represent a potential set of concentrations for each gas at a given point during the reaction?

	CO	O_2	CO_2
(A)	0.5	0.5	0
(B)	0	0.25	0.5
(C)	0.25	0.25	0.5
(D)	0.25	0.38	0.25

13. A mixture of helium and neon gases has a total pressure of 1.2 atm. If the mixture contains twice as many moles of helium as neon, what is the partial pressure due to neon?

   (A) 0.2 atm
   (B) 0.3 atm
   (C) 0.4 atm
   (D) 0.8 atm

14. Nitrogen gas was collected over water at 25°C. If the vapor pressure of water at 25°C is 23 mmHg, and the total pressure in the container is measured at 781 mmHg, what is the partial pressure of the nitrogen gas?

   (A) 46 mmHg
   (B) 551 mmHg
   (C) 735 mmHg
   (D) 758 mmHg

15. A 22.0 gram sample of an unknown gas occupies 11.2 liters at standard temperature and pressure. Which of the following could be the identity of the gas?

   (A) CO_2
   (B) SO_3
   (C) O_2
   (D) He

16.

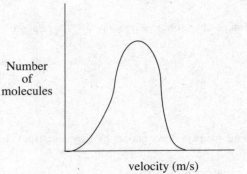

The diagram above shows the speed distribution of molecules in a gas held at 200 K. Which of the following representations would best represent the gas at a higher temperature? (Note: The original line is shown as a dashed line in the answer options.)

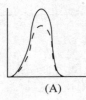

(A)

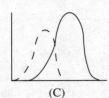

(C)

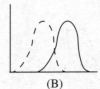

(B)

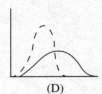

(D)

17. A sample of oxygen gas at 50°C is heated, reaching a final temperature of 100°C. Which statement best describes the behavior of the gas molecules?

 (A) Their velocity increases by a factor of two.
 (B) Their velocity increases by a factor of four.
 (C) Their kinetic energy increases by a factor of 2.
 (D) Their kinetic energy increases by a factor of less than 2.

Use the following information to answer questions 18–20.

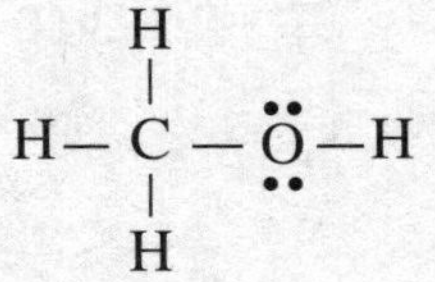

Methanol 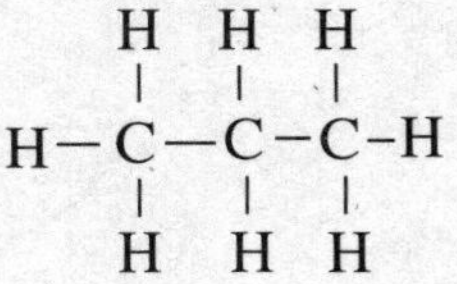   Propane

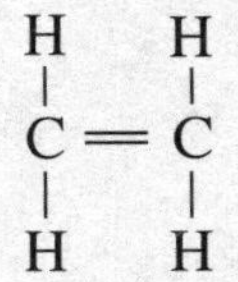

Ethene 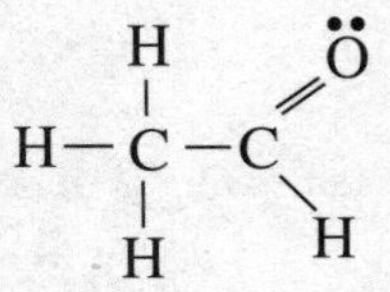   Ethanal

18. Based on the strength of the intermolecular forces in each substance, estimate from greatest to smallest the vapor pressures of each substance in liquid state at the same temperature.

 (A) Propane > Ethanal > Ethene > Methanol
 (B) Ethene > Propane > Ethanal > Methanol
 (C) Ethanal > Methanol > Ethene > Propane
 (D) Methanol > Ethanal > Propane > Ethene

19. When in liquid state, which two substances are most likely to be miscible with water?

 (A) Propane and ethene
 (B) Methanol and propane
 (C) Ethene and ethanal
 (D) Methanol and ethanal

20. Between propane and ethene, which will likely have the higher boiling point and why?

 (A) Propane, because it has a greater molar mass
 (B) Propane, because it has a more polarizable electron cloud
 (C) Ethene, because of the double bond
 (D) Ethene, because it is smaller in size

Use the following information to answer questions 21–23.

The diagram below shows three identical 1.0 L containers filled with the indicated amounts of gas. The stopcocks connecting the containers are originally closed and the gases are all at 25°C. Assume ideal behavior.

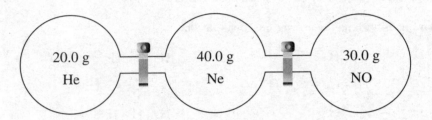

21. Which gas exerts the greatest pressure?

    (A) He
    (B) Ne
    (C) NO
    (D) All gases exert the same amount of pressure.

22. Which gas has the strongest IMFs?

    (A) He
    (B) Ne
    (C) NO
    (D) All gases have identical IMFs.

23. The stopcocks are opened. If the tubing connecting the containers has negligible volume, by what percentage will the pressure exerted by the neon gas decrease?

    (A) 25%
    (B) 33%
    (C) 50%
    (D) 67%

Use the following information to answer questions 24–26.

The outermost electron of an atom has a binding energy of 2.5 eV. The atom is exposed to light of a high enough frequency to cause exactly one electron to be ejected. The ejected electron is found to have a kinetic energy of 2.0 eV.

24. How much energy did photons of the incoming light contain?

    (A) 0.50 eV
    (B) 0.80 eV
    (C) 4.5 eV
    (D) 5.0 eV

25. If the wavelength of the light were to be shortened, how would that affect the kinetic energy of the ejected electron?

    (A) A shorter wavelength would increase the kinetic energy.
    (B) A shorter wavelength would decrease the kinetic energy.
    (C) A shorter wavelength would stop all electron emissions completely.
    (D) A shorter wavelength would have no effect on the kinetic energy of the ejected electrons.

26. If the intensity of the light were to be decreased (that is, if the light is made dimmer), how would that affect the kinetic energy of the ejected electron?

    (A) The decreased intensity would increase the kinetic energy.
    (B) The decreased intensity would decrease the kinetic energy.
    (C) The decreased intensity would stop all electron emissions completely.
    (D) The decreased intensity would have no effect.

27. A solution of Co^{2+} ions appears red when viewed under white light. Which of the following statements is true about the solution?

    (A) A spectrophotometer set to the wavelength of red light would read a high absorbance.
    (B) If the solution is diluted, the amount of light reflected by the solution will decrease.
    (C) All light with a frequency that is lower than that of red light will be absorbed by it.
    (D) Electronic transmissions within the solution match the wavelength of red light.

Free-Response Questions

1. Consider the Lewis structures for the following four molecules:

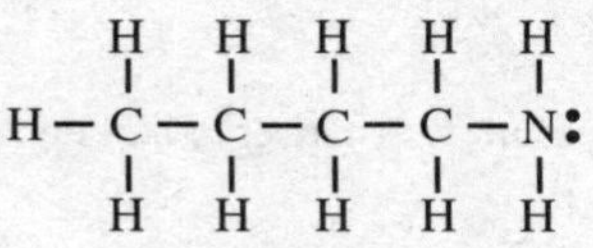

    n-Butylamine

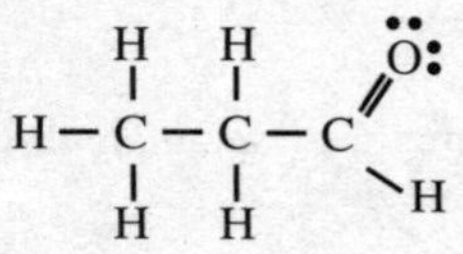

    Propanal

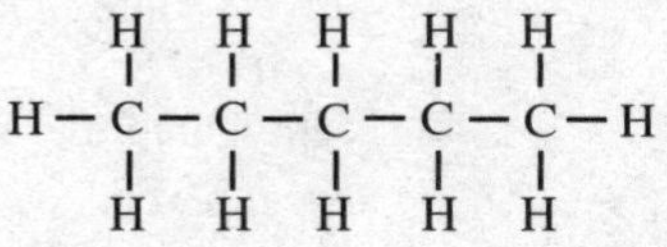

    Pentane

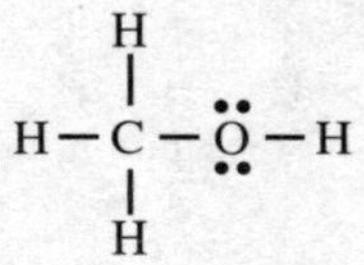

    Methanol

    (a) All of the substances are liquids at room temperature. Organize them from high to low in terms of boiling points, clearly differentiating between the intermolecular forces in each substance.

    (b) On the methanol diagram reproduced below, draw the locations of all partial charges.

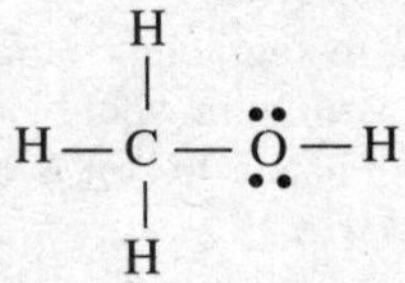

    Methanol

    (c) *n*-Butylamine is found to have the lowest vapor pressure at room temperature out of the four liquids. Justify this observation in terms of intermolecular forces.

2.

Substance	Boiling Point (°C)	Bond Length (Å)	Bond Strength (kcal/mol)
H_2	−253	0.75	104.2
N_2	−196	1.10	226.8
O_2	−182	1.21	118.9
Cl_2	−34	1.99	58.0

(a) Explain the differences in the properties given in the table above for each of the following pairs.
   (i) The bond strengths of N_2 and O_2
   (ii) The bond lengths of H_2 and Cl_2
   (iii) The boiling points of O_2 and Cl_2

(b) Use the principles of molecular bonding to explain why H_2 and O_2 are gases at room temperature, while H_2O is a liquid at room temperature.

3. A student has a mixture containing three different organic substances. The Lewis diagrams of the substances are below:

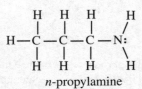

n-butanol   ethyl chloride

n-propylamine

(a) If the mixture was dabbed onto chromatography paper that was then placed into a nonpolar solvent, rank the R_f values for each component of the mixture from high to low after the solvent has saturated the paper. Justify your answer.

(b) If the mixture is poured into a chromatography column and then eluted with a very polar substance, which component of the mixture would leave the column first, and why?

(c) (i) The mixture is heated until it begins to boil. Which substance would be the easiest to separate via distillation, and why?
   (ii) After the substance begins boiling, it continues to be heated at the same rate. Compared to the rate at which it was changing prior to boiling, will the temperature increase faster, slower, or at the same rate? Explain.

(d) (i) After the components of the mixture have been separated, they are returned to room temperature. Of the three substances, which would have the highest vapor pressure at room temperature? Justify your answer.
   (ii) If the substances were heated (but not boiled), explain what would happen to their vapor pressures.

4. A rigid, sealed 12.00 L container is filled with 10.00 g each of three different gases: CO_2, NO, and NH_3. The temperature of the gases is held constant at 35.0°C. Assume ideal behavior for all gases.

    (a)   (i)   What is the mole fraction of each gas?

          (ii)   What is the partial pressure of each gas?

    (b)   Out of the three gases, molecules of which gas will have the highest velocity? Why?

    (c)   Name one circumstance in which the gases might deviate from ideal behavior, and clearly explain the reason for the deviation.

5. $$2\ KClO_3(s) \rightarrow 2\ KCl(s) +\ 3\ O_2(g)$$

The reaction above took place, and 1.45 liters of oxygen gas were collected over water at a temperature of 29°C and a pressure of 755 millimeters of mercury. The vapor pressure of water at 29°C is 30.0 millimeters of mercury.

    (a)   What is the partial pressure of the oxygen gas collected?

    (b)   How many moles of oxygen gas were collected?

    (c)   What would be the dry volume of the oxygen gas at a pressure of 760 millimeters of mercury and a temperature of 273 K?

    (d)   What was the mass of the $KClO_3$ consumed in the reaction?

6. Equal molar quantities of two gases, O_2 and H_2O, are confined in a closed vessel at constant temperature.

    (a)   Which gas, if either, has the greater partial pressure?

    (b)   Which gas, if either, has the greater density?

    (c)   Which gas, if either, has the greater concentration?

    (d)   Which gas, if either, has the greater average kinetic energy?

    (e)   Which gas, if either, will show the greater deviation from ideal behavior?

7. A student performs an experiment in which a butane lighter is held underwater directly beneath a 100-mL graduated cylinder which has been filled with water as shown in the diagram below.

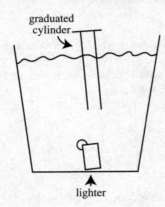

The switch on the lighter is pressed, and butane gas is released into the graduated cylinder. The student's data table for this lab is as follows:

Mass of lighter before gas release	20.432 g
Mass of lighter after gas release	20.296 g
Volume of gas collected	68.40 mL
Water Temperature	19.0°C
Atmospheric Pressure	745 mmHg

(a) Given that the vapor pressure of water at 19.0°C is 16.5 mmHg, determine the partial pressure of the butane gas collected in atmospheres.

(b) Calculate the molar mass of butane gas from the experimental data given.

(c) If the formula of butane is C_4H_{10}, determine the percent error for the student's results.

(d) The following are common potential error sources that occur during this lab. Explain whether or not each error could have been responsible for the error in the student's results.

  (i) The lighter was not sufficiently dried before massing it after the gas was released.

  (ii) The gas in the lighter was not held underwater long enough to sufficiently cool it to the same temperature of the water and was actually at a higher temperature than the water.

  (iii) Not all of the butane gas released was collected in the graduated cylinder.

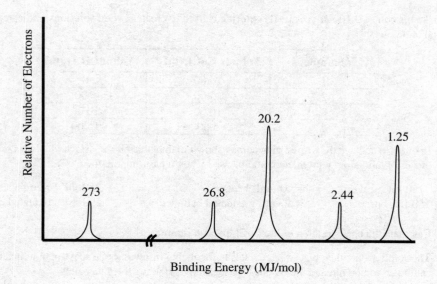

8. The above PES belongs to a neutral chlorine atom.

   (a) What wavelength of light would be required to eject a 3*s* electron from chlorine?
   (b) For the PES of a chloride ion, how would the following variables compare to the peaks on the PES above? Justify your answers.
      (i)   Number of peaks
      (ii) Height of the peaks

9. A stock solution of 0.100 M cobalt (II) chloride is used to create several solutions, indicated in the data table below:

Sample	Volume $CoCl_2$ (mL)	Volume H_2O (mL)
1	20.00	0
2	15.00	5.00
3	10.00	10.00
4	5.00	15.00

(a) In order to achieve the degree of accuracy shown in the table above, select which of the following pieces of laboratory equipment could be used when measuring out the $CoCl_2$:

150-mL beaker   400-mL beaker   250-mL Erlenmeyer flask

50-mL buret   50-mL graduated cylinder   100-mL graduated cylinder

(b) Calculate the concentration of the $CoCl_2$ in each sample.

The solutions are then placed in cuvettes before being inserted into a spectrophotometer calibrated to 560 nm and their values are measured, yielding the data below:

Sample	Absorbance
1	0.485
2	0.367
3	0.249
4	0.131

(c) If gloves are not worn when handling the cuvettes, how might this affect the absorbance values gathered?

(d) If the path length of the cuvette is 1.00 cm, what is the molar absorptivity value for $CoCl_2$ at 560 nm?

(e) On the axes below, plot a graph of absorbance vs. concentration. The y-axis scale is set, and be sure to scale the x-axis appropriately.

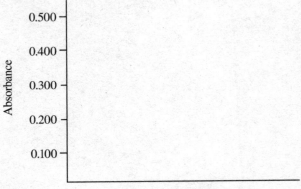

(f) What would the absorbance values be for $CoCl_2$ solutions at the following concentrations?

     (i)   0.067 M

     (ii)   0.180 M

UNIT 3 ANSWERS AND EXPLANATIONS

Multiple-Choice

1. **C** There are 1 mole of N_2, 1 mole of Ar, and 2 moles of water in the container. The mole fraction of nitrogen is $\frac{1}{4} = 0.25$.

    $P_{N_2} = (X_{N_2})(P_{total})$

    $P_{N_2} = (0.25)(2.0) = 0.50$ atm

2. **A** Via $c = \lambda v$, you can see that there is an inverse relationship between wavelength and frequency (as one goes up, the other goes down). So, ultraviolet radiation has a higher frequency, and via $E = hv$, also has more energy.

3. **B** Moles of $H_2 = \dfrac{2.00 \text{ g}}{2.00 \text{ g/mol}} = 1.00$ mol (remember, hydrogen is a diatomic)

    Moles of $Ne = \dfrac{10.00 \text{ g}}{20.0 \text{ g/mol}} = 0.500$ mol

    Total moles = 1.50 moles. $X_{Ne} = \dfrac{\text{moles Ne}}{\text{total moles}} = \dfrac{0.500}{1.500} = 0.33$

4. **C** At STP, 1 mole of gas takes up 22.4 L of space: $(1.5)(22.4) = 33.5$ L

5. **A** The gas molecules have the same amount of kinetic energy due to their temperature being the same. Via $KE = \frac{1}{2}mv^2$, if KE is the same, then the molecule with less mass must correspondingly have a higher velocity.

6. **D** When a covalent substance undergoes a phase change, the bonds between the various molecules inside the substance break apart.

7. **A** In a polar solvent, polar molecules will be the most soluble (like dissolves like). Of the four options, methanol and acetone would both have dipoles, but those of methanol would be significantly stronger due to the H-bonding.

8. **B** In phase I, an increased temperature means the molecules are moving faster and will spread out more, leading to an increased volume. For the volume to remain constant in phase II, either both pressure and temperature have to remain constant, or they both have to increase or decrease together, as they have inverse effects on volume.

9. **C** To determine the number of moles of SO_3 created, stoichiometry must be used.

$$SO_2: \quad 4 \text{ mol } SO_2 \times \frac{2 \text{ mol } SO_3}{2 \text{ mol } SO_2} = 4 \text{ mol } SO_3$$

$$O_2: \quad 6 \text{ mol } O_2 \times \frac{2 \text{ mol } SO_3}{1 \text{ mol } O_2} = 12 \text{ mol } SO_3$$

The oxygen is in excess, and only 2.0 mol of it will react. (As every 2.0 mol of SO_2 react with 1 mol of O_2.) Thus, 4.0 mol of SO_3 are created and 4.0 mol of O_2 remain.

10. **B** If 4.0 mol of SO_3 are created and 4.0 mol of O_2 remain, there are 8.0 mol of gas present after the reaction.

Prior to the reaction there were 10.0 mol of gas. If there are 8/10 = 80% as many moles after the reaction, there is also 80% as much pressure.

11. **C** One of the assumptions of kinetic molecular theory is that the amount of intermolecular forces between the gas molecules is negligible. If the molecules are moving very slowly, as happens when the temperature is lowered, the IMFs between them are more likely to cause deviations from ideal behavior.

12. **C** For every two moles of CO that react, one mole of O_2 will also react. If one mole of CO reacts, half a mole of O_2 will react. At that point, the concentration of CO will be (1.0 mol/4.0 L = 0.25 M) and the concentration of the O_2 will be (1.5 mol/4.0 L = 0.38 M), rendering (C) impossible.

13. **C** From Dalton's Law, the partial pressure of a gas depends on the number of moles of the gas that are present. If the mixture has twice as many moles of helium as neon, then the mixture must be $\frac{1}{3}$ neon. So $\frac{1}{3}$ of the pressure must be due to neon.

$$\left(\frac{1}{3}\right)(1.2 \text{ atm}) = 0.4 \text{ atm}$$

14. **D** From Dalton's Law, the partial pressures of nitrogen and water vapor must add up to the total pressure in the container. The partial pressure of water vapor in a closed container will be equal to the vapor pressure of water, so the partial pressure of nitrogen is

$$781 \text{ mmHg} - 23 \text{ mmHg} = 758 \text{ mmHg}$$

15. **A** Use the following relationship:

$$\text{Moles} = \frac{\text{liters}}{22.4 \text{ L/mol}}$$

$$\text{Moles of unknown gas} = \frac{11.2 \text{ L}}{22.4 \text{ L/mol}} = 0.500 \text{ mole}$$

$$\text{MW} = \frac{\text{grams}}{\text{mole}}$$

$$\text{MW of unknown gas} = \frac{22.0 \text{ g}}{0.500 \text{ mole}} = 44.0 \text{ grams/mole}$$

That's the molecular weight of CO_2.

16. **D** At a higher temperature, the average velocity of the gas molecules would be greater. Additionally, they would have a greater spread of potential velocities, which would lead to a wider curve.

17. **D** The absolute temperature of a gas, measured in the Kelvin scale, is proportional to the average kinetic energy of its particles. Doubling the temperature in the Celsius scale is a much smaller change than doubling the absolute temperature, so you'd expect the average kinetic energy to change by some factor less than two.

18. **B** Vapor pressure is dependent on intermolecular forces. The weaker the IMFs are, the easier it is for molecules to escape from the surface of the liquid. To begin, polar molecules have stronger IMFs than nonpolar molecules. Methanol and ethanal are both polar, but methanol has hydrogen bonding, meaning it has stronger IMFs (and thus a lower vapor pressure) than ethanal. Ethene and propane are both nonpolar, but propane is larger, meaning it is more polarizable than ethene and thus has stronger IMFs and lower vapor pressure.

19. **D** Water is polar, and using "like dissolves like," only polar solutes will be able to fully mix with it to create a homogenous solution.

20. **B** Both are nonpolar, but propane has a lot more electrons and thus is more polarizable than ethene.

21. **A** Pressure is directly dependent on the number of moles. In their respective containers, there are 5 moles of He, 2 moles of Ne, and 1 mole of NO. As there are the most moles of He, the He must exert the greatest pressure.

22. **D** One of the precepts of kinetic molecular theory is that gas molecules exert no forces on each other; thus, in all containers there are no IMFs present.

23. **D** $P_1V_1 = P_2V_2$

   $P_1(1.0 \text{ L}) = P_2(3.0 \text{ L})$

   $P_2/P_1 = 1.0/3.0$

   Thus, the pressure of the neon gas is 33% of what it was originally, meaning a decrease of 67%. Note that the same calculation could be used for any of the gases; each gas is expanding to take up three times as much space as it has originally, and thus exerts one-third as much pressure.

24. **C** Radiation Energy = Binding Energy + Kinetic Energy

   Radiation energy (photon) = 2.5 eV + 2.0 eV

   Radiation energy (photon) = 4.5 eV

25. **A** If the wavelength is decreased, that increases the frequency via $c = \lambda v$ (remember, the speed of light is constant). An increased frequency increases the amount of photon energy via $\Delta E = hv$. As the binding energy of the electron would not change, the excess radiation energy would turn into kinetic energy, increasing the velocity of the electron.

26. **D** The intensity of the light is independent of the amount of energy that the light has. Energy is entirely based on frequency and wavelength, and the brightness of the light would not change the amount of radiation energy. Thus, the amount of kinetic energy of the ejected electrons would not change either.

27. **D** A spectrophotometer works by reading light absorbance. A red solution appears to be red because it transmits red light. The amount of light reflected by the solution does not change as it dilutes; the amount of light that is transmitted and absorbed does. The solution has a red color because the distance between the energy levels within the cobalt ions corresponds with the wavelength of red light.

Free-Response

1. (a)   The strongest type of intermolecular force present here is hydrogen bonding, which both *n*-butylamine and methanol exhibit. Of the two, *n*-butylamine would have the stronger London dispersion forces because it has more electrons (is more polarizable), and so it has stronger IMFs than methanol. For the remaining two structures, propanal has permanent dipoles, while pentane is completely nonpolar. Therefore, propanal would have stronger IMFs than pentane.

    So, from high to low boiling point: *n*-butylamine > methanol > propanal > pentane.

    (b)

    (c)   Vapor pressure arises from molecules overcoming the intermolecular forces to other molecules and escaping the surface of the liquid to become a gas. Due to its hydrogen bonding and large, highly polarizable electron cloud, *n*-butylamine would have the strongest intermolecular forces and thus its molecules would have the hardest time escaping the surface, leading to a low vapor pressure.

2. (a)   (i)   The bond strength of N_2 is larger than the bond strength of O_2 because N_2 molecules have triple bonds and O_2 molecules have double bonds. Triple bonds are stronger and shorter than double bonds.

         (ii) The bond length of H_2 is smaller than the bond length of Cl_2 because hydrogen is a smaller atom than chlorine, allowing the hydrogen nuclei to be closer together.

         (iii) Liquid oxygen and liquid chlorine are both nonpolar substances that experience only London dispersion forces of attraction. These forces are greater for Cl_2 because it has more electrons (which makes it more polarizable), so Cl_2 has a higher boiling point than O_2.

    (b) H_2 and O_2 are both nonpolar molecules that experience only London dispersion forces, which are too weak for a substance to be liquid at room temperature.

    H_2O is a polar substance whose molecules form hydrogen bonds with each other. Hydrogen bonds are strong enough for water to be a liquid at room temperature.

3. (a) All three components are polar, but ethyl chloride has no hydrogen bonding and thus has the weakest dipoles, meaning it would travel the furthest and have the highest R_f value.

    Between *n*-butanol and *n*-propylamine, both have H-bonds but the butanol has more electrons, meaning its London dispersion forces are stronger. It thus has the highest polarity and would have the smallest R_f value. So: ethyl chloride > *n*-propylamine > *n*-butanol.

    (b) *n*-butanol is the most polar and would be most attracted to a polar eluent, and thus would leave the column first.

    (c) (i) Ethyl chloride has the weakest overall IMFs, and thus would have the lowest boiling point and be the easiest to separate out.

    (ii) Prior to boiling, molecules in all three substances were speeding up as heat was added. The increased velocity caused the temperature increase. Once the ethyl chloride starts boiling, though, the heat that would ordinarily be causing the molecules to speed up is instead breaking the IMFs. While the molecules of both the *n*-butanol and *n*-propylamine would still be experiencing a velocity increase, those of the ethyl chloride would not. Thus, the overall rate of the temperature change would be less than it was prior to the ethyl chloride starting to boil.

    (d) (i) The substance with the weakest IMFs would allow the largest number of molecules to escape the liquid phase and turn into a gas, which is the cause of vapor pressure. As such, the ethyl chloride should have the highest vapor pressure of the three substances.

    (ii) As temperature increases, so does the molecular velocity. The faster the molecules are going, the more energy they have, and the more likely they are to be able to overcome the IMFs and escape the liquid phase. For all substances, as temperature increases, so does vapor pressure.

4. (a) (i) First, the moles of each gas need to be calculated:

    CO_2 = 10.00 g × 1 mol/44.01 g = 0.2272 mol
    NO   = 10.00 g × 1 mol/30.01 g = 0.3332 mol
    NH_3 = 10.00 g × 1 mol/17.03 g = 0.5872 mol

    Total moles of gas = 1.1476 moles

    X_{CO_2} = 0.2272/1.1476 = 0.1980
    X_{NO} = 0.3332/1.1476 = 0.2903
    X_{NH_3} = 0.5872/1.1476 = 0.5117

(ii) Using the Ideal Gas Law, calculate the total pressure in the container:

$PV = nRT$

$P(12.00 \text{ L}) = (1.1476 \text{ mol})(0.08206 \text{ atm·L/mol·K})(308 \text{ K})$

$P = 2.415 \text{ atm}$

The partial pressure of a gas is equal to the total pressure times the mole fraction of that gas.

$P_{CO_2} = (2.415 \text{ atm})(0.1980) = 0.4782 \text{ atm}$

$P_{NO} = (2.415 \text{ atm})(0.2903) = 0.7011 \text{ atm}$

$P_{NH_3} = (2.415 \text{ atm})(0.5117) = 1.236 \text{ atm}$

(b) If all gases are at the same temperature, they have the same amount of kinetic energy. *KE* has aspects of both mass and velocity, so the gas with the lowest mass would have the highest velocity. Thus, the NH_3 molecules have the highest velocity.

(c) The most common reason for deviation from ideal behavior is that the intermolecular forces of the gas molecules are acting upon one another. This would occur when the molecules are very close together and/or moving very slowly, so deviations would occur at high pressures and/or low temperatures.

5. (a) Use Dalton's Law.

$$P_{Total} = P_{Oxygen} + P_{Water}$$

$$(755 \text{ mmHg}) = (P_{Oxygen}) + (30.0 \text{ mmHg})$$

$$P_{Oxygen} = 725 \text{ mmHg}$$

(b) Use the Ideal Gas Law. Don't forget to convert to the proper units.

$$n = \frac{PV}{RT} = \frac{\left(\frac{725}{760} \text{ atm}\right)(1.45 \text{ L})}{(0.08206 \text{ L} \cdot \text{atm/mol} \cdot \text{K})(302 \text{ K})} = 0.056 \text{ mole}$$

(c) At STP, moles of gas and volume are directly related.

Volume = (moles)(22.4 L/mol)

Volume of O_2 = (0.056 mol)(22.4 L/mol) = 1.3 L

(d) We know that 0.056 mole of O_2 was produced in the reaction.

From the balanced equation, you know that for every 3 moles of O_2 produced, 2 moles of $KClO_3$ are consumed. So there are $\dfrac{2}{3}$ as many moles of $KClO_3$ as O_2.

$$\text{Moles of } KClO_3 = \left(\frac{2}{3}\right)(\text{moles of } O_2)$$

$$\text{Moles of } KClO_3 = \left(\frac{2}{3}\right)(0.056 \text{ mol}) = 0.037 \text{ mole}$$

$$\text{Grams} = (\text{moles})(MW)$$

$$\text{Grams of } KClO_3 = (0.037 \text{ mol})(123 \text{ g/mol}) = 4.6 \text{ g}$$

6. (a) The partial pressures depend on the number of moles of gas present. Because the number of moles of the two gases are the same, the partial pressures are the same.

(b) O_2 has the greater density. Density is mass per unit volume. Both gases have the same number of moles in the same volume, but oxygen has heavier molecules, so it has greater density.

(c) Concentration is moles per volume. Both gases have the same number of moles in the same volume, so their concentrations are the same.

(d) According to kinetic-molecular theory, the average kinetic energy of a gas depends only on the temperature. Both gases are at the same temperature, so they have the same average kinetic energy.

(e) H_2O will deviate most from ideal behavior. Ideal behavior for gas molecules assumes that there will be no intermolecular interactions.

H_2O is polar, and O_2 is not. H_2O undergoes hydrogen bonding, while O_2 does not. So H_2O has stronger intermolecular interactions, which will cause it to deviate more from ideal behavior.

7. (a) $745 \text{ mmHg} - 16.5 \text{ mmHg} = 729 \text{ mmHg}$

$$\frac{729 \text{ mmHg}}{760 \text{ mmHg}} = 0.959 \text{ atm}$$

(b) To determine the mass of the butane, subtract the mass of the lighter after the butane was released from the mass of the lighter before the butane was released.

$$20.432 \text{ g} - 20.296 \text{ g} = 0.136 \text{ g}$$

To determine the moles of butane, use the Ideal Gas Law, making any necessary conversions first.

$$PV = nRT$$
$$(0.959 \text{ atm})(0.06840 \text{ L}) = n(0.08206 \text{ atm·L/mol·K})(292 \text{ K})$$
$$n = 2.74 \times 10^{-3} \text{ mol}$$

Molar mass is defined as grams per mole, so
$0.136 \text{ g}/2.74 \times 10^{-3} \text{ mol} = 49.6 \text{ g/mol}$

(c)   Actual molar mass of butane:

$(12.01 \text{ g/mol} \times 4) + (1.01 \text{ g/mol} \times 10) = 58.14 \text{ g/mol}$

Percent error is:

$$\frac{|\text{Actual value} - \text{experimental value}|}{\text{Actual value}} \times 100\%$$

So:

$$\frac{|58.14 - 49.6|}{58.14} \times 100\% = 15\% \text{ error}$$

(d)   (i)   If the lighter is not sufficiently dried, then the mass of the butane calculated will be artificially low. That means the numerator in the molar mass calculation will be too low, which would lead to an experimental molar mass that is too low. This is consistent with the student's error.

(ii)   If the temperature of the butane is higher than the water temperature, the calculated moles of butane will be artificially high. This means the denominator in the molar mass calculation will be too high, which would lead to an experimental molar mass that is too low. This is consistent with the student's error.

(iii) If some butane gas escaped without going into the graduated cylinder, the volume of butane gas collected will be artificially low. That will make the calculation for moles of butane too low, which in turns means the denominator of the molar mass calculation will be too low. This would lead to an experimental molar mass that is too high. This is NOT consistent with the student's error.

8. (a) The 3s electron belongs to the peak located at 2.44 MJ/mol. First, you need to calculate the amount of energy needed to remove a single 3s electron (rather than a mole of them):

$$\frac{2.44 \text{ MJ}}{1 \text{ mol}} \times \frac{1 \text{ mol}}{6.022 \times 10^{23} \text{ electrons}} = 4.05 \times 10^{-24} \text{ MJ} \times \frac{1 \times 10^6 \text{ J}}{1 \text{ MJ}} = 4.05 \times 10^{-18} \text{ J}$$

Then, you can use $E = hc/\lambda$ to calculate the wavelength of light that would have sufficient energy:

$$4.05 \times 10^{-18} \text{ J} = \frac{(6.626 \times 10^{-34} \text{ J·s})(2.998 \times 10^8 \text{ m/s})}{\lambda}$$

$$\lambda = 4.91 \times 10^{-8} \text{ m}$$

(b) (i) The chloride ion would have one more electron, which would enter the 3p energy level. This is represented by the rightmost peak, and so the PES of chloride would have the same number of peaks as that of the chlorine atom.

(ii) The first four peaks would have the same height, but the last peak would have an additional electron. That would give it six total electrons, making it the same height as the 2p (middle) peak shown in the PES of the neutral chlorine atom.

9. (a) The 50-mL buret is what is needed here in order to get measurements that are accurate to the hundredths place. Graduated cylinders are generally accurate to the tenths, and using flasks or beakers to measure out volume is highly inaccurate.

(b) Sample 1: 0.100 M (no dilution occurred)

For samples 2–4, the first step is to calculate the number of moles of $CoCl_2$, using the Molarity = moles/volume formula. Then, divide the number of moles by the new volume (volume of $CoCl_2$ + volume H_2O) to determine the new concentration.

Sample 2: 0.100 M = n/0.01500 L
n = 0.00150 mol
0.00150 mol/0.02000 L = 0.0750 M

Sample 3: 0.100 M = n/0.01000 L
n = 0.00100 mol
0.00100 mol/0.02000 L = 0.0500 M

Sample 4: 0.100 M = n/0.00500 L
n = 0.000500 mol
0.000500 mol/0.02000 L = 0.0250 M

(c) If gloves are not worn when handling the cuvettes, fingertip oils may be deposited on the surface of the cuvette. These oils may absorb some light, increasing the overall absorbance values.

(d)   Using Beer's Law: $A = abc$ and the data from Sample 1:

0.485 = a(1.00 cm)(0.100 M) a = 4.85 $M^{-1}cm^{-1}$

Values from any sample can be used with identical results.

(e)

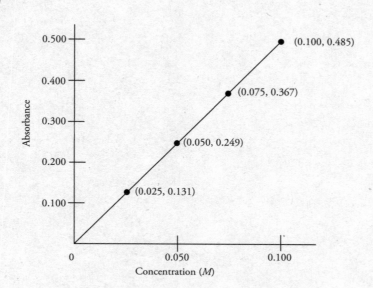

(f)   While the absorbance values for (i) can be estimated from a correctly drawn graph, the absorbance value for (ii) would be off the chart. The more accurate method to determine the absorbances of both samples would be to determine the slope of the line from the graph in (e) and then plug in the appropriate values. (Note: Estimating the value for (i) is still an acceptable solution.)

Any two points can be taken to determine the slope of the graph, but by examining the Beer's Law calculation in (d), you can see it is already in slope-intercept form (the y-intercept would be zero, as at 0 concentration there would be no absorbance). So, the slope of the line is equal to 4.85 $M^{-1}cm^{-1}$ × 1.00 cm = 4.85 M^{-1}.

(i)   0.067 M

    A = (4.85 M^{-1})(0.067 M) = 0.32

(ii)   0.180 M

    A = (4.85 M^{-1})(0.180 M) = 0.873

Chapter 6
Unit 4: Chemical Reactions

TYPES OF REACTIONS

Reactions may be classified into several categories.

1. Synthesis Reactions: When elements or simple compounds are combined to form a single, more complex compound.

    $$2\ Mg(s) + O_2(g) \rightarrow 2\ MgO(s)$$

2. Decomposition: The opposite of a synthesis. A reaction in which a single compound is split into two or more elements or simple compounds, usually in the presence of heat.

    $$HgO(s) + Heat \rightarrow Hg(s) + \frac{1}{2}\ O_2(g)$$

3. Acid-Base Reaction: A reaction when an acid (i.e., H^+) reacts with a base (i.e., OH^-) to form water and a salt.

    $$HCl(aq) + NaOH(aq) \rightarrow NaCl(aq) + H_2O(l)$$

4. Oxidation-Reduction (Redox) Reaction: A reaction that results in the change of the oxidation states of some participating species.

    $$Cu^{2+}(aq) + 2\ e^- \rightarrow Cu(s)$$

5. Hydrocarbon Combustion: When a covalent substance containing carbon and hydrogen (and sometimes oxygen) is ignited, it reacts with the oxygen in the atmosphere. The products of a hydrocarbon combustion are always CO_2 and H_2O. The combustion of butane, a fuel found in many lighters, is an example of a combustion reaction:

    $$C_4H_{10} + \frac{13}{2}\ O_2 \rightarrow 4\ CO_2 + 5\ H_2O$$

    If elements other than carbon and hydrogen are present in the substance being combusted, they, too, often combine with oxygen to form various gases.

    $$2\ C_2H_5S + \frac{17}{2}\ O_2 \rightarrow 4\ CO_2 + 5\ H_2O + 2\ SO_2$$

    It is not always possible to predict the exact formula of the non-carbon or hydrogen containing compounds that are created, but carbon will always lead to CO_2, and hydrogen to H_2O.

6. Precipitation: When two aqueous solutions mix, sometimes a new cation/anion pairing can create an insoluble salt. This type of reaction is called a precipitation reaction. When potassium carbonate and magnesium nitrate mix, a precipitate of magnesium carbonate will form as follows:

    $$K_2CO_3(aq) + Mg(NO_3)_2(aq) \rightarrow 2\ KNO_3(aq) + MgCO_3(s)$$

    That can also be written as a net ionic equation. In solution, the potassium and nitrate ions do not actually take part in the reaction. They start out as free ions and end up as free ions. We call those ions **spectator ions**. The only thing that is changing is that carbonate and magnesium ions are bonding to form magnesium carbonate. The net ionic equation is:

    $$CO_3^{2-}(aq) + Mg^{2+}(aq) \rightarrow MgCO_3(s)$$

Finally, this can be represented using particulate diagrams, as shown below:

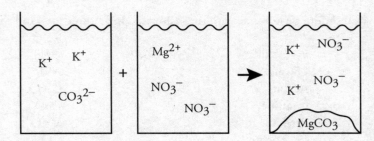

You may have heard that you need to memorize solubility rules—that is, what ions will form insoluble precipitates when you combine them. You DO NOT need to memorize most of these; for the most part, the AP Exam will provide you with them as needed. The only ones you need to know are:

1.   Compounds with an alkali metal cation (Na^+, Li^+, K^+, etc.) or an ammonium cation (NH_4^+) are always soluble.
2.   Compounds with a nitrate (NO_3^-) anion are always soluble.

That's it! You should know how to interpret any solubility rules that you are given on the test, but you need not memorize anything beyond those two.

Ionic substances that dissolve in water do so because the attraction of the ions to the dipoles of the water molecules exceeds the attraction of the ions to each other. Let's take NaBr for example:

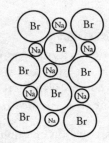

When it dissolves in water, the positive Na^+ cations are attracted to the negative (oxygen) ends of the water molecules. The negative Br^- anions are attracted to the positive (hydrogen) ends of the water molecules. Thus, when the substance dissolves, it looks like this on the particulate level:

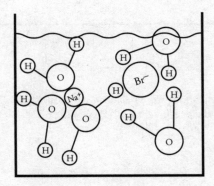

Are You a Visual Learner?

If you're getting over-whelmed by all of the concepts for an AP course, consider looking at our *Fast Track* or *ASAP* books, both available for AP Chemistry. These handy guides focus on the most-tested content or present it in a friendly, illustrated fashion.

You should be familiar with the following polyatomic ions and their charges.

Hydroxide	OH^-
Nitrate	NO_3^-
Acetate	$C_2H_3O_2^-$
Cyanide	CN^-
Permanganate	MnO_4^-
Carbonate	CO_3^{2-}
Sulfate	SO_4^{2-}
Dichromate	$Cr_2O_7^{2-}$
Phosphate	PO_4^{3-}
Ammonium	NH_4^+
Chromate	CrO_4^{2-}
Peroxide	O_2^{2-}
Oxalate	$C_2O_4^{2-}$
Thiosulfate	$S_2O_3^{2-}$

CHEMICAL EQUATIONS

Balancing Chemical Equations

Normally, balancing a chemical equation is a trial-and-error process. You start with the most complicated-looking compound in the equation and work from there. There is, however, an old Princeton Review SAT trick that you may want to try if you see a balancing equation question on the multiple-choice section. The trick is called **backsolving.**

It works like this. To make a balancing equation question work in a multiple-choice format, one of the answer choices is the correct coefficient for one of the species in the reaction. So instead of starting blind in the trial-and-error process, you can insert the answer choices one by one to see which one works. You probably won't have to try all four, and if you start in the middle and the number doesn't work, it might be obviously too small or large, eliminating other choices before you have to try them. Let's try it.

$$...NH_3 + ...O_2 \rightarrow ...N_2 + ...H_2O$$

1. If the equation above were balanced with lowest whole number coefficients, the coefficient for NH_3 would be

    (A) 1
    (B) 2
    (C) 3
    (D) 4

Here's How to Crack It

Start at (C) because it's a middle number. If there are 3 NH_3's, then there can't be a whole number coefficient for N_2, so (C) is wrong, and so is the other odd number answer, (A).

Try (D).

If there are 4 NH_3's, then there must be 2 N_2's and 6 H_2O's.

If there are 6 H_2O's, then there must be 3 O_2's, and the equation is balanced with lowest whole number coefficients. Thus, (D) is correct.

───────────○───────────

Chemical Equations and Calculations

Because of the math involved, stoichiometry like this will most likely appear only as a free-response question. In those instances, expect the following: you will be given a balanced chemical equation and told that you have some number of grams (or liters of gas, or molar concentration, and so on) of reactant. Then you will be asked what number of grams (or liters of gas, or molar concentration, and so on) of products are generated.

> In these cases, follow this simple series of steps.
>
> 1. Convert whatever quantity you are given into moles.
> 2. If you are given information about two reactants, you may have to use the equation coefficients to determine which one is the limiting reagent. Remember, the limiting reagent is not necessarily the reactant that you have the least of; it is the reactant that runs out first.
> 3. Use the balanced equation to determine how many moles of the desired product are generated.
> 4. Convert moles of product to the desired unit.

Let's try one.

───────────○───────────

$$2\ HBr(aq) + Zn(s) \rightarrow ZnBr_2(aq) + H_2(g)$$

2. A piece of solid zinc weighing 98 grams was added to a solution containing 324 grams of HBr. What is the volume of H_2 produced at standard temperature and pressure if the reaction above runs to completion?

A Note About Backsolving

Backsolving is more efficient than the methods that you're accustomed to. If you use the answer choices that you're given, you streamline the trial-and-error process and allow yourself to use POE as you work on the problem.

Here's How to Crack It

1. Convert whatever quantity you are given into moles.

$$\text{Moles of Zn} = \frac{\text{grams}}{\text{molar mass}} = \frac{(98 \text{ g})}{(65.4 \text{ g/mol})} = 1.5 \text{ mol}$$

$$\text{Moles of HBr} = \frac{\text{grams}}{\text{molar mass}} = \frac{(324 \text{ g})}{(80.9 \text{ g/mol})} = 4.0 \text{ mol}$$

2. Use the balanced equation to find the limiting reagent. From the balanced equation, 2 moles of HBr are used for every mole of Zn that reacts, so when 1.5 moles of Zn react, 3 moles of HBr are consumed, and there will be HBr left over when all of the Zn is gone. That makes Zn the limiting reagent.

3. Use the balanced equation to determine how many moles of the desired product are generated.
    1 mole of H_2 is produced for every mole of Zn consumed, so if 1.5 moles of Zn are consumed, then 1.5 moles of H_2 are produced.

4. Convert moles of product to the desired unit.
    The H_2 gas is at STP, so you can convert directly from moles to volume.
    Volume of H_2 = (moles)(22.4 L/mol) = (1.5 mol)(22.4 L/mol) = 33.6 L ≈ 34 L

Let's try another one using the same reaction.

$$2 \text{ HBr}(aq) + \text{Zn}(s) \rightarrow \text{ZnBr}_2(aq) + \text{H}_2(g)$$

3. A piece of solid zinc weighing 13.1 grams was placed in a container. A 0.10-molar solution of HBr was slowly added to the container until the zinc was completely dissolved. What was the volume of HBr solution required to completely dissolve the solid zinc?

Here's How to Crack It

1. Convert whatever quantity you are given into moles.

$$\text{Moles of Zn} = \frac{\text{grams}}{\text{MW}} = \frac{(13.1 \text{ g})}{(65.4 \text{ g/mol})} = 0.200 \text{ mol}$$

2. Use the balanced equation to find the limiting reagent.

3. Use the balanced equation to determine how many moles of the desired product are generated.
    In this case, use the balanced reaction to find out how much of one reactant is required to consume the other reactant.
    You can see from the balanced equation that it takes 2 moles of HBr to react completely with 1 mole of Zn, so it will take 0.400 mole of HBr to react completely with 0.200 mole of Zn.

4. Convert moles of product to the desired unit.
   Moles of HBr = (molarity)(volume)

$$\text{Volume of HBr} = \frac{\text{moles}}{\text{molarity}} = \frac{(0.400\,\text{mol})}{(0.10\,\text{mol/L})} = 4.0\ \text{L}$$

───○───

When you perform calculations, always include units. Including units in your calculations will help you (and the person scoring your test) keep track of what you are doing. While using the proper units in an answer is not typically worth any points in and of itself (with a few exceptions), AP graders may take off points if your units are missing or incorrect.

Percent Error

The 33.6 L of H_2 we calculated in the first example is the theoretical yield, the maximum amount the reaction could produce. What if a researcher actually only collects 25.8 L of H_2 in that experiment? In that case, the percent error can be calculated by comparing the difference between the calculated and experimentally observed values.

$$\% \text{ error} = \frac{|\text{experimental value} - \text{expected value}|}{\text{expected value}} \times 100\% = \frac{|25.8\,\text{L} - 33.6\,\text{L}|}{33.6\,\text{L}} \times 100\% = 23.2\% \text{ error}$$

Combustion Analysis

When a hydrocarbon is combusted, producing carbon dioxide and water, the formula of that hydrocarbon can often be determined using a little math. One of the most important concepts in chemistry is the **Law of Conservation of Mass**. This states that matter can be neither created nor destroyed during a chemical reaction. Thus, when a hydrocarbon is combusted, all of the carbon in that hydrocarbon will end up in CO_2, and all of the hydrogen will end up in H_2O. With that in mind, you can often determine the empirical formula of a hydrocarbon by determining the mass of the products it forms.

───○───

4. A compound containing hydrogen, carbon, and oxygen is combusted. 44.0 g of CO_2 and 27.0 g of H_2O are produced. Which of the following is a possible empirical formula for that compound?

   (A) CH_4O
   (B) C_2H_6O
   (C) $C_2H_3O_2$
   (D) $C_3H_5O_2$

Here's How to Crack It

If 44.0 grams of CO_2 are created, that is 1.00 mole of CO_2, and thus, 1.00 mole of C as well (as there is one mole of carbon in each mole of CO_2). Through conservation of mass, this means there was 1.00 mole of C in the original compound.

27.0 g of H_2O means 1.5 moles of H_2O were created. Because there are two hydrogen atoms in every H_2O, that means there were 3.0 moles of hydrogen in the water that was created, and thus, 3.00 moles of H in the original compound. The C:H ratio in the original compound must be 1:3, which is consistent with (B).

GRAVIMETRIC ANALYSIS

A common way to use precipitation reactions in order to make quantitative determinations about the identity of an unknown sample is through gravimetric analysis. In this process, a soluble sample of an unknown compound is mixed with another solution that will cause one of the ions from the unknown sample to fully precipitate out. If the identity of the precipitate is known, stoichiometric conversions will reveal the mass of the ion in the unknown sample, which allows for a mass percent calculation.

Due to the large amount of math involved, it's easiest to demonstrate.

Example:

A 4.33 g sample of an unknown alkali hydroxide compound is dissolved completely in water. A sufficient solution of copper (II) nitrate is added to the hydroxide solution such that it will fully precipitate copper (II) hydroxide via the following reaction:

$$Cu^{2+}(aq) + 2\ OH^-\ (aq) \rightarrow Cu(OH)_2(s)$$

After the precipitate is filtered and dried, its mass is found to be 3.81 g. Is the original alkali hydroxide sample most likely LiOH, NaOH, or KOH?

First, you convert the mass of the precipitate to moles:

$$3.81 \text{ g Cu(OH)}_2 \times \frac{1 \text{ mol Cu(OH)}_2}{97.57 \text{ g Cu(OH)}_2} = 0.0390 \text{ mol Cu(OH)}_2$$

Next, you need to determine the moles of hydroxide in the precipitate and convert that to grams.

$$0.0390 \text{ mol Cu(OH)}_2 \times \frac{2 \text{ mol OH}^-}{1 \text{ mol Cu(OH)}_2} \times \frac{17.01 \text{ g OH}^-}{1 \text{ mol OH}^-} = 1.33 \text{ g OH}^-$$

If there are 1.33 g of OH^- present in the precipitate, then there were also 1.33 g of OH^- in the original sample. The final calculation is to determine the mass percent of the OH^- in the original sample:

$$\frac{1.33 \text{ g OH}^-}{4.33 \text{ g sample}} \times 100\% = 30.7\%$$

To determine the identity of the sample, you have to know the mass percent of hydroxide in all of the possible compounds. To do that, you divide the mass of one mole of hydroxide by the molar mass of the compound.

$$\text{LiOH: } \frac{17.01 \text{ g}}{23.95 \text{ g}} \times 100\% = 71.0\%$$

$$\text{NaOH: } \frac{17.01 \text{ g}}{40.00 \text{ g}} \times 100\% = 42.5\%$$

$$\text{KOH: } \frac{17.01 \text{ g}}{56.11 \text{ g}} \times 100\% = 30.3\%$$

The hydroxide mass percent in KOH is closest to the mass percent determined via gravimetric analysis, so that is the most likely identity of the original unknown.

OXIDATION STATES

In many chemical reactions, electrons are transferred between the reactants. In order to determine which reactants are gaining electrons and which are losing electrons, we use something called an **oxidation state**. There are several rules when it comes to assigning oxidation states.

1) Any neutral atom that is not bonded to atoms of any other element has an oxidation state of zero. Examples: In $Cu(s)$, the oxidation state on copper is zero. In $O_2(g)$, the oxidation state on both oxygen atoms is zero.

2) Any monoatomic ion has an oxidation state equal to the charge on that ion. This includes ions bonded to other ions in any kind of ionic compound. Examples: In $Zn^{2+}(aq)$, zinc has an oxidation state of +2. In $FeCl_3$, iron is +3 and chlorine is –1.

3) In most compounds, oxygen is –2. An exception worth knowing is that in hydrogen peroxide (H_2O_2), oxygen is –1. Example: In $C_6H_{12}O_6$, oxygen is –2.

4) When bonded to a nonmetal, hydrogen is +1. When bonded to a metal, hydrogen is –1, Examples: In CH_4, hydrogen is +1. In NaH, hydrogen is –1.

5) In the absence of oxygen, the most electronegative element in a compound will take on an oxidation state equal to its most common charge. Examples: In CF_4, fluorine is –1. In CS_2, sulfur is –2.

6) The combined oxidation states on all elements in a neutral compound must add up to zero. The combined oxidation states on all elements in a polyatomic ion must add up to the charge on that ion.

Rule 6 is only applied for covalent compounds to determine the oxidation state on an element which does not fall under the first five rules. Nonmetals with low electronegativity values (such as carbon, nitrogen, sulfur, and phosphorus) have frequently varying oxidation states.

Example 1: Determine the oxidation state on every element in H_2SO_4.

We know that H is +1 and O is –2, and that the total charge on the compound must be 0. Thus,

$$+1(2) + S + -2(4) = 0.$$

Solve for S, and you will find S = +6.

Example 2: Determine the oxidation state on every element in the phosphate ion, PO_4^{3-}.

$$P + -2(4) = -3$$
$$P = +5$$

OXIDATION-REDUCTION REACTIONS

In an oxidation-reduction (or redox, for short) reaction, electrons are exchanged by the reactants, and the oxidation states of some of the reactants are changed over the course of the reaction. Look at the following reaction:

$$Fe + 2\ HCl \rightarrow FeCl_2 + H_2$$

The oxidation state of Fe changes from 0 to +2.

The oxidation state of H changes from +1 to 0.

- *When an atom gains electrons, its oxidation number decreases, and it is said to have been reduced.*

In the reaction above, H was reduced.

- *When an atom loses electrons, its oxidation number increases, and it is said to have been oxidized.*

Here's a mnemonic device that might be useful.

LEO the lion says GER
LEO: you **L**ose **E**lectrons in **O**xidation
GER: you **G**ain **E**lectrons in **R**eduction

Or, if lions don't appeal to you, another way to memorize this concept is **OIL RIG**:
Oxidation **I**s **L**oss
Reduction **I**s **G**ain

In the reaction above, Fe was oxidized.

Oxidation and reduction go hand in hand. If one atom is losing electrons, another atom must be gaining them.

An oxidation-reduction reaction can be written as two **half-reactions**: one for the reduction and one for the oxidation. For example, the reaction

$$Fe + 2\ HCl \rightarrow FeCl_2 + H_2$$

can be written as

$$Fe \rightarrow Fe^{2+} + 2\ e^- \qquad \text{Oxidation}$$
$$2\ H^+ + 2\ e^- \rightarrow H_2 \qquad \text{Reduction}$$

Redox Titrations

A titration involves the slow addition of a solution at a known concentration to another solution of unknown concentration in order to determine the concentration of the unknown solution. To determine the endpoint of a titration, a color change is often used. Titrations are frequently used in acid-base reactions (more on that in Unit 4), but redox reactions can also be used in titration situations.

Potassium permanganate, $KMnO_4$, is frequently used in redox titrations. The manganese ion has an oxidation state of +7 in the permanganate ion (MnO_4^-), and a solution containing permanganate ions is a deep purple color. The manganese in MnO_4^- reduces to Mn^{2+} (thus changing its oxidation state to +2) when mixed with compounds that can be oxidized. Mn^{2+} is a colorless ion.

Frequently, when potassium permanganate is titrated into a colorless solution that contains a compound that can be oxidized, the end of the titration is marked by the solution turning pink. Initially, the permanganate ions take electrons from the oxidized compound and reduce to Mn^{2+} However, once the compound that is being oxidized runs out, there are no electrons left for the MnO_4^- to take, and thus any extra permanganate ion that is added remains unreduced and retains its purple color, which when diluted sufficiently appears pink.

Example:

A 50.0 mL solution of sodium oxalate, $Na_2C_2O_4$, is poured into an Erlenmeyer flask. An acidified solution of 0.135 M potassium permanganate is titrated into the flask while the solution is swirled on a stir station and heated. The solution in the flask is originally colorless, but turns pink after 14.56 mL of potassium permanganate is added. Given the following half-reactions, what is the concentration of the oxalate solution?

Reduction: $8\ H^+(aq) + MnO_4^-(aq) + 5\ e^- \rightarrow Mn^{2+}(aq) + 4\ H_2O(l)$

Oxidation: $C_2O_4^{2-}(aq) \rightarrow 2\ CO_2(g) + 2\ e^-$

The first step to solving this problem is to balance the half-reactions. In this case, the oxidation half-reaction must be multiplied by 5 in order to make the electrons balance with those in the reduction half-reaction, and the reduction half-reaction must be multiplied by 2. Thus, the full reaction (which omits the sodium and potassium spectator ions) is:

$16\ H^+(aq) + 5\ C_2O_4^{2-}(aq) + 2\ MnO_4^-(aq) \rightarrow 2\ Mn^{2+}(aq) + 8\ H_2O(l) + 10\ CO_2(g)$

So, for every two moles of permanganate that reacts, five moles of oxalate are required. The number of moles of permanganate can be easily determined:

$$M = \frac{n}{V} \qquad 0.135\ M\ MnO_4^- = \frac{n}{0.01456\ L\ MnO_4^-} \qquad n = 0.00197\ MnO_4^-$$

That can then be converted to moles of oxalate:

$$0.00197\ MnO_4^- \times \frac{5\ mol\ C_2O_4^{2-}}{2\ mol\ MnO_4^-} = 0.00493\ mol\ C_2O_4^{2-}$$

Finally, the concentration of the oxalate solution can be determined:

$$\frac{0.00493 \text{ mol } C_2O_4^{2-}}{0.0500 \text{ L } C_2O_4^{2-}} = 0.0986 \; M = [C_2O_4^{2-}] = [Na_2C_2O_4]$$

You will not be expected to memorize any of the various color changes that can occur throughout a redox titration. However, you should understand the concept of using a color change to determine an endpoint if you are provided with the colors of the various compounds that are present during a titration.

ACIDS AND BASES

J. N. Brønsted and T. M. Lowry defined an acid as a substance that is capable of donating a proton, which is the same as donating an H^+ ion, and they defined a base as a substance that is capable of accepting a proton. This definition is the one that will be used most frequently on the exam.

Look at the reversible reaction below.

$$HC_2H_3O_2 + H_2O \rightleftharpoons C_2H_3O_2^- + H_3O^+$$

According to Brønsted-Lowry

$HC_2H_3O_2$ and H_3O^+ are acids.

$C_2H_3O_2^-$ and H_2O are bases.

Now look at this reversible reaction.

$$NH_3 + H_2O \rightleftharpoons NH_4^+ + OH^-$$

According to Brønsted-Lowry

NH_3 and OH^- are bases.

H_2O and NH_4^+ are acids.

So in each case, the species with the H^+ ion is the acid, and the same species without the H^+ ion is the base; the two species are called a **conjugate pair.** The following are the acid-base conjugate pairs in the reactions above:

$HC_2H_3O_2$ and $C_2H_3O_2^-$

NH_4^+ and NH_3

H_3O^+ and H_2O

H_2O and OH^-

Notice that water can act either as an acid or a base. Any substance which has that ability is called **amphoteric.**

UNIT 4 QUESTIONS

Multiple-Choice Questions

Use the following solubility rules to answer questions 1–4.

Salts containing halide anions are soluble except for those containing Ag^+, Pb^{2+}, and Hg_2^{2+}.

Salts containing carbonate anions are insoluble except for those containing alkali metals or ammonium.

1. If solutions of iron (III) nitrate and sodium carbonate are mixed, what would be the formula of the precipitate?

    (A) Fe_3CO_3
    (B) $Fe_2(CO_3)_3$
    (C) $NaNO_3$
    (D) No precipitate would form.

2. If solutions containing equal amounts of $AgNO_3$ and KCl are mixed, what is the identity of the spectator ions?

    (A) Ag^+, NO_3^-, K^+, and Cl^-
    (B) Ag^+ and Cl^-
    (C) K^+ and Ag^+
    (D) K^+ and NO_3^-

3. If equimolar solutions of $Pb(NO_3)_2$ and $NaCl$ are mixed, which ion will NOT be present in significant amounts in the resulting solution after equilibrium is established?

    (A) Pb^{2+}
    (B) NO_3^-
    (C) Na^+
    (D) Cl^-

4. Choose the correct net ionic equation representing the reaction that occurs when solutions of potassium carbonate and copper (I) chloride are mixed.

    (A) $K_2CO_3(aq) + 2\ CuCl(aq) \rightarrow 2\ KCl(aq) + Cu_2CO_3(s)$
    (B) $K_2CO_3(aq) + 2\ CuCl(aq) \rightarrow 2\ KCl(s) + Cu_2CO_3(aq)$
    (C) $CO_3^{2-}(aq) + 2\ Cu^+(aq) \rightarrow Cu_2CO_3(s)$
    (D) $CO_3^{2-}(aq) + Cu^{2+}(aq) \rightarrow CuCO_3(s)$

---

5. In which of the following compounds is the oxidation number of chromium the greatest?

    (A) CrO_4^{2-}
    (B) CrO
    (C) Cr^{3+}
    (D) $Cr(s)$

6. What is the mass of oxygen in 148 grams of calcium hydroxide ($Ca(OH)_2$)?

    (A) 24 grams
    (B) 32 grams
    (C) 48 grams
    (D) 64 grams

7. A sample of a compound known to consist of only carbon, hydrogen, and oxygen is found to have a total mass of 29.05 g. If the mass of the carbon is 18.02 g and the mass of the hydrogen is 3.03 g, what is the empirical formula of the compound?

(A) C_2H_4O
(B) C_3H_6O
(C) $C_2H_6O_3$
(D) $C_3H_8O_2$

Use the following information to answer questions 8–10.

When heated in a closed container in the presence of a catalyst, potassium chlorate decomposes into potassium chloride and oxygen gas via the following reaction:

$$2 \ KClO_3(s) \rightarrow 2 \ KCl(s) + 3 \ O_2(g)$$

8. If 12.25 g of potassium chlorate decomposes, how many grams of oxygen gas will be generated?

(A) 1.60 g
(B) 3.20 g
(C) 4.80 g
(D) 18.37 g

9. Approximately how many liters of oxygen gas will be evolved at STP?

(A) 2.24 L
(B) 3.36 L
(C) 4.48 L
(D) 22.4 L

10. If the temperature of the gas is doubled while the volume is held constant, what will happen to the pressure exerted by the gas and why?

(A) It will also double, because the gas molecules will be moving faster.
(B) It will also double, because the gas molecules are exerting a greater force on each other.
(C) It will be cut in half, because the molecules will lose more energy when colliding.
(D) It will increase by a factor of 4, because the kinetic energy will be four times greater.

11. A sample of a hydrate of $CuSO_4$ with a mass of 250 grams was heated until all the water was removed. The sample was then weighed and found to have a mass of 160 grams. What is the formula for the hydrate?

(A) $CuSO_4 \cdot 10 \ H_2O$
(B) $CuSO_4 \cdot 7 \ H_2O$
(C) $CuSO_4 \cdot 5 \ H_2O$
(D) $CuSO_4 \cdot 2 \ H_2O$

12. A gaseous mixture at 25°C contained 1 mole of CH_4 and 2 moles of O_2 and the pressure was measured at 2 atm. The gases then underwent the reaction shown below.

$$CH_4(g) + 2\, O_2(g) \rightarrow CO_2(g) + 2\, H_2O(g)$$

What was the pressure in the container after the reaction had gone to completion and the temperature was allowed to return to 25°C?

(A) 1 atm
(B) 2 atm
(C) 3 atm
(D) 4 atm

13. During a chemical reaction, $NO(g)$ gets reduced and no nitrogen-containing compound is oxidized. Which of the following is a possible product of this reaction?

(A) $NO_2(g)$
(B) $N_2(g)$
(C) $NO_3^-(aq)$
(D) $NO_2^-(aq)$

14. Solutions of potassium carbonate and calcium chloride are mixed, and the particulate representation below shows which are present in significant amounts after the reaction has gone to completion.

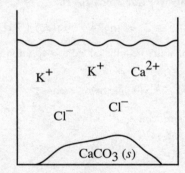

Which of the two original solutions is the limiting reagent and why?

(A) The potassium carbonate, because of the polyatomic anion
(B) The potassium carbonate, because there is no carbonate left after the reaction
(C) The calcium chloride, because there is an excess of calcium ions post-reaction
(D) The calcium chloride, because the component ions are smaller than those in potassium carbonate

15. A student mixes equimolar amounts of KOH and $Cu(NO_3)_2$ in a beaker. Which of the following particulate diagrams correctly shows all species present after the reaction occurs?

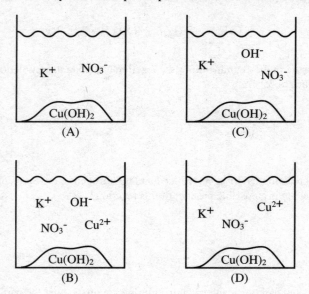

Use the following information to answer questions 16–18.

$$14\ H^+(aq) + Cr_2O_7^{2-}(aq) + 3\ Ni(s) \rightarrow 2\ Cr^{3+}(aq) + 3\ Ni^{2+}(aq) + 7\ H_2O(l)$$

In the above reaction, a piece of solid nickel is added to a solution of potassium dichromate.

16. Which species is being oxidized and which is being reduced?

	Oxidized	Reduced
(A)	$Cr_2O_7^{2-}(aq)$	$Ni(s)$
(B)	$Cr^{3+}(aq)$	$Ni^{2+}(aq)$
(C)	$Ni(s)$	$Cr_2O_7^{2-}(aq)$
(D)	$Ni^{2+}(aq)$	$Cr^{3+}(aq)$

17. How many moles of electrons are transferred when 1 mole of potassium dichromate is mixed with 3 moles of nickel?

(A) 2 moles of electrons
(B) 3 moles of electrons
(C) 5 moles of electrons
(D) 6 moles of electrons

18. How does the pH of the solution change as the reaction progresses?

(A) It increases until the solution becomes basic.
(B) It increases, but the solution remains acidic.
(C) It decreases until the solution becomes basic.
(D) It decreases, but the solution remains acidic.

Use the following information to answer questions 19–21.

20.0 mL of 1.0 M Na_2CO_3 is placed in a beaker and titrated with a solution of 1.0 M $Ca(NO_3)_2$, resulting in the creation of a precipitate.

19. How much $Ca(NO_3)_2$ must be added to reach the equivalence point?

    (A) 10.0 mL
    (B) 20.0 mL
    (C) 30.0 mL
    (D) 40.0 mL

20. Which of the following diagrams correctly shows the species present in the solution in significant amounts at the equivalence point?

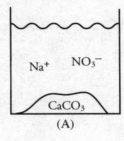
(A)

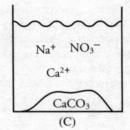
(C)

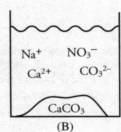
(B)

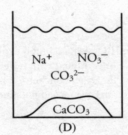
(D)

21. If the experiment were repeated and the Na_2CO_3 was diluted to 40.0 mL with distilled water prior to the titration, how would that affect the volume of $Ca(NO_3)_2$ needed to reach the equivalence point?

    (A) It would be cut in half.
    (B) It would decrease by a factor of 1.5.
    (C) It would double.
    (D) It would not change.

22. A sample of an unknown chloride compound was dissolved in water, and then titrated with excess $Pb(NO_3)_2$ to create a precipitate. After drying, it is determined there are 0.0050 mol of precipitate present. What mass of chloride is present in the original sample?

    (A) 0.177 g
    (B) 0.355 g
    (C) 0.522 g
    (D) 0.710 g

Free-Response Questions

1. 2.54 g of beryllium chloride are completely dissolved into 50.00 mL of water inside a beaker.

    (a) Draw a particulate representation of all species in the beaker after the solute has dissolved. Your diagram should include at least one beryllium ion, one chloride ion, and four water molecules. Make sure the atoms and ions are correctly sized and oriented relative to each other.
    (b) What is the concentration of beryllium and chloride ions in the beaker?

    A solution of 0.850 M lead nitrate is then titrated into the beaker, causing a precipitate of lead (II) chloride to form.

    (c) Identify the net ionic reaction occurring in the beaker.
    (d) What volume of lead nitrate must be added to the beaker to cause the maximum precipitate formation?
    (e) What is the theoretical yield of precipitate?
    (f) Students performing this experiment suggested the following techniques to separate the precipitate from the water. Their teacher rejected each idea. Explain why the teacher may have done so, and name the inherent errors of
       (i) boiling off the water
       (ii) decanting (pouring off) the water

2. Hydrogen peroxide, H_2O_2, is a common disinfectant. Pure hydrogen peroxide is a very strong oxidizer, and as such, it is diluted with water to low percentages before being bottled and sold. One method to determine the exact concentration of H_2O_2 in a bottle of hydrogen peroxide is to titrate a sample with a solution of acidified potassium permanganate. This causes the following redox reactions to occur:

    Reduction: $8 H^+(aq) + MnO_4^-(aq) + 5 e^- \rightarrow Mn^{2+}(aq) + 4 H_2O(l)$

    Oxidation: $H_2O_2(aq) \rightarrow 2 H^+(aq) + O_2(g) + 2 e^-$

    During a titration, a student measures out 5.0 mL of hydrogen peroxide solution into a graduated cylinder, and he pours it into a flask, diluting it to 50.0 mL with water. The student then titrates 0.150 M potassium permanganate solution into the flask with constant stirring.

    (a) Write out the full, balanced redox reaction that is taking place during the titration.
    (b) List two observations that the student will see as the titration progresses that are indicative of chemical reactions.

Diagrams of the permanganate in the buret at the start and end of the titration are as follows:

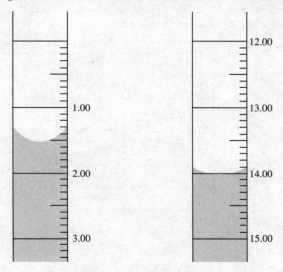

(c) (i)   What volume of KMnO₄ was titrated?
     (ii) What is the concentration of hydrogen peroxide in the original sample?
(d) How would the precision of the student's results have changed if the hydrogen peroxide sample were measured out in a 50 mL beaker instead of a graduated cylinder?
(e) How would each of the following errors affect the student's final calculated hydrogen peroxide concentration?
     (i)   Not filling the buret tip with solution prior to the titration
     (ii) Not rinsing down the sides of the flask during titration

3. The **<u>unbalanced</u>** reaction between potassium permanganate and acidified iron (II) sulfate is a redox reaction that proceeds as follows:

$$H^+(aq) + Fe^{2+}(aq) + MnO_4^-(aq) \rightarrow Mn^{2+}(aq) + Fe^{3+}(aq) + H_2O(l)$$

(a) Provide the equations for both half-reactions that occur below:
     (i)   Oxidation half-reaction
     (ii) Reduction half-reaction
(b) What is the balanced net ionic equation?

A solution of 0.150 *M* potassium permanganate is placed in a buret before being titrated into a flask containing 50.00 mL of iron (II) sulfate solution of unknown concentration. The following data describes the colors of the various ions in solution:

Ion	Color in solution
H^+	Colorless
Fe^{2+}	Pale Green
MnO_4^-	Dark Purple
Mn^{2+}	Colorless
Fe^{3+}	Yellow
K^+	Colorless
SO_4^{2-}	Colorless

(c) Describe the color of the solution in the flask at the following points:
   (i)   Before titration begins
   (ii) During titration prior to the endpoint
   (iii) At the endpoint of the titration

(d) (i)   If 15.55 mL of permanganate are added to reach the endpoint, what is the initial concentration of the iron (II) sulfate?
   (ii) The actual concentration of the $FeSO_4$ is 0.250 *M*. Calculate the percent error.

(e) Could the following errors have led to the experimental result deviating in the direction that it did? You must justify your answers quantitatively.
   (i)   55.0 mL of $FeSO_4$ was added to the flask prior to titration instead of 50.0 mL.
   (ii) The concentration of the potassium permanganate was actually 0.160 *M* instead of 0.150 *M*.

4. $2 Mg(s) + 2 CuSO_4(aq) + H_2O(l) \rightarrow 2 MgSO_4(aq) + Cu_2O(s) + H_2(g)$

(a) If 1.46 grams of Mg(*s*) are added to 500. milliliters of a 0.200-molar solution of $CuSO_4$, what is the maximum molar yield of $H_2(g)$?

(b) When all of the limiting reagent has been consumed in (a), how many moles of the other reactant (not water) remain?

(c) What is the mass of the Cu_2O produced in (a)?

(d) What is the value of $[Mg^{2+}]$ in the solution at the end of the experiment? (Assume that the volume of the solution remains unchanged.)

UNIT 4 ANSWERS AND EXPLANATIONS

Multiple-Choice

1. **B** Carbonates are insoluble when paired with iron. Iron has a charge of +3 and carbonate has a charge of −2. To cancel out, both charges need to have a magnitude of 6, requiring two iron atoms and three carbonate ions. The best representation of that is (B).

2. **D** When those solutions mix, a precipitate of AgCl will form, removing those ions from the solution. The remaining ions (K^+ and NO_3^-) do not react and remain the same as they were when they started.

3. **D** The precipitate that would form is $PbCl_2$, meaning that both NO_3^- and Na^+ are spectator ions. The precipitate would require twice as many chloride ions as lead ions. Therefore, if equal moles of both are used, the NaCl would be the limiting reagent, and almost all of the chloride ions would be present in the solid, with very few left in the solution. There would, however, still be significant Pb^{2+} remaining in solution, as it is in excess.

4. **C** Via the information given from the solubility rules, you can determine that the precipitate would be copper (I) carbonate, which has a formula of Cu_2CO_3. In the net ionic equation, spectator ions (in this case, K^+ and Cl^-) cancel out and do not appear in the final reaction.

5. **A** In (D), chromium is a pure element and has an oxidation number of 0. In (C), chromium's oxidation number is equal to its charge of +3, and in (B), it must balance the −2 on the oxygen, so it has a charge of +2. In (A), the total charge on the ion is −2, and each oxygen is −2. Solving the following, where X is the oxidation number on chromium: $X + -2(4) = -2$. So, the oxidation number is +6.

6. **D** $\text{Moles} = \dfrac{\text{grams}}{\text{MW}}$

   $\text{Moles of calcium hydroxide} = \dfrac{(148\ g)}{(74\ g/mol)} = 2\ \text{moles}$

   Every mole of $Ca(OH)_2$ contains 2 moles of oxygen.

   So there are $(2)(2) = 4$ moles of oxygen

   $\text{Grams} = (\text{moles})(\text{MW})$

   So grams of oxygen = $(4\ mol)(16\ g/mol) = 64$ grams

7. **B** First, the mass of the oxygen must be calculated: 29.05 g − 18.02 g − 3.03 g = 8.00 g.

Converting each of those to moles yields 0.5 mole of oxygen, 1.5 moles of carbon, and 3.0 moles of hydrogen.

Thus, for every one oxygen atom there are three carbon atoms and six hydrogen atoms.

8. **C** $12.25 \text{ g KClO}_3 \times \dfrac{1 \text{ mol KClO}_3}{122.55 \text{ g KClO}_3} = 0.1000 \text{ mol KClO}_3$

$0.1000 \text{ mol KClO}_3 \times \dfrac{3 \text{ mol O}_2}{2 \text{ mol KClO}_3} = 0.1500 \text{ mol O}_2$

$0.1500 \text{ mol O}_2 \times \dfrac{32.00 \text{ g O}_2}{1 \text{ mol O}_2} = 4.800 \text{ g O}_2$

9. **B** $0.1500 \text{ mol O}_2 \times \dfrac{22.4 \text{ L}}{1 \text{ mol O}_2} = 3.36 \text{ L O}_2$

10. **A** Pressure is a measure of the amount of force with which the gas particles are hitting the container walls. An increase in temperature is indicative of increased molecular velocity. If the molecules are moving faster, they will not only have more energy when they hit the container walls, but they will also hit those walls more often. Both of those factors contribute to the increased pressure.

11. **C** The molecular weight of $CuSO_4$ is 160 g/mol, so you have only 1 mole of the hydrate. The lost mass was due to water, so 1 mole of the hydrate must have contained 90 grams of H_2O.

$\text{Moles} = \dfrac{\text{grams}}{\text{MW}}$

$\text{Moles of water} = \dfrac{(90 \text{ g})}{(18 \text{ g/mol})} = 5 \text{ moles}$

So if 1 mole of hydrate contains 5 moles of H_2O, then the formula for the hydrate must be $CuSO_4 \cdot 5 H_2O$.

12. **B** All of the reactants are consumed in the reaction and the temperature doesn't change, so the pressure will change only if the number of moles of gas changes over the course of the reaction. The number of moles of gas (3 moles) doesn't change in the balanced equation, so the pressure will remain the same (2 atm) at the end of the reaction as at the beginning.

13. **B** The oxidation state of nitrogen in NO is +2. If the nitrogen is reduced, that value must become more negative. The oxidation state of nitrogen in N_2 is 0, so that fits the bill. The oxidation state on the nitrogen in the other choices is greater than +2.

14. **B** If extra Ca^{2+} ions are in solution, that means there were not enough CO_3^{2-} ions present for the Ca^{2+} ions to fully react.

15. **D** The overall reaction (excluding spectator ions) is $2\ OH^-(aq)\ +\ Cu^{2+}(aq) \rightarrow Cu(OH)_2(s)$. Both the K^+ and the NO_3^- are spectator ions which are present in the solution both before and after the reaction. Additionally, if equimolar amounts of the two reactants are initially present, the OH^- will run out before the Cu^{2+}, meaning that some Cu^{2+} ions will also be present in the final solution.

16. **C** The $Cr_2O_7^{2-}(aq)$ gains electrons to change the oxidation state on Cr from +6 to +3. The $Ni(s)$ loses electrons to change the oxidation state on Ni from 0 to +2.

17. **D** In one mole of dichromate, two moles of chromium are reduced, each requiring three moles of electrons. Three moles of nickel are being oxidized, each requiring two moles of electrons. Thus, six moles of electrons are being transferred.

18. **B** As the reaction progresses, the H^+ ions are converting to H_2O molecules. This would raise the pH of the solution. As the reaction progresses, the solution will become less acidic, but there is no mechanism for it to become basic.

19. **B** As the formula of the precipitate is $CaCO_3$, for every mole of sodium carbonate in the beaker, one mole of calcium nitrate will be needed. As both solutions have equal molarity, equal volumes of each would be necessary to ensure the moles are equal.

20. **A** At the equivalence point, the moles of Ca^{2+} and CO_3^{2-} in solution will be negligible.

21. **D** The equivalence point is reached when there are equal moles of each reactant present. Diluting the sodium carbonate solution will not change the number of moles of Na_2CO_3 present in solution.

22. **B** The precipitate is $PbCl_2$, and in every mole of precipitate there are two moles of chlorine ions. Thus, there are 0.010 mol of chlorine present in the precipitate, and through conservation of mass, the same number of moles of chlorine were present in the original sample: $0.010\ mol \times 35.5\ g = 0.355\ g$.

Free-Response

1. (a) The beryllium ions only have one full energy level, while chloride ions have three and would thus be much larger. The positive beryllium ions would be attracted to the negative partial charges of the water molecules, and the negative chloride ions would be attracted to the positive partial charges of the water molecules.

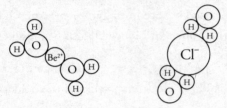

(b) First, calculate the concentration of the $BeCl_2$.

$$2.54 \text{ g } BeCl_2 \times \frac{1 \text{ mol } BeCl_2}{79.91 \text{ g } BeCl_2} = 0.0318 \text{ mol } BeCl_2$$

$$\frac{0.0318 \text{ mol } BeCl_2}{0.05000 \text{ L}} = 0.636 \ M \ BeCl_2$$

For every mole of $BeCl_2$ that dissociates, one Be^{2+} and two Cl^- ions are produced. Thus, the concentration of the Be^{2+} will be the same as the base concentration of the $BeCl_2$, and the concentration of the Cl^- will be twice that. So, $[Be^{2+}] = 0.636 \ M$ and $[Cl^-] = 1.27 \ M$.

(c) $Pb^{2+} (aq) + 2 \ Cl^- (aq) \rightarrow PbCl_2(s)$

(d) There must be one lead ion present for every two chloride ions in order to create the maximum amount of precipitate. Figure out the number of moles of chloride from the concentration and volume, which you already have, and then convert:

$$1.27 \ M \ Cl^- = \frac{n}{0.0500 \text{ L}}$$

$$n = 0.0635 \text{ mol } Cl^- \times \frac{1 \text{ mol } Pb^2}{2 \text{ mol } Cl^-} = 0.0318 \text{ mol } Pb^{2+}$$

From there, you can use the given concentration of the lead nitrate solution to determine how much of it is required.

$$0.850 \ M = \frac{0.0318 \text{ mol}}{V} \qquad V = 0.0374 \text{ L} = 37.4 \text{ mL}$$

(e) If both reactants are mixed in equal amounts, either value can be used to calculate the amount of precipitate that will be created.

$$0.0318 \text{ mol } Pb^{2+} \times \frac{1 \text{ mol } PbCl_2}{1 \text{ mol } Pb^{2+}} \times \frac{278.1 \text{ g}}{1 \text{ mol } PbCl_2} = 8.84 \text{ g}$$

(f) (i) Boiling the water would evaporate all the water, but it would also cause the spectator ions to no longer be dissolved in water. In addition to any $PbCl_2$ that was created, the excess beryllium and nitrate ions would adhere to the solid, giving a falsely high molar mass and an impure product.

(ii) While decanting the solution, smaller particles of precipitate may get poured out with the water. Decanting the solution into a filtered funnel setup would be a much better option, as the filter paper would then catch all of the precipitate while allowing the water (which would carry the spectator ions) through.

2. (a) In a full redox reaction, the electrons from each half-reaction must cancel. To do that, the reduction reaction must be multiplied by two, and the oxidation reaction multiplied by five. When combining the reactions after doing so, this yields:

$$6\ H^+ (aq) + 2\ MnO_4^- (aq) + 5\ H_2O_2(aq) \rightarrow 2\ Mn^{2+} (aq) + 8\ H_2O(l) + 5\ O_2(g)$$

Note that 10 of the hydrogen ions cancel out when the reactions combine.

(b) A gas will be evolved as the hydrogen peroxide oxidizes into oxygen gas. Additionally, there will be a color change as the purple permanganate solution reduces into colorless manganese ions.

(c) (i) 14.03 mL – 1.52 mL = 12.51 mL

(ii) First, calculate the moles of permanganate that were titrated.

$$0.150\ M = \frac{n}{0.01251\ L} \qquad n = 0.00188\ mol\ MnO_4^-$$

Then use the stoichiometric ratios from the balanced equation to convert to moles of hydrogen peroxide.

$$0.00188\ mol\ MnO_4^- \times \frac{5\ mol\ H_2O_2}{2\ mol\ MnO_4^-} = 0.00470\ mol\ H_2O_2$$

Finally, convert that to the concentration of H_2O_2. We will use a volume of 5.00 mL and not 50.0 mL in our calculations because we are interested in the concentration of the solution before it was diluted with water.

$$\frac{0.00470\ mol\ H_2O_2}{0.00500\ L} = 0.940\ M\ H_2O_2$$

(d) Beakers are precise to the ones places; you cannot accurately measure out a volume in a beaker to any number of decimal places. Graduated cylinders are accurate to one decimal place, so using a beaker would reduce the number of significant figures in your answer by one.

(e) (i) If the tip of the buret is not filled prior to dispensing the permanganate solution from it, the first milliliter or so that would be dispensed is air instead of solution. Thus, the experimental volume of permanganate would be artificially high (instead of 12.51 mL of permanganate, it would have really been about 11.5 mL of permanganate with the rest being air). If the recorded volume of the permanganate is too high, that will eventually lead to a calculated $[H_2O_2]$ that is also too high.

(ii) During a titration, it is best practice to regularly rinse down the sides of the flask you are titrating into, just in case any of the titrated solution splashes onto it and does not mix with the solution you are titrating into. If you do not do this, some of the permanganate that leaves the buret might stick to the sides of the flask and not react with the hydrogen peroxide. This would mean the recorded volume of permanganate is too high (not all of the measured permanganate actually reacted), which would again lead to an artificially high calculated concentration of hydrogen peroxide.

3. (a) (i) Oxidation half-reaction:

Oxidation: $Fe^{2+}(aq) \rightarrow Fe^{3+}(aq) + e^-$

(ii) Reduction half-reaction:

Reduction: $5\ e^- + 8\ H^+(aq) + MnO_4^-(aq) \rightarrow Mn^{2+}(aq) + 4\ H_2O(l)$

(b) The oxidation reaction must be multiplied by a factor of 5 in order for the electrons to balance out. So:

$5\ Fe^{2+}(aq) + 8\ H^+(aq) + MnO_4^-(aq) \rightarrow 5\ Fe^{3+}(aq) + Mn^{2+}(aq) + 4\ H_2O(l)$

(c) (i) Before titration begins

The only ions present in the flask are Fe^{2+}, SO_4^{2-}, and H^+. The latter two are colorless, so the solution would be pale green.

(ii) During titration prior to the endpoint

The MnO_4^- is reduced to Mn^{2+} upon entering the flask, and the Fe^{2+} ions are oxidized into Fe^{3+} ions. The solution would become less green and more yellow as more Fe^{3+} ions are formed, as all other ions present are colorless.

(iii) At the endpoint of the titration

After the Fe^{2+} ions have all been oxidized, there is nothing left to donate electrons to the MnO_4^- ions. Therefore, they will no longer be reduced upon entering the flask, and the solution will take on a light purplish/yellow hue due to the mixture of MnO_4^- and Fe^{3+} ions.

(d) (i) First, the moles of permanganate added must be calculated:

$$0.150\ M = \frac{n}{0.0155\ L} \qquad n = 2.33 \times 10^{-3}\ mol\ MnO_4^-$$

Then, the moles of iron (II) can be determined via stoichiometry:

$$2.33 \times 10^{-3}\ mol\ MnO_4^- \times \frac{5\ mol\ Fe^{2+}}{1\ mol\ MnO_4} = 0.0117\ mol\ Fe^{2+}$$

Finally, the concentration of the $FeSO_4$ can be determined:

$$Molarity = \frac{0.0117\ mol}{0.05000\ L} = 0.234\ M$$

(ii) $$Percent\ error = \frac{actual\ value - experimental\ value}{actual\ value} \times 100$$

$$Percent\ error = \frac{0.250 - 0.234}{0.250} \times 100 = 6.40\%\ error$$

(e) (i) If the volume of $FeSO_4$ was artificially low in the calculations, that would lead to the experimental value for the concentration of $FeSO_4$ being artificially high. As the calculated value for the concentration of $FeSO_4$ was too low, this error source is not supported by data.

(ii) If the molarity of the permanganate was artificially low in the calculations, the moles of permanganate, and by extension, the moles of Fe^{2+} would also be artificially low. This would lead to an artificially low value for the concentration of $FeSO_4$. This matches with the experimental results and is thus supported by the data.

4. (a) You need to find the limiting reagent. There's plenty of water, so it must be one of the other two reactants.

$$\text{Moles} = \frac{\text{grams}}{\text{MW}}$$

$$\text{Moles of Mg} = \frac{(1.46 \text{ g})}{(24.3 \text{ g/mol})} = 0.0601 \text{ mole}$$

Moles = (molarity)(volume)

Moles of $CuSO_4$ = (0.200 M)(0.500 L) = 0.100 mole

From the balanced equation, Mg and $CuSO_4$ are consumed in a 1:1 ratio, so Mg runs out first. Mg is the limiting reagent, so it is used to calculate the yield of H_2.

From the balanced equation, 1 mole of H_2 is produced for every 2 moles of Mg consumed, so the number of moles of H_2 produced will be half the number of moles of Mg consumed.

$$\text{Moles of } H_2 = \frac{1}{2} (0.0601 \text{ mol}) = 0.0301 \text{ mole}$$

(b) Mg is the limiting reagent, so some $CuSO_4$ will remain. From the balanced equation, Mg and $CuSO_4$ are consumed in a 1:1 ratio, so when 0.060 mole of Mg are consumed, 0.060 mole of $CuSO_4$ are also consumed.

Moles of $CuSO_4$ remaining = (0.100 mol) − (0.060 mol) = 0.040 mole

(c) From the balanced equation, 1 mole of Cu_2O is produced for every 2 moles of Mg consumed, so the number of moles of Cu_2O produced will be half the number of moles of Mg consumed.

$$\text{Moles of } Cu_2O = \frac{1}{2} (0.0601 \text{ mol}) = 0.0301 \text{ mole}$$

Grams = (moles)(MW)

Grams of Cu_2O = (0.0301 mol)(143 g/mol) = 4.30 grams

(d) All of the Mg consumed ends up as Mg^{2+} ions in the solution.

$$\text{Molarity} = \frac{\text{moles}}{\text{liters}} \qquad [Mg^{2+}] = \frac{(0.0601 \text{ mol})}{(0.500 \text{ L})} = 0.120 \text{ } M$$

Chapter 7
Unit 5: Kinetics

RATE LAW USING INITIAL CONCENTRATIONS

The rate law for a reaction describes the dependence of the initial rate of a reaction on the concentrations of its reactants. It includes the Arrhenius constant, k, which takes into account the activation energy for the reaction and the temperature at which the reaction occurs. The rate of a reaction is described in terms of the rate of appearance of a product or the rate of disappearance of a reactant. The rate law for a reaction cannot be determined from a balanced equation; it must be determined from experimental data, which is presented on the test in table form.

Here's How It's Done

The data below was collected for the following hypothetical reaction:

$$A + 2B + C \rightarrow D$$

Experiment	Initial Concentration of Reactants (M)			Initial Rate of Formation of D (M/sec)
	[A]	[B]	[C]	
1	0.10	0.10	0.10	0.01
2	0.10	0.10	0.20	0.01
3	0.10	0.20	0.10	0.02
4	0.20	0.20	0.10	0.08

The rate law always takes the following form, using the concentrations of the reactants:

$$\text{Rate} = k[A]^x[B]^y[C]^z$$

The greater the value of a reactant's exponent, the more a change in the concentration of that reactant will affect the rate of the reaction. To find the values for the exponents x, y, and z, we need to examine how changes in the individual reactants affect the rate. The easiest way to find the exponents is to see what happens to the rate when the concentration of an individual reactant is doubled.

Let's Look at [A]

From experiment 3 to experiment 4, [A] doubles while the other reactant concentrations remain constant. For this reason, it is useful to use the rate values from these two experiments to calculate x (the order of the reaction with respect to reactant A).

As you can see from the table, the rate quadruples from experiment 3 to experiment 4, going from 0.02 M/sec to 0.08 M/sec.

We need to find a value for the exponent x that relates the doubling of the concentration to the quadrupling of the rate. The value of x can be calculated as follows:

$$(2)^x = 4, \text{ so } x = 2$$

Because the value of x is 2, the reaction is said to be second order with respect to A.

$$\text{Rate} = k[A]^2[B]^y[C]^z$$

Let's Look at [B]

From experiment 1 to experiment 3, [B] doubles while the other reactant concentrations remain constant. For this reason it is useful to use the rate values from these two experiments to calculate y (the order of the reaction with respect to reactant B).

As you can see from the table, the rate doubles from experiment 1 to experiment 3, going from 0.01 M/sec to 0.02 M/sec.

We need to find a value for the exponent y that relates the doubling of the concentration to the doubling of the rate. The value of y can be calculated as follows:

$$(2)^y = 2, \text{ so } y = 1$$

Because the value of y is 1, the reaction is said to be first order with respect to B.

$$\text{Rate} = k[A]^2[B][C]^z$$

Let's Look at [C]

From experiment 1 to experiment 2, [C] doubles while the other reactant concentrations remain constant.

The rate remains the same at 0.01 M.

The rate change is $(2)^z = 1$, so $z = 0$.

Because the value of z is 0, the reaction is said to be zero order with respect to C.

$$\text{Rate} = k[A]^2[B]$$

Because the sum of the exponents is 3, the reaction is said to be third order overall.

Once the rate law has been determined, the value of the rate constant can be calculated using any of the lines of data on the table. The units of the rate constant are dependent on the order of the reaction, so it's important to carry along units throughout all rate constant calculations.

Let's use experiment 3.

$$k = \frac{\text{Rate}}{[A]^2[B]} = \frac{(0.02\,M\,/\,\sec)}{(0.10\,M)^2\,(0.20\,M)} = 10\left(\frac{(M)}{(M)^3\,(\sec)}\right) = 10\,M^{-2}\cdot\sec^{-1}$$

You should note that we can tell from the coefficients in the original balanced equation that the rate of appearance of D is equal to the rate of disappearance of A and C because the coefficients of all three are the same. The coefficient of D is half as large as the coefficient of B, however, so the rate at which D appears is half the rate at which B disappears.

RATE LAW USING CONCENTRATION AND TIME

It's also useful to have rate laws that relate the rate constant k to the way that concentrations change over time. The rate laws will be different depending on whether the reaction is first, second, or zero order, but each rate law can be expressed as a graph that relates the rate constant, the concentration of a reactant, and the elapsed time.

Zero-Order Rate Laws

The rate of a zero-order reaction does not depend on the concentration of reactants at all, so the rate of a zero-order reaction will always be the same at a given temperature.

$$\text{Rate} = k$$

The graph of the change in concentration of a reactant of a zero-order reaction versus time will be a straight line with a slope equal to $-k$.

First-Order Rate Laws

The rate of a first-order reaction depends on the concentration of a single reactant raised to the first power.

$$\text{Rate} = k[A]$$

The rate law for a first-order reaction uses natural logarithms. This means that the use of natural logarithms in the rate law creates a linear graph comparing concentration and time. The slope of the line is given by $-k$ and the y-intercept is given by $\ln[A]_0$.

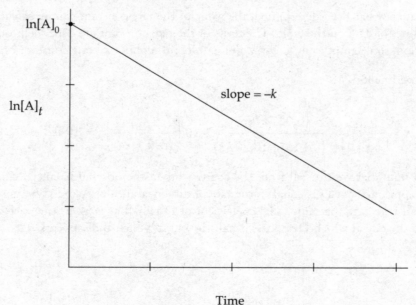

Using slope-intercept form, we can interpret this graph to come up with a useful equation.

$$y = mx + b$$

$y = \ln[A]_t$
$m = -k$
$x = $ time
$b = \ln[A]_0$

After substitution and rearrangement, the slope-intercept equation becomes:

First-Order Rate Law

$$\ln[A]_t = -kt + \ln[A]_0$$

$[A]_t$ = concentration of reactant A at time t
$[A]_0$ = initial concentration of reactant A
k = rate constant
t = time elapsed

Second-Order Rate Laws

The rate of a second-order reaction depends on the concentration of a single reactant raised to the second power.

$$\text{Rate} = k[A]^2$$

The rate law for a second-order reaction uses the inverses of the concentrations.

Second-Order Rate Law

$$\frac{1}{[A]_t} = kt + \frac{1}{[A]_0}$$

$[A]_t$ = concentration of reactant A at time t
$[A]_0$ = initial concentration of reactant A
k = rate constant
t = time elapsed

The use of inverses in the rate law creates a linear graph comparing concentration and time.

Notice that the line moves upward as the concentration decreases. The slope of the line is given by k and the y-intercept is given by $\frac{1}{[A]_0}$.

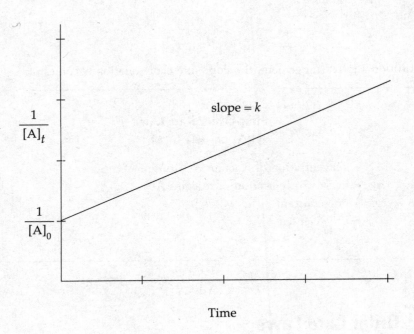

Half-Life

The half-life of a reactant in a chemical reaction is the time it takes for half of the substance to react. Most half-life-problems can be solved by using a simple chart:

Time	Sample
0	100%
1 half-life	50%
2 half-lives	25%
3 half-lives	12.5%

So a sample with a mass of 120 grams and a half-life of 3 years will decay as follows:

Time	Sample
0	120 g
3 years	60 g
6 years	30 g
9 years	15 g

For a first-order reactant, half-life remains constant. That means if it takes 30 seconds for 50% of a reactant to decay (from 100% to 50%), then in the next 30 seconds, 50% more will decay, leaving 25% of the original amount remaining. Thirty seconds later, only 12.5% will remain, etc.

Reminder
Don't forget that the chart should start with the time at zero.

This can be demonstrated graphically as shown below:

$$\text{Half-life} = \frac{\ln 2}{k} = \frac{0.693}{k}$$

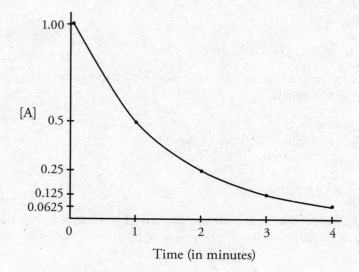

Note that for zero-order or second-order reactants, half-life is NOT constant. You cannot use the half-life equation for anything other than a first-order reactant. Zero-order and second-order reactants have half-lives that vary over the course of the reaction. In other words, the time it takes for the amount of a zero- or second-order reactant to decrease from 100% to 50% will NOT be the same as it takes for the reactant to decrease from 50% to 25%.

Let's do an example. As substance A decomposes, its concentration is monitored as time passes. A student looks at the data and concludes the decomposition reaction is first order due to the constant half-life.

[A] (M)	Time (min)
1.52	0.0
0.76	5.0
0.38	10.0
0.19	15.0

(a) What is the half-life of the decomposition?

As the concentration is cut in half every 5.0 minutes, the half-life of this reaction is exactly 5.0 minutes.

(b)   Determine the value for the rate constant, k. Include units!

Given that this is a first-order reaction, we can use the half-life equation to solve for the rate constant.

$t_{1/2} = \ln 2/k$
$5.0 \text{ min} = 0.693/k$
$k = 0.14 \text{ min}^{-1}$

(c)   What will [A] be at $t = 20.0$ min?

Now that we know k we can solve for the concentration at any given time by using the first-order rate law.

$\ln[A]_{20} = -kt + \ln[A]_0$
$\ln[A]_{20} = -(0.14 \text{ min}^{-1})(20.0 \text{ min}) + \ln(1.52)$
$\ln[A]_{20} = -2.8 + 0.419$
$\ln[A]_{20} = -2.4$
$[A]_{20} = 0.091 \ M$

Another use of half-life is to examine the rate of decay of a radioactive substance. A radioactive substance is one that will slowly decay into a more stable form as time goes on.

COLLISION THEORY

According to collision theory, chemical reactions occur because reactants are constantly moving around and colliding with one another.

When reactants collide with sufficient energy (**activation energy, E_a**), a reaction occurs. These collisions are referred to as effective collisions, because they lead to a chemical reaction. Ineffective collisions do not produce a chemical reaction. At any given time during a reaction, a certain fraction of the reactant molecules will collide with sufficient energy to cause a reaction between them.

For gaseous or aqueous reactants, an increased concentration will increase the rate of reaction. This is because with more molecules moving around in a given volume, they are more likely to collide. If one of the reactants is a solid, the reaction will proceed faster if the surface area available to react is larger. For example, dropping a large chunk of metal into acid will cause the metal to dissolve, but slowly, as the acid can react only with the metal atoms on the surface. If you were to grind the metal into a powder before dropping it into the acid, the acid can react with a lot more metal atoms, and the metal would dissolve faster.

One other physical factor that can affect a reaction rate is stirring. It's certainly true that if you were to drop sugar into water, stirring the water will make the sugar dissolve faster. To explore why stirring a mixture doesn't always speed up a reaction, though, we need to talk about the concept of homogeneity. A **heterogeneous mixture** is one in which all parts of a mixture would not be identical. Before the sugar dissolves, some portions of the water will contain more sugar molecules than others, and the mixture would be heterogeneous. After the sugar fully dissolves, though, each portion of water will contain the same number of aqueous sugar molecules. A mixture in which all portions are identical is called **homogeneous**.

Stirring only increases the reaction rate in a heterogeneous mixture. Causing the solid sugar molecules to move around will make them collide with the liquid water molecules more often, which increases the likelihood that a reaction will occur. However, after the sugar dissolves completely and the mixture is homogenous, any additional sugar/water molecule collisions caused by stirring it are negligible compared to the number of collisions that are already happening due to the inherent motion of the molecules in their aqueous state.

Reaction rate increases with increasing temperature because increasing temperature means that the molecules are moving faster, which means that the molecules have greater average kinetic energy. The higher the temperature, the greater the number of reactant molecules colliding with each other with enough energy (E_a) to cause a reaction.

This can be demonstrated visually on a Maxwell-Boltzmann diagram.

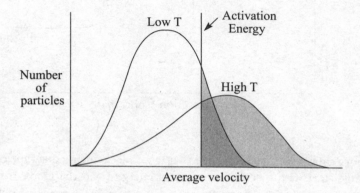

On the above graph, you can see that for the reaction at the higher temperature, a larger fraction of the reactant molecules have sufficient energy to exceed the activation energy barrier.

In addition, reactions only occur if the reactants collide with the correct orientation. For example, in the reaction $2\ NO_2F \rightarrow 2\ NO_2 + F_2$, there are many possible collision orientations. Two of them are drawn below:

The reaction will only occur if the two NO_2F molecules collide in such a fashion that the N–F bonds can break and the F–F bonds can form. While there is no simple way to quantify collision orientation, it is an important part of collision theory, and you should be familiar with the basic concept underlying it.

Reaction Energy Profile

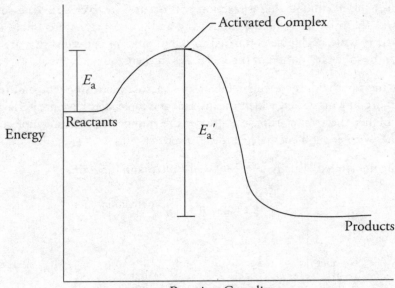

The diagram above shows the energy change that takes place during one type of chemical reaction. The reactants start with a certain amount of energy (read the graph from left to right). For the reaction to proceed, the reactants must have enough energy to reach the transition state, where they are part of an activated complex. This is the highest point on the graph above. The amount of energy needed to reach this point is the activation energy, E_a. At this point, all reactant bonds have been broken, but no products have been formed, so this is the point in the reaction with the highest energy and lowest stability. The energy needed for the reverse reaction is shown as line E_a'. Moving to the right past the activated complex, product bonds start to form, and we eventually reach the energy level of the products.

REACTION MECHANISMS

Many chemical reactions are not one-step processes. Rather, the balanced equation is the sum of a series of individual steps, called elementary steps.

For instance, the hypothetical reaction

$$2A + 2B \rightarrow C + D \qquad\qquad \text{Rate} = k[A]^2[B]$$

could take place by the following three-step mechanism:

I. $\qquad$ $A + A \rightleftharpoons X$ $\qquad\qquad$ (fast)

II. $\qquad$ $X + B \rightarrow C + Y$ $\qquad\qquad$ (slow)

III. $\qquad$ $Y + B \rightarrow D$ $\qquad\qquad$ (fast)

Species X and Y are called reaction **intermediates**, because they are produced but also fully consumed over the course of the reaction. Intermediates will always cancel out when adding up the various elementary steps in a reaction, as shown below.

I. $A \rightarrow B + C$

II. $B + D \rightarrow C$

III. $C + C \rightarrow E$

$$\overline{}$$

$A + B + D + 2C \rightarrow B + 2C + E$

Cancel the species that appear on both sides.

$A + D \rightarrow E$

By adding up all the steps, we get the balanced equation for the overall reaction, so this mechanism is consistent with the balanced equation. All elementary steps have either one or two reactants. Any elementary step that has two reactants (even if they are the same!) is called **bimolecular**. Any elementary step that only has one reactant is called **unimolecular**. In the above mechanism, step I is unimolecular and steps II and III are bimolecular.

As in any process in which many steps are involved, the speed of the whole process can't be faster than the speed of the slowest step in the process, so the slowest step of a reaction is called the **rate-determining step.** Since the slowest step is the most important step in determining the rate of a reaction, the slowest step and the steps leading up to it are used to see if the mechanism is consistent with the rate law for the overall reaction.

The rate for an elementary step can be determined by taking the concentration of the reactants in that step and raising them to the power of any coefficient attached to that reactant. Let's look at an example:

I. $A + A \rightleftharpoons X$ (so $2A \rightleftharpoons X$)   Rate = $k[A]^2$ (fast)

II. $X + B \rightarrow Y$ Rate = $k[X][B]$ (slow)

III. $Y + B \rightarrow D$ Rate = $k[Y][B]$ (fast)

The rate law for the entire reaction is equal to that of the slowest elementary step, which is step II. However, step II has an intermediate (X) present in it. Even though intermediates can appear in rate laws, if they can be replaced with a reactant via an earlier step, we should try to do so. Step I is an equilibrium reaction, meaning that both the forward and reverse reactions are proceeding. We'll see some more details in Unit 7 when we discuss equilibrium, but [X] will be proportional to $[A]^2$, so we can substitute $[A]^2$ for [X] in the rate law for step II. The rate law then becomes rate law = $k[A]^2[B]$, which is the overall rate law for this reaction.

It is important to emphasize that you can only use the coefficient method to determine the rate law of elementary steps. You cannot use the coefficients from the overall balanced equation in a similar fashion to determine the overall rate law. (For instance, the rate law for the above example is not rate = $k[A]^2[B]^2$). The only way to use the coefficient method to determine the rate law for a full reaction is by knowing which elementary step is the slowest and applying the above method to that step. If you do not know the relative speed of the elementary steps, you must use experimental data to determine the rate law.

As discussed previously, an energy diagram is a graphical representation that shows the energy level of the products and the reactants, as well as the required activation energy for a reaction to occur. As many reactions take place over several steps, an energy diagram can be broken down to examine the energy change in each individual step. Take the diagram below:

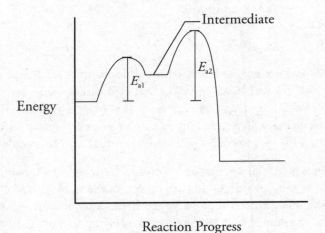

The presence of two activated complexes means that this reaction takes place through two elementary steps, or molecular collisions. A reaction intermediate is produced in the first step and consumed in the second. In both steps, the activation energy is measured from the level of the reactants. As such, the second step would be slower because it has the higher activation energy.

CATALYSTS

A catalyst increases the rate of a chemical reaction without being consumed in the process; catalysts do not appear in the balanced equation. In some cases, a catalyst is a necessary part of a reaction because in its absence, the reaction would proceed at too slow of a rate to be at all useful.

When looking at elementary steps, a catalyst is present both before and after the overall reaction. Catalysts cancel out of the overall reaction but are present in elementary steps.

Example: $A + B \rightarrow C$

    I. $A + X \rightarrow Y$

    II. $B + Y \rightarrow C + X$

> A catalyst increases the rate of a chemical reaction by providing an alternative reaction pathway with a lower activation energy or by lowering the activation energy of an existing reaction pathway.

In the above example, X is a catalyst and Y is a reaction intermediate.

Note that catalysts and (very rarely) intermediates can appear in rate laws. Let's look at the decomposition of H_2O_2 in the presence of the iodide ion as an example. The two elementary steps are shown below:

    I. $H_2O_2(aq) + I^-(aq) \rightleftharpoons H_2O(l) + OI^-(aq)$ (slow)

    II. $H_2O_2(aq) + OI^-(aq) \rightleftharpoons H_2O(l) + O_2(g) + I^-(aq)$ (fast)

If you combine those steps, the overall reaction becomes $H_2O_2(aq) \rightleftharpoons 2\,H_2O(l) + O_2(g)$. In the given mechanism, I^- is a catalyst, and OI^- is an intermediate. As I^- appears as a reactant in the slow step, it would be part of the overall rate law, which would be rate = $k[H_2O_2][I^-]$.

When a catalyst is introduced to a reaction, the ensuing reaction is said to undergo catalysis. There are many types of catalysis. One of the most common is surface catalysis, in which a reaction intermediate is formed as in the example above. Another is enzyme catalysis, in which the catalyst binds to the reactants in a way to reduce the overall activation energy of the reaction. Enzymes are very common in biological applications. Finally, in acid-base catalysis, reactants will gain or lose protons in order to change the rate of reaction. Acids and bases will be studied in more depth in Unit 8.

UNIT 5 QUESTIONS

Multiple-Choice Questions

Use the following information to answer questions 1–4.

A multi-step reaction takes place with the following elementary steps:

Step I $\qquad$ $2\,NO(g) \rightleftharpoons N_2O_2(g)$

Step II $\qquad$ $N_2O_2(g) + H_2(g) \rightarrow N_2O(g) + H_2O(g)$

Step III $\qquad$ $N_2O(g) + H_2(g) \rightarrow N_2(g) + H_2O(g)$

1. What is the overall balanced equation for this reaction?

   (A) $N_2O_2(g) + N_2O(g) + 2\,H_2(g) + 2\,NO(g) \rightarrow N_2O_2(g) + N_2O(g) + N_2(g) + 2\,H_2O(g)$
   (B) $2\,NO(g) \rightarrow N_2O(g) + H_2O(g)$
   (C) $2\,NO(g) + N_2O_2(g) + N_2O(g) + H_2(g) \rightarrow N_2O_2(g) + N_2(g) + H_2O(g)$
   (D) $2\,H_2(g) + 2\,NO(g) \rightarrow N_2(g) + 2\,H_2O(g)$

2. What is the role of N_2O_2 in the overall reaction?

   (A) It is a reactant.
   (B) It is a reaction intermediate.
   (C) It is a catalyst.
   (D) It is a product.

3. If step II is the slow step, what is the rate law for the overall reaction?

   (A) Rate $= k[NO]^2$
   (B) Rate $= k[NO]^2[H_2]$
   (C) Rate $= k[N_2O_2][H_2]$
   (D) Rate $= k[NO]^2[H_2]^2$

4. Why would increasing the temperature make the reaction rate go up?

   (A) It is an endothermic reaction that needs an outside energy source to function.
   (B) The various molecules in the reactions will move faster and collide more often.
   (C) The overall activation energy of the reaction will be lowered.
   (D) A higher fraction of molecules will have the same activation energy.

5.
$$SO_2Cl_2(g) \rightarrow SO_2(g) + Cl_2(g)$$

At 600 K, SO_2Cl_2 will decompose to form sulfur dioxide and chlorine gas via the above equation. If the reaction is found to be first order overall, which of the following will cause an increase in the half-life of SO_2Cl_2?

(A) Increasing the initial concentration of SO_2Cl_2
(B) Increasing the temperature at which the reaction occurs
(C) Decreasing the overall pressure in the container
(D) None of these will increase the half-life.

6. $A + B \rightarrow C + D$ rate $= k[A][B]^2$

What are the potential units for the rate constant for the above reaction?

(A) s^{-1}
(B) $s^{-1}M^{-1}$
(C) $s^{-1}M^{-2}$
(D) $s^{-1}M^{-3}$

7. The following mechanism is proposed for a reaction:

$2A \rightleftharpoons B$ (fast equilibrium)
$C + B \rightarrow D$ (slow)
$D + A \rightarrow E$ (fast)

Which of the following is the correct rate law for the complete reaction?

(A) Rate $= k[C]^2[B]$
(B) Rate $= k[C][A]^2$
(C) Rate $= k[C][A]^3$
(D) Rate $= k[D][A]$

8. $2\ NOCl \rightarrow 2\ NO + Cl_2$

The reaction above takes place with all of the reactants and products in the gaseous phase. Which of the following is true of the relative rates of disappearance of the reactants and appearance of the products?

(A) NO appears at twice the rate that NOCl disappears.
(B) NO appears at the same rate that NOCl disappears.
(C) NO appears at half the rate that NOCl disappears.
(D) Cl_2 appears at the same rate that NOCl disappears.

9.
$$H_2(g) + I_2(g) \rightarrow 2\,HI(g)$$

When the reaction given above takes place in a sealed isothermal container, the rate law is

$$\text{Rate} = k[H_2][I_2]$$

If a mole of H_2 gas is added to the reaction chamber and the temperature remains constant, which of the following will be true?

(A) The rate of reaction and the rate constant will increase.
(B) The rate of reaction and the rate constant will not change.
(C) The rate of reaction will increase and the rate constant will decrease.
(D) The rate of reaction will increase and the rate constant will not change.

10.
$$4\,NH_3(g) + 5\,O_2(g) \rightarrow 4\,NO(g) + 6\,H_2O(g)$$

The above reaction will experience a rate increase by the addition of a catalyst. Which of the following best explains why?

(A) The catalyst causes the value for ΔG to become more negative.
(B) The catalyst decreases the bond energy in the products.
(C) The catalyst introduces a new reaction mechanism for the reaction.
(D) The catalyst increases the activation energy for the reaction.

11.
$$A + B \rightarrow C$$

Based on the following experimental data, what is the rate law for the hypothetical reaction given above?

Experiment	[A] (M)	[B] (M)	Initial Rate of Formation of C (mol/L·sec)
1	0.20	0.10	3×10^{-2}
2	0.20	0.20	6×10^{-2}
3	0.40	0.20	6×10^{-2}

(A) Rate = $k[A]$
(B) Rate = $k[A]^2$
(C) Rate = $k[B]$
(D) Rate = $k[A][B]$

12.
$$NO_2 + O_3 \rightarrow NO_3 + O_2 \qquad \text{slow}$$
$$NO_3 + NO_2 \rightarrow N_2O_5 \qquad \text{fast}$$

A proposed reaction mechanism for the reaction of nitrogen dioxide and ozone is detailed above. Which of the following is the rate law for the reaction?

(A) Rate = $k[NO_2][O_3]$
(B) Rate = $k[NO_3][NO_2]$
(C) Rate = $k[NO_2]^2[O_3]$
(D) Rate = $k[NO_3][O_2]$

13.

Time (Hours)	[A] M
0	0.40
1	0.20
2	0.10
3	0.05

Reactant A underwent a decomposition reaction. The concentration of A was measured periodically and recorded in the chart above. Based on the data in the chart, which of the following is the rate law for the reaction?

(A) Rate = $k[A]$

(B) Rate = $k[A]^2$

(C) Rate = $2k[A]$

(D) Rate = $\frac{1}{2}k[A]$

14.
$$A \rightarrow B + C \qquad \text{rate} = k[A]^2$$

Which of the following graphs may have been created using data gathered from the above reaction?

(A)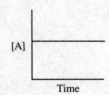

(C)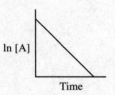

(B)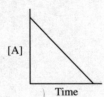

(D)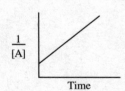

15. After 44 minutes, a sample of $^{44}_{19}K$ is found to have decayed to 25 percent of the original amount present. What is the half-life of $^{44}_{19}K$?

   (A) 11 minutes
   (B) 22 minutes
   (C) 44 minutes
   (D) 66 minutes

16. $A + B \rightarrow C$

Based on the following experimental data, what is the rate law for the hypothetical reaction given above?

Experiment	[A] (M)	[B] (M)	Initial Rate of Formation of C (M/sec)
1	0.20	0.10	5.0×10^{-3}
2	0.20	0.30	1.5×10^{-2}
3	0.60	0.10	4.5×10^{-2}

   (A) Rate = $k[A][B]^2$
   (B) Rate = $k[A]^2[B]^2$
   (C) Rate = $k[A][B]^2$
   (D) Rate = $k[A]^2[B]$

Free-Response Questions

1.
$$2 NO(g) + Br_2(g) \rightarrow 2 NOBr(g)$$

The following results were obtained in experiments designed to study the rate of the reaction above:

Experiment	Initial Concentration (mol/L)		Initial Rate of Appearance of NOBr (M/sec)
	[NO]	[Br_2]	
1	0.02	0.02	9.6×10^{-2}
2	0.04	0.02	3.8×10^{-1}
3	0.02	0.04	1.9×10^{-1}

(a) Write the rate law for the reaction.
(b) Calculate the value of the rate constant, k, for the reaction. Include the units.
(c) In experiment 2, what was the concentration of NO remaining when half of the original amount of Br_2 was consumed?
(d) Which of the following reaction mechanisms is consistent with the rate law established in (a)? Explain your choice.

   I. $NO + NO \rightleftharpoons N_2O_2$ (fast)

   $N_2O_2 + Br_2 \rightarrow 2NOBr$ (slow)

   II. $Br_2 \rightarrow Br + Br$ (slow)

   $2(NO + Br \rightarrow NOBr)$ (fast)

2.
$$2 N_2O_5(g) \rightarrow 4 NO_2(g) + O_2(g)$$

Dinitrogen pentoxide gas decomposes according to the equation above. The first-order reaction was allowed to proceed at 40°C and the data below was collected.

[N_2O_5] (M)	Time (min)
0.400	0.0
0.289	20.0
0.209	40.0
0.151	60.0
0.109	80.0

(a) Calculate the rate constant for the reaction using the values for concentration and time given in the table. Include units with your answer.
(b) After how many minutes will [N_2O_5] be equal to 0.350 M?
(c) What will be the concentration of N_2O_5 after 100. minutes have elapsed?
(d) Calculate the initial rate of the reaction. Include units with your answer.
(e) What is the half-life of the reaction?

3. The decomposition of phosphine occurs via the pathway below:

$$4\ PH_3(g) \rightarrow P_4(g) + 6\ H_2(g)$$

A scientist observing this reaction at 250 K plots the following data:

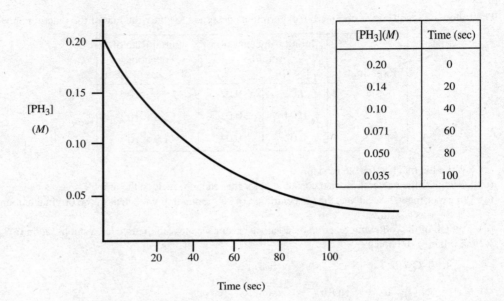

$[PH_3](M)$	Time (sec)
0.20	0
0.14	20
0.10	40
0.071	60
0.050	80
0.035	100

(a) (i)   What order is this reaction? Why?
     (ii) What would the concentration of the PH_3 gas be after 120 sec?
(b) If the rate of disappearance of PH_3 is 2.5×10^{-3} M/s at $t = 20$ s:
     (i)   What is the rate of appearance of P_4 at the same point in time?
     (ii) How will the rate of appearance of P_4 change as the reaction progresses forward?
(c) The experiment is repeated with the same initial concentration of phosphine, but this time the temperature is set at 500 K. How and why would the following values change, if at all?
     (i)   The half-life of the phosphine
     (ii) The rate law
     (iii) The value of the rate constant

4. $$A(g) + B(g) \rightarrow C(g)$$

The reaction above is second order with respect to A and zero order with respect to B. Reactants A and B are present in a closed container. Predict how each of the following changes to the reaction system will affect the rate and rate constant, and explain why.

(a) More gas A is added to the container.
(b) More gas B is added to the container.
(c) The temperature is increased.
(d) An inert gas D is added to the container.
(e) The volume of the container is decreased.

5. Use your knowledge of kinetics to answer the following questions. Justify your answers.

(a)

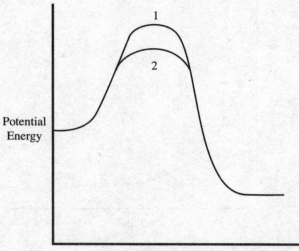

The two lines in the diagram above show different reaction pathways for the same reaction. Which of the two lines shows the reaction when a catalyst has been added?

(b)

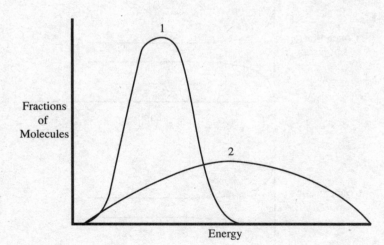

Which of the two lines in the energy distribution diagram shows the conditions at a higher temperature?

(c)

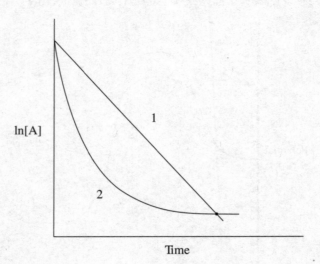

Which of the two lines in the diagram above shows the relationship of ln[A] to time for a first-order reaction with the following rate law?

$$Rate = k[A]$$

(d)

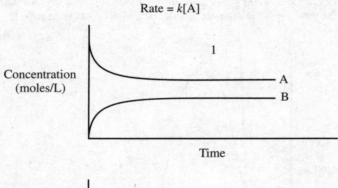

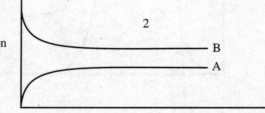

Which of the two graphs above shows the changes in concentration over time for the following reaction?

$$A \rightarrow B$$

6. Hydrazine (N_2H_4) can be produced commercially via the Raschig process. The following is a proposed mechanism:

$$Step\ 1:\quad NH_3(aq) + OCl^-(aq) \rightarrow NH_2Cl(aq) + OH^-(aq)$$
$$Step\ 2:\quad NH_2Cl(aq) + NH_3(aq) \rightarrow N_2H_5^+(aq) + Cl^-(aq)$$
$$Step\ 3:\quad N_2H_5^+(aq) + OH^-(aq) \rightarrow N_2H_4(aq) + H_2O(l)$$

(a) (i)   What is the equation for the overall reaction?
     (ii) Identify any catalysts or intermediates from the reaction mechanism.
(b) The rate law for the reaction is determined to be rate = $k[NH_3][OCl^-]$.
     (i)   Which elementary step is the slowest one? Justify your answer.
     (ii) If the reaction is measured over the course of several minutes, what would the units of the rate constant be?

7. The reaction between crystal violet (a complex organic molecule represented by CV$^+$) and sodium hydroxide is as follows:

$$CV^+ \quad + \quad OH^- \quad \rightarrow \quad CVOH$$
$$(violet) \qquad (colorless) \qquad (colorless)$$

As the crystal violet is the only colored species in the reaction, a spectrophotometer calibrated to a specific wavelength can be used to determine its concentration over time. The following data was gathered:

[CV$^+$] (M)	Time (s)
5.5×10^{-5}	0
3.8×10^{-5}	60
2.6×10^{-5}	120
1.8×10^{-5}	180

(a) (i)   What is the rate of disappearance for crystal violet from $t = 60$ s to $t = 120$ s?
     (ii) If the solution is placed in the spectrophotometer 30 s after mixing instead of immediately after mixing, how would that affect the calculated rate of disappearance for crystal violet in part (i)?
(b) Given the path length of the cuvette is 1.00 cm and the molar absorptivity of the solution is 26,000 cm$^{-1}M^{-1}$ at the wavelength of the spectrophotometer, what would the absorbance reading on the spectrophotometer be at $t = 60$ s?

(c) This reaction is known to be first order with respect to crystal violet. On the provided axes, graph a function of $[CV^+]$ vs. time that will provide you with a straight line graph.

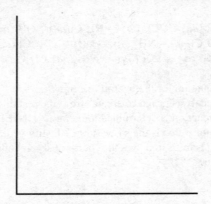

The following data was also gathered over the course of three experiments:

Experiment	$[CV^+]_{init}$ (M)	$[OH^-]_{init}$ (M)	Initial rate of formation of CVOH (M/s)
1	5.5×10^{-5}	0.12	3.60×10^{-7}
2	5.5×10^{-5}	0.24	7.20×10^{-7}
3	4.1×10^{-5}	0.18	?

(d) Write the rate law for this reaction.
(e) What is the rate constant, k, for this reaction? Include units in your answer.
(f) Determine the initial rate of formation of CVOH for experiment 3.

UNIT 5 ANSWERS AND EXPLANATIONS

Multiple-Choice

1. **D**   Option A has every single species present in the overall reaction. However, anything that appears in both the reactants and products (in this case, N_2O_2 and N_2O) will be eliminated from the full reaction.

2. **B** N_2O_2 is produced in Step I, and then consumed in Step II. This makes it an intermediate.

3. **B**   In general, the coefficients on the reactants in the slow step become the order for those reactants in the overall rate law. However, N_2O_2 is an intermediate, and as such it should not appear in the overall rate law. As N_2O_2 is in equilibrium with 2 NO in step I, the N_2O_2 can/should be replaced with 2 NO for the purpose of determining the overall rate law.

4. **B**   Increasing the temperature increases the speed of the molecules. This will increase how often those molecules collide, meaning that a successful collision is more likely to happen. Additionally, there will also be more energy within each collision, making it more likely that the collision will exceed the activation energy barrier.

5. **D**   None of the options would decrease the rate of reaction, which would be required for the half-life of the reactant to increase.

6. **C**   To answer this, you can do a unitless dimensional analysis:

    $$M/s = k(M)(M)^2$$
    $$M/s = k(M)^3$$
    $$k = s^{-1}M^{-2}$$

7. **B**   The coefficients on the reactants in the slowest elementary step match up with the order of those reactants in the overall rate law. However, B is an intermediate that is not present at the start of the reaction and as such cannot be part of the rate law. B is in equilibrium with 2A in the first step, though, so the 2A can be substituted for B in the slow step, which then yields (B) when the rate law is determined.

8. **B**   For every two NO molecules that form, two NOCl molecules must disappear, so NO is appearing at the same rate that NOCl is disappearing. Choice (D) is wrong because for every mole of Cl_2 that forms, two moles of NOCl are disappearing, so Cl_2 is appearing at *half* the rate that NOCl is disappearing.

9. **D**   From the rate law given in the question (rate = $k[H_2][I_2]$), you can see that increasing the concentration of H_2 will increase the rate of reaction. The rate constant, k, is not affected by changes in the concentration of the reactants.

10. **C** Catalysts work by creating a new reaction pathway with a lower activation energy than the original pathway.

11. **C** From a comparison of experiments 1 and 2, when [B] is doubled while [A] is held constant, the rate doubles. That means that the reaction is first order with respect to B.

From a comparison of experiments 2 and 3, when [A] is doubled while [B] is held constant, the rate doesn't change. That means that the reaction is zero order with respect to A and that A will not appear in the rate law.

So the rate law is rate = k[B].

12. **A** The overall rate law is always equal to the rate law of the slowest elementary step. The rate law of any elementary step can be determined using the coefficients of the reactants in that step.

13. **A** The key to this question is to recognize that reactant A is disappearing with a characteristic half-life. This is a signal that the reaction is first order with respect to A. So the rate law must be rate = k[A].

14. **D** The reaction is a second-order reaction, of which a graph of the inverse of concentration always produces a straight line with a slope equal to k, the rate constant.

15. **B** Make a chart. Always start at time = 0.

Half-Lives	Time	Stuff
0	0	100%
1	X	50%
2	44 min.	25%

It takes two half-lives for the amount of $^{44}_{19}K$ to decrease to 25 percent. If the time it takes for two half-lives is 44 minutes, one half-life must be 22 minutes.

16. **D** Between trial 1 and 3, the concentration of A tripled while the concentration of B was held constant, and the rate went up by a factor of nine. $3^2 = 9$, so the reaction is second order with respect to A. Between trial 1 and 2, the concentration of A was held constant while the concentration of B tripled, and the rate also went up by a factor of three. Thus, the reaction is first order with respect to B.

Free-Response

1. (a) Comparing the results of experiments 1 and 2, you can see that when [NO] doubles, the rate quadruples, so the reaction is second order with respect to NO.

   Comparing the results of experiments 1 and 3, you can see that when [Br$_2$] doubles, the rate doubles, so the reaction is first order with respect to Br$_2$.

   Rate = $k[NO]^2[Br_2]$

   (b) Using the values from experiment 1, we first need to realize that the rate of reaction would be half of the rate of appearance of NOBr, because of the "2" coefficient in front of the NOBr. So, the rate of reaction for experiment 1 would be (0.096 M/s)/2 = 0.048 M/s. Plugging that into the rate law yields:

   Rate = $k[NO]^2[Br_2]$ 0.048 M/s = $k(0.02\ M)^2(0.02\ M)$ $k = 6 \times 10^3\ M^{-2} \cdot s^{-1}$

   (c) In experiment 2, you started with [Br$_2$] = 0.02 M, so 0.01 M was consumed.

   From the balanced equation, 2 moles of NO are consumed for every mole of Br$_2$ consumed. So 0.02 M of NO are consumed.

   [NO] remaining = 0.04 M – 0.02 M = 0.02 M

   (d) Choice (I) agrees with the rate law.

   NO + NO $\rightleftharpoons$ N$_2$O$_2$ (fast)

   N$_2$O$_2$ + Br$_2$ $\rightarrow$ 2 NOBr   (slow)

   The slow step is the rate-determining step, with the following rate law:

   Rate = $k[N_2O_2][Br_2]$

   Looking at the first step in the proposed mechanism, you can replace the N$_2$O$_2$ with 2 NO. Doing so gives you:

   2 NO + Br$_2$ $\rightarrow$ 2 NOBr

   The rate law for that elementary step matches the overall rate law determined previously.

2. (a) Use the First-Order Rate Law and insert the first two lines from the table.

$\ln[N_2O_5]_t - \ln[N_2O_5]_0 = -kt$

$\ln(0.289) - \ln(0.400) = -k(20.0 \text{ min})$

$-.325 = -k(20.0 \text{ min})$

$k = 0.0163 \text{ min}^{-1}$

(b) Use the First-Order Rate Law.

$\ln[N_2O_5]_t - \ln[N_2O_5]_0 = -kt$

$\ln(0.350) - \ln(0.400) = -(0.0163 \text{ min}^{-1})t$

$-0.134 = -(0.0163 \text{ min}^{-1})t$

$t = 8.22 \text{ min}$

(c) $\ln[N_2O_5] = \ln[N_2O_5]_0 - kt$

$\ln[N_2O_5] = \ln(0.400) - (0.0163 \text{ min}^{-1})(100. \text{ min})$

$\ln[N_2O_5] = -2.55$

$[N_2O_5] = e^{-2.55} = 0.0781 \; M$

(d) For a first-order reaction, rate = $k[N_2O_5]$ = $(0.0163 \text{ min}^{-1})(0.400 \; M)$ = $0.00652 \; M/\text{min}$.

(e) You can see from the numbers on the table that the half-life is slightly over 40 min. To calculate it exactly, use the formula

$$\text{Half-life} = \frac{0.693}{k} = \frac{0.693}{0.0163 \text{ min}^{-1}} = 42.5 \text{ min}$$

3. (a) (i) The reaction is first order. The half-life is constant; it takes 40 seconds for the concentration to drop from 0.20 M to 0.10 M, and another 40 seconds for it to drop from 0.10 M to 0.050 M. This type of exponential decay is typical of a first-order reaction.

(ii) After $t = 80$ s, another 40 seconds is an additional half-life, and it would bring the concentration down to 0.025 M (half of 0.050 M).

(b) (i) Use the stoichiometric ratios to determine the rate of appearance of P_4:

$$2.5 \times 10^{-3} \ M/s \ PH_3 \times \frac{1 \ mol \ P_4}{4 \ mol \ PH_3} = \ 6.3 \ \times \ 10^{-4} \ M/s \ P_4$$

(ii) As the reaction progresses, the rate at which the phosphine disappears will decrease, as indicated by the decreasing slope of the line. Thus, the rate of appearance of P_4 will also decrease.

(c) (i) The half-life of the phosphine will decrease. This is because at a higher temperature, the reaction will proceed faster. It will thus take less time for the phosphine to decay.

(ii) The rate law is unaffected by outside conditions and will remain unchanged.

(iii) The rate constant will increase. There is a directly proportional relationship between the rate constant and temperature. As the temperature increases, so will the rate constant.

4. (a) The rate of the reaction will increase because the rate depends on the concentration of A as given in the rate law: Rate = $k[A]^2$.

The rate constant is independent of the concentration of the reactants and will not change.

(b) The rate of the reaction will not change. If the reaction is zero order with respect to B, then the rate is independent of the concentration of B.

The rate constant is independent of the concentration of the reactants and will not change.

(c) The rate of the reaction will increase with increasing temperature because the rate constant increases with increasing temperature.

The rate constant increases with increasing temperature because at a higher temperature more gas molecules will collide with enough energy to overcome the activation energy for the reaction.

(d) Neither the rate nor the rate constant will be affected by the addition of an inert gas.

(e) The rate of the reaction will increase because decreasing the volume of the container will increase the concentration of A: Rate = $k[A]^2$.

The rate constant is independent of the concentration of the reactants and will not change.

5. (a) Line 2 is the catalyzed reaction. Adding a catalyst lowers the activation energy of the reaction, making it easier for the reaction to occur.

    (b) Line 2 shows the higher temperature distribution. At a higher temperature, more of the molecules will be at higher energies, causing the distribution to flatten out and shift to the right.

    (c) Line 1 is correct. The ln[reactant] for a first-order reaction changes in a linear fashion over time, as shown in the following equation.

$$\ln[A]_t = -kt + \ln[A]_0$$

$$y = mx + b$$

    Notice the similarity to the slope-intercept form for a linear equation.

    (d) Graph 1 is correct, showing a decrease in the concentration of A as it is consumed in the reaction and a corresponding increase in the concentration of B as it is produced.

6. (a) (i) To determine the net equation, all three equations must be added together, and species that appear on both sides of the arrow can be eliminated.

$$NH_3(aq) + OCl^-(aq) + \cancel{NH_2Cl(aq)} + NH_3(aq) + \cancel{N_2H_5^+(aq)} + \cancel{OH^-(aq)} \rightarrow$$
$$\cancel{NH_2Cl(aq)} + \cancel{OH^-(aq)} + \cancel{N_2H_5^+(aq)} + Cl^-(aq) + N_2H_4(aq) + H_2O(l)$$

$$2\,NH_3(aq) + OCl^-(aq) \rightarrow N_2H_4(aq) + H_2O(l) + Cl^-(aq)$$

        (ii) There are no catalysts present, but NH_2Cl, $N_2H_5^+$, and OH^- are all intermediates in the process.

    (b) (i) The overall rate law will match the rate law of the slowest step. The rate law of an elementary step can be determined by the reactants present, and in this case, the rate law for Step 1 matches the overall rate law. Therefore, Step 1 is the slowest step.

        (ii) Using unit analysis:

Rate = $k[NH_3][OCl^-]$
M/min = $k(M)(M)$
$k = M^{-1}\text{min}^{-1}$

7. (a) (i) $\dfrac{\left(3.8\times10^{-5}\ M\right)\ -\ \left(2.6\times10^{-5}\ M\right)}{60\ s\ -\ 120\ s}=2.0\times10^{-7}\ M/s$

(ii) The solution starts reacting (and thus, fading) immediately after it is mixed. If the student waited 30 seconds before putting the cuvette in, the calculated rate of disappearance would thus decrease.

(b) $A = abc$

$A = (26{,}000\ cm^{-1}M^{-1})(1.00\ cm)(3.8\times10^{-5}\ M)$

$A = 0.99$

(c) A reaction that is first order with respect to [CV⁺] will create a straight line in a graph of ln [CV⁺] vs. time.

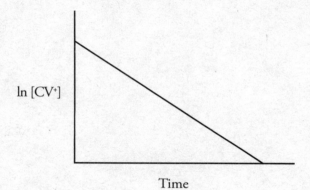

(d) Between experiments 1 and 2, the value of the [CV⁺] remained constant, while the [OH⁻] doubled. At the same time, the rate of reaction also doubled, meaning there is a direct relationship between [OH⁻] and the rate, so the reaction is first order with respect to [OH⁻]. The rate with respect to [CV⁺] has already been established as first order, so:

Rate = k[CV⁺][OH⁻]

(e) Either of the experiments can be used for this. Taking experiment 2:

$7.20\times10^{-7}\ M/s = k(5.5\times10^{-5}\ M)(0.24\ M)$

$k = 0.055\ M^{-1}s^{-1}$

(f) Rate = k[CV⁺][OH⁻]

Rate = $(0.055\ M^{-1}s^{-1})(4.1\times10^{-5}\ M)(0.18\ M)$

Rate = $4.1\times10^{-7}s^{-1}$

Chapter 8
Unit 6:
Thermodynamics

HEAT AND TEMPERATURE

Many people do not understand that heat and temperature are not, in fact, synonyms. The fact is, while the concepts are certainly related, they represent different things. **Temperature** represents the average amount of kinetic energy due to molecular motion in a given substance. **Heat** represents energy flow between two substances at different temperatures.

Energy can be neither created nor destroyed; the amount of energy in the universe is constant. Energy can only transfer and change forms. This is called the **First Law of Thermodynamics.** Energy transfers because of molecular collisions. As faster-moving molecules collide with slower-moving ones, they transfer some of their energy, changing the speed of both molecules.

Anytime energy is transferred, we classify it according to the direction of energy transfer. In an **exothermic** process, energy is transferred from the system to its surroundings. If you dissolve a salt in water and the temperature of the water increases, it is because the system itself (the salt dissolving) emitted energy. The surroundings (in this case, the water), absorbed it. For an **endothermic** process, the opposite is true. Energy transfers from the surroundings into the system. When an ice cube melts in your hand, the energy leaves your hand (the surroundings) and goes into the ice cube (the system). This is an endothermic process. Both physical and chemical changes can be classified as exothermic or endothermic.

ENTHALPY

Enthalpy Change, ΔH

The enthalpy of a substance is a measure of the energy that is released or absorbed by the substance when bonds are broken and formed during a reaction.

> **The Basic Rules of Enthalpy**
> When bonds are *formed*, energy is *released*.
> When bonds are *broken,* energy is *absorbed*.

All substances are the most stable in their lowest energy state. This means that, in general, exothermic processes are more likely to be favorable than endothermic processes.

If more energy is released when the bonds in the products form than is necessary to break the bonds in the reactants, the overall reaction releases energy into the surroundings and the reaction is called **exothermic**. Exothermic reactions have a negative enthalpy change.

If more energy is required to break the bonds in the reactants than is released when the bonds in the products form, the overall reaction requires energy from the surroundings and is called **endothermic**. Endothermic reactions have a positive enthalpy change.

ENERGY DIAGRAMS

Exothermic and Endothermic Reactions

This diagram represents an exothermic reaction, as the products are at a lower energy level than the reactants and ΔH is negative.

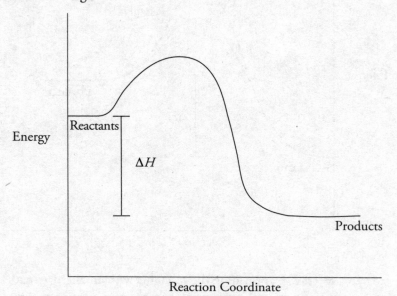

EXOTHERMIC REACTION

This diagram represents an endothermic reaction, as the products are at a higher energy level than the reactants and ΔH is positive.

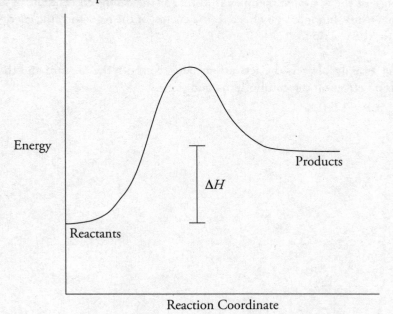

ENDOTHERMIC REACTION

Reaction diagrams can be read in both directions, so the reverse reaction for an exothermic reaction is endothermic, and vice versa.

CATALYSTS AND ENERGY DIAGRAMS

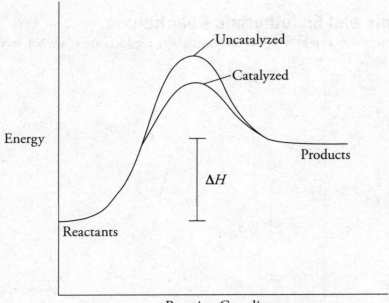

A catalyst speeds up a reaction by providing the reactants with an alternate pathway that has a lower activation energy or lowering the activation energy of an existing pathway, as shown in the diagram above.

Notice that the only difference between the catalyzed reaction and the uncatalyzed reaction is that the energy of the activated complex is lower for the catalyzed reaction. A catalyst lowers the activation energy, but it has no effect on the energy of the reactants, the energy of the products, or ΔH for the reaction.

Also note that a catalyst lowers the activation energy for both the forward and the reverse reaction, so it has no effect on the equilibrium conditions.

ENTHALPY OF FORMATION, ΔH_f°

Enthalpy of formation (also known as heat of formation) is the change in energy that takes place when one mole of a compound is formed from its component pure elements under standard state conditions. Enthalpy of formation is almost always calculated at a temperature of 25°C (298 K).

An example of an enthalpy of formation equation is below.

$$C(s) + \frac{1}{2}\,O_2(g) + 2\,H_2(g) \rightarrow CH_3OH(l)$$

Note that in any enthalpy of formation equation, there is always exactly one mole of product. When that product (in this case, CH_3OH, also known as methanol) contains an odd number of atoms from an element that is diatomic in nature, it is thus common practice to use halves to balance the equation. This practice of using fractions to balance equations is only used when dealing with diatomics and even then the only fractions that can be used are halves, so don't start getting creative when balancing other equations!

An easy mnemonic to remember the diatomic elements is Dr. HOFBrINCl! (Hydrogen, oxygen, fluorine, bromine, iodine, nitrogen, and chlorine).

ΔH_f° for a pure element is defined as zero. This is even true for elements that, in their pure state, appear as diatomic molecules (such as oxygen and hydrogen).

- If ΔH_f° for a compound is negative, energy is released when the compound is formed from pure elements, and the product is *more* stable than its constituent elements. That is, the process is exothermic.

- If ΔH_f° for a compound is positive, energy is absorbed when the compound is formed from pure elements, and the product is *less* stable than its constituent elements. That is, the process is endothermic.

If the ΔH_f° values of the products and reactants are known, ΔH for a reaction can be calculated.

$$\Delta H^\circ = \Sigma\,\Delta H_f^\circ\ products - \Sigma\,\Delta H_f^\circ\ reactants$$

The ΔH for a reaction describes the energy change in that reaction when it goes to completion as written. On the AP Exam, the units for reaction enthalpy are kJ/mol$_{rxn}$. These units are often simply just written as kJ/mol, but either way they indicate the same quantity—the amount of heat released (negative ΔH) or absorbed (positive ΔH) when the reaction occurs with the mole quantities represented in the balanced equation. Let's find ΔH° for the reaction below.

$$2\,CH_3OH(g) + 3\,O_2(g) \rightarrow 2\,CO_2(g) + 4\,H_2O(g)$$

Compound	ΔH_f° (kJ/mol)
$CH_3OH(g)$	−201
$O_2(g)$	0
$CO_2(g)$	−394
$H_2O(g)$	−242

$\Delta H° = \Sigma \Delta H_f°$ products $- \Sigma \Delta H_f°$ reactants

$\Delta H° = [(2)(\Delta H_f° \ CO_2) + (4)(\Delta H_f° \ H_2O)] - [(2)(\Delta H_f° \ CH_3OH) + (3)(\Delta H_f° \ O_2)]$

$\Delta H° = [(2)(-394 \text{ kJ}) + (4)(-242 \text{ kJ})] - [(2)(-201 \text{ kJ}) + (3)(0 \text{ kJ})]$

$\Delta H° = (-1{,}756 \text{ kJ}) - (-402 \text{ kJ})$

$\Delta H° = -1{,}354 \text{ kJ/mol}_{rxn}$

In the example above, when 2 moles of CH_3OH react with 3 moles of oxygen, exactly 1,354 kJ of energy is released. On a related note, if we were to look at HALF of the above overall reaction, we'd get:

$$CH_3OH(l) + \frac{3}{2} O_2(g) \rightarrow CO_2(g) + 2 H_2O(l)$$

$$\Delta H° = -677 \text{ kJ/mol}$$

This represents the **enthalpy of combustion** for methanol. Combustion is always an exothermic process, and the enthalpy of combustion specifically describes the amount of energy released when ONE mole of a hydrocarbon combusts. Thus, when talking about enthalpy of combustion, we again often see halves used to balance the diatomic oxygen molecule that is always part of a combustion reaction. Note that the enthalpy change is exactly half of the calculated enthalpy change from above, as we are combusting exactly half as much methanol. However, when reactions occur, they are rarely in exact quantities like that. Instead, you have to combine stoichiometry concepts with thermodynamics to determine the energy change.

Example:

How much heat is released when 5.00 g of CH_3OH is combusted in excess oxygen?

In this case, the prompt tells us that CH_3OH is the limiting reactant. Otherwise, we would have to determine which reactant was limiting. The limiting reactant not only limits the amount of products formed, but it also limits the amount of heat change that occurs during the reaction.

Since we know the CH_3OH is limiting, we can set up some stoichiometry to get to our answer:

$$5.00 \text{ g } CH_3OH \times \frac{1 \text{ mol } CH_3OH}{32.04 \text{ g } CH_3OH} \times \frac{-677 \text{ kJ}}{1 \text{ mol } CH_3OH} = -106 \text{ kJ}$$

So, 106 kJ of energy is released (indicated by the negative sign). Don't forget to take mole ratios into account when doing thermodynamics problems that involve stoichiometry. As a general rule, whenever you are given specific amounts of reactants (either in grams or moles) and asked about the energy change for a reaction, you will have to use stoichiometry to get the correct answer.

BOND ENERGY

Bond energy is the energy required to break a bond. Because the breaking of a bond is an endothermic process, bond energy is always a positive number. When a bond is formed, energy equal to the bond energy is released.

$$\Delta H° = \Sigma \text{ Bond energies of bonds broken} - \Sigma \text{ Bond energies of bonds formed}$$

The bonds broken will be the reactant bonds, and the bonds formed will be the product bonds.

The number of bonds broken and formed is affected by the number of that particular type of bond within a molecule as well as how many of those molecules there are in a balanced reaction. For instance, in the equation below, there are four O–H bonds being formed because each water molecule has two O–H bonds, and there are two water molecules present.

Let's find $\Delta H°$ for the following reaction.

$$2 \text{ H}_2(g) + \text{O}_2(g) \rightarrow 2 \text{ H}_2\text{O}(g)$$

Bond	Bond Energy (kJ/mol)
H–H	436
O=O	499
O–H	463

$\Delta H° = \Sigma$ Bond energies of bonds broken $- \Sigma$ Bond energies of bonds formed
$\Delta H° = [(2)(\text{H–H}) + (1)(\text{O=O})] - [(4)(\text{O–H})]$
$\Delta H° = [(2)(436 \text{ kJ}) + (1)(499 \text{ kJ})] - [(4)(463 \text{ kJ})]$
$\Delta H° = (1{,}371 \text{ kJ}) - (1{,}852 \text{ kJ})$
$\Delta H° = -481 \text{ kJ/mol}$

HESS'S LAW

Hess's Law states that if a reaction can be described as a series of steps, then ΔH for the overall reaction is simply the sum of the ΔH values for all the steps.

When manipulating equations for use in enthalpy calculations, there are three rules:

1. If you flip the equation, flip the sign on the enthalpy value.
2. If you multiply or divide an equation by an integer, also multiply/divide the enthalpy value by that same integer.
3. If several equations, when summed up, create a new equation, you can also add the enthalpy values of those component equations to get the enthalpy value for the new equation.

For example, let's say you want to calculate the enthalpy change for the following reaction:

$$4\,NH_3(g) + 5\,O_2(g) \rightarrow 4\,NO(g) + 6\,H_2O(g)$$

Given:

Equation A: $N_2(g) + O_2(g) \rightarrow 2\,NO(g)$ $\Delta H = 180.6$ kJ/mol

Equation B: $N_2(g) + 3\,H_2(g) \rightarrow 2\,NH_3(g)$ $\Delta H = -91.8$ kJ/mol

Equation C: $2\,H_2(g) + O_2(g) \rightarrow 2\,H_2O(g)$ $\Delta H = -483.7$ kJ/mol

First, we want to make sure all of the equations have the products and reactants on the correct side. A quick glance shows us the NH_3 should end up on the left but is given to us on the right, so equation B needs to be flipped, which will change the ΔH value to 91.8 kJ/mol. All other species appear to be on the side they need to be on, so we'll leave the other two reactions alone.

Next, we're going to see what coefficients we need to get to. The NO needs to have a 4 but is only 2 in equation A, so we'll multiply that equation by 2, which will change the ΔH value to 180.6 kJ/mol $\times$ 2 = 361.2 kJ/mol. The NH_3 also needs to have a 4 but is only 2 in equation B, so we'll also multiply that equation by 2, changing ΔH to 91.8 kJ/mol $\times$ 2 = 184 kJ/mol. Finally, the H_2O needs to have a 6 but only has a 2 in equation C, so we multiply that equation by 3, changing ΔH to -483.7 kJ/mol $\times$ 3 = $-1,451$ kJ/mol. We now have:

$$2\,N_2(g) + 2\,O_2(g) \rightarrow 4\,NO(g) \quad \Delta H = 361.2 \text{ kJ/mol}$$

$$4\,NH_3(g) \rightarrow 2\,N_2(g) + 6\,H_2(g) \quad \Delta H = 184 \text{ kJ/mol}$$

$$6\,H_2(g) + 3\,O_2(g) \rightarrow 6\,H_2O(g) \quad \Delta H = -1,451 \text{ kJ/mol}$$

Fortunately, the two O_2 on the products side add up to 5, and both the N_2 and H_2 cancel out completely, giving us the final equation that we wanted. So, our final value for the enthalpy will be 361.2 kJ/mol + 184 kJ/mol + ($-1,451$ kJ/mol) = -906 kJ/mol.

ENTHALPY OF SOLUTION

When an ionic substance dissolves in water, a certain amount of heat is emitted or absorbed. The bond between the cation and anion is breaking, which requires energy, but energy is released when those ions form new attractions to the dipoles of the water molecules. Typically, this process can be broken down into three steps. We will use NaCl dissolving in water for this example.

Step 1: Breaking the Solute Bonds

The bonds between the Na^+ and Cl^- ions must be broken. The amount of energy needed to break this bond is equal to the lattice energy (see page 94 for a more detailed explanation on lattice energy). As energy is required to break the bonds, this step always has a positive ΔH.

Step 2: Separating the Solvent Molecules

The water molecules must spread out to make room for the Na^+ and Cl^- ions. This requires energy to weaken (but not break) the intermolecular forces between them, so this step always has a positive ΔH.

Step 3: Creating New Attractions

The last step involves the free-floating ions being attracted to the dipoles of the water molecules. Note that while bonds are not being formed (no electrons are transferring or being shared), energy is still released in this process. As such, this step always has a negative ΔH.

The energy values from step 2 and step 3 combined are often called the **hydration energy.** This value is always negative, as the magnitude of the energy change in step 3 exceeds the magnitude of the energy change in step 2. As with lattice energy, hydration energy is a Coulombic energy and thus its magnitude increases as the ions either increase their charge magnitude (e.g., $|Mg^{2+}| > |Na^+|$ and $|O^{2-}| > |F^-|$) or decrease in size (e.g., $|Na^+| > |K^+|$).

If you add the energy values for all three steps together, you can determine the enthalpy of solution for that compound. Thus, if the magnitude of the hydration energy exceeds the magnitude of the lattice energy, the enthalpy of solution is negative. However, if the magnitude of the lattice energy exceeds the magnitude of the hydration energy, the enthalpy of solution is positive.

THERMODYNAMICS OF PHASE CHANGE

Naming the Phase Changes

Phase changes occur because of changes in temperature and/or pressure. Phase changes are always physical changes, as they do not involve breaking any bonds or the creation of new substances.

Solid to *liquid*	—	**Melting**
Liquid to *solid*	—	**Freezing**
Liquid to *gas*	—	**Vaporization**
Gas to *liquid*	—	**Condensation**
Solid to *gas*	—	**Sublimation**
Gas to *solid*	—	**Deposition**

Some particles in a liquid or solid will have enough energy to break away from the surface and become gaseous. The pressure exerted by these molecules as they escape from the surface is called the **vapor pressure.** When the liquid or solid phase of a substance is in equilibrium with the gas phase, the pressure of the gas will be equal to the vapor pressure of the substance. As temperature increases, the vapor pressure of a liquid will increase. When the vapor pressure of a liquid increases to the point where it is equal to the surrounding atmospheric pressure, the liquid boils. This also means that at higher elevations, where the atmospheric pressure is lower, the boiling point of liquids will also be lower.

Enthalpy of Fusion

The **enthalpy of fusion** is the energy that must be put into a solid to melt it. This energy is needed to overcome the forces holding the solid together. Alternatively, the heat of fusion is the heat given off by a substance when it freezes. The intermolecular forces within a solid are more stable and therefore have lower energy than the forces within a liquid, so energy is released in the freezing process.

Enthalpy of Vaporization

The **enthalpy of vaporization** is the energy that must be put into a liquid to turn it into a gas. This energy is needed to overcome the forces holding the liquid together. Alternatively, the heat of vaporization is the heat given off by a substance when it condenses. Intermolecular forces become stronger when a gas condenses; the gas becomes a liquid, which is more stable, and energy is released.

As heat is added to a substance, the temperature of the substance can increase or the substance can change phases, but both changes cannot occur simultaneously. Therefore, when a substance is changing phases, the temperature of that substance remains constant.

CALORIMETRY

The specific heat of a substance is the amount of heat required to raise the temperature of one gram of that substance by one degree Celsius (or one Kelvin). An object with a large **specific heat** can absorb a lot of heat without undergoing much of a temperature change, while a substance with a low specific heat will experience much greater temperature changes.

The easiest way to think about this is by considering how it feels to be outside when the air temperature is 50°F (10°C), compared to how it would feel to be in water of the same temperature. In both cases, since you are warmer than the surroundings, they would absorb heat from you and increase their temperature (meaning you lose heat and decrease your temperature). Water has a very high specific heat, so would absorb a lot of heat to change its temperature. Air, with its much lower specific heat, would absorb quite a bit less heat. That's partially why swimming in water at that temperature would feel extremely uncomfortable and rapidly cause hypothermia, but walking around outside at that temperature does not.

Specific Heat

$$q = mc\Delta T$$

q = heat added (J or cal)
m = mass of the substance (g or kg)
c = specific heat
ΔT = temperature change (K or °C)

Calorimetry is the measurement of heat changes during chemical reactions, and it is frequently accomplished via the equation above. Determining the amount of heat transfer in a reaction can lead directly to determining the enthalpy for that reaction, as shown in the example below.

$$H^+(aq) + OH^-(aq) \rightarrow H_2O(l)$$

25.0 mL of 1.5 M HCl and 30.0 mL of 2.0 M NaOH are mixed together in a Styrofoam cup and the reaction above occurs. The temperature of the reaction rises from 23.00°C to 31.60°C over the course of the reaction. Assuming the density of the solutions is 1.0 g/mL and the specific heat of the mixture is 4.18 J/g·°C, calculate the enthalpy of the reaction.

To solve this, we first need to determine the amount of heat released during the reaction. The mass of the final solution can be determined by taking the total volume of the solution (55.0 mL) and multiplying that by the density, yielding 55.0 g. The temperature change is 31.60 − 23.00 = 8.60°C. So:

$$q = mc\Delta T$$
$$q = (55.0\text{g})(4.18 \text{ J/g·°C})(8.60°C)$$
$$q = 1{,}980 \text{ J}$$

From here, we need to determine how many moles of product are formed in this reaction. Looking at the two reactants, we can see that there is going to be less of the HCl, which has a lower molarity and volume. As everything here is in a 1:1 ratio, that means the HCl will be limiting, and we can use it to determine how many moles of product will form.

$$\text{Molarity} = \text{mol/volume}$$

$$1.5 \ M = n/0.025 \ \text{L}$$

$$n = 0.038 \ \text{mol HCl} = 0.038 \ \text{mol } H_2O \text{ formed}$$

The calculated heat gain from earlier (1,980 J) was the heat gained by the water. Due to conservation of energy, this is also the heat lost by the reaction itself. To calculate the enthalpy of reaction, we have to flip that sign to indicate heat is lost, and then divide that value by the number of moles.

$$\Delta H = -1,980 \ \text{J}/0.038 \ \text{mol} = -52,000 \ \text{J/mol} = -52 \ \text{kJ/mol}$$

It is important to emphasize that ΔH is always measured in joules per mole (or kilojoules per mole). Enthalpy is not just heat; it's the amount of heat given per mole of product that is created. Limiting reagent calculations can be done in situations where it is not easy to determine which reactant is limiting by inspection, as we did above.

Also, bear in mind that if you completely dissolve a solid into water, you must add the mass of that solid to the mass of the water when doing your calculations. So, if you were to dissolve 5.0 g of NaCl into 200.0 mL of water (remember, the density of water is 1.0 g/mL), the final mass of the solution would be 205.0 g.

HEATING CURVES

A heating (or cooling) curve shows what happens to the temperature of a substance as heat is added. If the substance is in a single phase, the temperature of the substance will increase. The amount of the temperature increase can be calculated using calorimetry. If the substance is in the process of a phase change, the temperature will remain constant, and the amount of heat required to fully cause the substance to change phases can be calculated using the heat of fusion or the heat of vaporization.

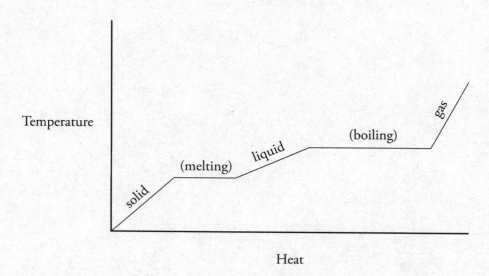

There are a lot of different units that can be used when talking about these heat versus temperature curves, so it is very important to pay extremely close attention to what your units are when doing calculations.

Example:

A 1.53 g piece of ice is in a freezer and initially at a temperature of −15.1°C. The ice is removed from the freezer and melts completely after reaching a temperature of 0.0°C. If the specific heat of ice is 2.03 J/g·°C and its molar heat of fusion is 6.01 kJ/mol, how much heat is required for the entire process to occur?

There are two parts to this problem. The first part involves the temperature change of the ice.

$$q = mc\Delta T$$
$$q = (1.53\text{g})(2.03\text{ J/g·°C})(15.1°C)$$
$$q = 46.9\text{ J}$$

The second part involves the ice melting. As this is a molar heat capacity, the number of moles of ice must first be determined.

$$1.53 \text{ g H}_2\text{O} \times \frac{1 \text{ mol H}_2\text{O}}{18.02 \text{ g}} = 0.0849 \text{ mol H}_2\text{O}$$

$$6.01 \text{ kJ/mol} \times 0.0849 \text{ mol} = 0.510 \text{ kJ} = 510. \text{ J}$$

Finally, the two values can be added together.

$$46.9 \text{ J} + 510. \text{ J} = 557 \text{ J}$$

UNIT 6 QUESTIONS

Multiple-Choice Questions

1. 1.50 g of $NaNO_3$ is dissolved into 25.0 mL of water, causing the temperature to increase by 2.2°C. The density of the final solution is found to be 1.02 g/mL. Which of the following expressions will correctly calculate the heat gained by the water as the $NaNO_3$ dissolves? Assume the volume of the solution remains unchanged.

   (A) (25.0)(4.18)(2.2)

   (B) $\dfrac{(26.5)(4.18)(2.2)}{1.02}$

   (C) $\dfrac{(1.02)(4.18)(2.2)}{1.50}$

   (D) (25.0)(1.02)(4.18)(2.2)

2. $$C(s) + 2\ S(s) \rightarrow CS_2(l) \qquad \Delta H = +92.0\ kJ/mol$$

   Which of the following energy level diagrams gives an accurate representation of the above reaction?

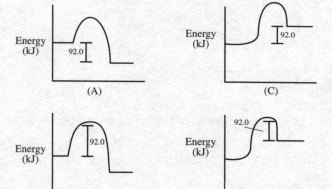

3. A student adds 100 mL of 25°C liquid water to 100 g of ice initially at 0°C, its melting point. What is the initial change observed in this system?

   (A) Some of the liquid water freezes.
   (B) The water increases in temperature.
   (C) The ice increases in temperature.
   (D) Some of the ice melts.

4.
$$2\ H_2(g) + O_2(g) \rightarrow 2\ H_2O(g)$$

Based on the information given in the table below, what is $\Delta H°$ for the above reaction?

Bond	Average bond energy (kJ/mol)
H–H	432
O=O	498
O–H	467

(A) −506 kJ
(B) −428 kJ
(C) 428 kJ
(D) 506 kJ

Use the following diagram to answer questions 5–7.

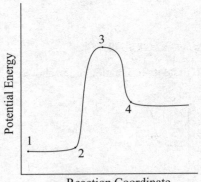

5. Which point on the graph shown above corresponds to an activated complex or transition state?

(A) 1
(B) 2
(C) 3
(D) 4

6. The distance between which two points is equal to the enthalpy change for this reaction?

(A) Points 1 and 2
(B) Points 1 and 3
(C) Points 1 and 4
(D) Points 2 and 3

7. What would happen to this graph if a catalyst were to be added?

(A) Point 3 would be lower.
(B) The distance between points 2 and 4 would be decreased.
(C) The slope of the line between points 2 and 3 would increase.
(D) Point 4 would be higher.

8.
$$C(s) + O_2(g) \rightarrow CO_2(g) \qquad \Delta H^\circ = -390 \text{ kJ/mol}$$

$$H_2(g) + \frac{1}{2} O_2(g) \rightarrow H_2O(l) \qquad \Delta H^\circ = -290 \text{ kJ/mol}$$

$$2 C(s) + H_2(g) \rightarrow C_2H_2(g) \qquad \Delta H^\circ = +230 \text{ kJ/mol}$$

Based on the information given above, what is ΔH° for the following reaction?

$$C_2H_2(g) + \frac{5}{2} O_2(g) \rightarrow 2 CO_2(g) + H_2O(l)$$

(A) −1,300 kJ
(B) −1,070 kJ
(C) −840 kJ
(D) −780 kJ

9. Consider the following reaction showing photosynthesis:

$$6 CO_2(g) + 6 H_2O(l) \rightarrow C_6H_{12}O_6(s) + 6 O_2(g) \qquad \Delta H = + 2{,}800 \text{ kJ/mol}$$

Which of the following is true regarding the thermal energy in this system?

(A) It is transferred from the surroundings to the reaction.
(B) It is transferred from the reaction to the surroundings.
(C) It is transferred from the reactants to the products.
(D) It is transferred from the products to the reactants.

10.
$$H_2(g) + F_2(g) \rightarrow 2 HF(g)$$

Gaseous hydrogen and fluorine combine in the reaction above to form hydrogen fluoride with an enthalpy change of −540 kJ. What is the value of the heat of formation of $HF(g)$?

(A) −1,080 kJ/mol
(B) −270 kJ/mol
(C) 270 kJ/mol
(D) 540 kJ/mol

Use the following information to answer questions 11–13.

When calcium chloride ($CaCl_2$) dissolves in water, the temperature of the water increases dramatically.

11. Which of the following must be true regarding the enthalpy of solution?

(A) The lattice energy in $CaCl_2$ exceeds the bond energy within the water molecules.
(B) The magnitude of the hydration energy between the water molecules and the solute ions exceeds the magnitude of the lattice energy within $CaCl_2$.
(C) The strength of the intermolecular forces between the solute ions and the dipoles on the water molecules must exceed the hydration energy.
(D) The hydration energy must exceed the strength of the intermolecular forces between the water molecules.

12. During this reaction, heat transfers from

(A) the reactants to the products
(B) the reactants to the system
(C) the system to the surroundings
(D) the products to the surroundings

13. Compared to $CaCl_2$, what must be true regarding the magnitude of the hydration energy of CaF_2?

(A) It would be greater because fluoride is smaller than chloride.
(B) It would be the same because the charges of fluoride and chloride are identical.
(C) It would be the same because hydration energy is only dependent on the IMFs present in water.
(D) It would be smaller because the molar mass of CaF_2 is smaller than that of $CaCl_2$.

Free-Response Questions

1.

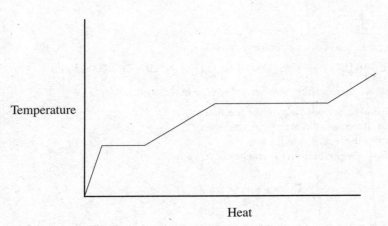

The diagram above shows how the temperature of a certain covalent substance changes as heat is added to it.

(a) Which is greater for the substance: the heat of fusion or the heat of vaporization? How do you know?

(b) If additional heat is added to the substance, the line would continue at its current slope and never become horizontal again. Why is this?

(c) Reading the graph above, a student theorizes that the specific heat capacity of the substance is greatest in the solid phase. Do you agree? Why or why not?

2. A student designs an experiment to determine the specific heat of aluminum. The student heats a piece of aluminum with a mass of 5.86 g to various temperatures, and then drops it into a calorimeter containing 25.0 mL of water. The following data is gathered during one of the trials:

Initial Temperature of Al (°C)	Initial Temperature of H_2O (°C)	Final Temperature of Al + H_2O (°C)
109.1	23.2	26.8

(a) Given that the specific heat of water is 4.18 J/g·°C and assuming its density is exactly 1.00 g/mL, calculate the heat gained by the water.

(b) Calculate the specific heat of aluminum from the experimental data given.

(c) Calculate the enthalpy change for the cooling of aluminum in water in kJ/mol.

(d) If the accepted specific heat of aluminum is 0.900 J/g·°C, calculate the percent error.

(e) Suggest two potential sources of error that would lead the student's experimental value to be different from the actual value. Be specific in your reasoning and make sure any identified error can be quantitatively tied to the student's results.

UNIT 6 ANSWERS AND EXPLANATIONS

Multiple-Choice

1. **D** Use the formula $q = mc\Delta T$. The mass is the solution's mass, which can be obtained by multiplying 25.0 mL by 1.02 g/mL.

2. **C** A positive H means the reaction is endothermic, so the products have more bond energy than the reactants. The difference between the energy levels of the products and reactants is equal to ΔH. (The difference between the energy level of the reactants and the top of the hump is the value for the activation energy.)

3. **D** Because the ice is colder, it will absorb heat from the liquid water. The liquid water is not initially at a phase transition temperature, so its temperature will decrease from this loss of heat. At its melting point, absorbing heat will cause the ice to undergo a phase change, not a change in temperature.

4. **A** The bond energy is the energy that must be put into a bond to break it. First, figure out how much energy must be put into the reactants to break their bonds.

   To break 2 moles of H–H bonds, it takes $(2)(432)$ kJ = 864 kJ.

   To break 1 mole of O=O bonds, it takes 498 kJ.

   So to break up the reactants, it takes 1,362 kJ.

   Energy is given off when a bond is formed; that's the negative of the bond energy. Now see how much energy is given off when 2 moles of H_2O are formed.

   2 moles of H_2O molecules contain 4 moles of O–H bonds, so $(4)(-467)$ kJ = $-1,868$ kJ are given off.

   So the value of $\Delta H°$ for the reaction is

   $(-1,868$ kJ, the energy given off$) + (1,362$ kJ, the energy put in$) = -506$ kJ.

5. **C** Point 3 represents the activated complex, which is the point of highest energy. This point is the transition state between the reactants and the products.

6. **C** The enthalpy change is the distance between the energy level of the reactants (on which points 1 and 2 lie) and the energy level of the products (on which point 4 lies).

7. **A** A catalyst lowers the activation energy of the reaction without changing the energy level of either the reactants or products.

8. **A** The equations given on top give the heats of formation of all the reactants and products (remember: the heat of formation of O_2, an element in its most stable form, is zero).

$\Delta H°$ for a reaction = ($\Delta H°$ for the products) – ($\Delta H°$ for the reactants).

First, the products:

From CO_2, you get (2)(–390 kJ) = –780 kJ.

From H_2O, you get –290 kJ.

So $\Delta H°$ for the products = (–780 kJ) + (–290 kJ) = –1,070 kJ.

Now the reactants:

From C_2H_2, you get +230 kJ. The heat of formation of O_2 is defined as zero, so that's it for the reactants.

$\Delta H°$ for the reaction = (–1,070 kJ) – (+230 kJ) = –1,300 kJ.

9. **A** In an endothermic reaction, heat is transferred into the reaction system.

10. **B** The reaction that forms 2 moles of HF(g) from its constituent elements has an enthalpy change of –540 kJ. The heat of formation is given by the reaction that forms 1 mole from these elements, so you can just divide –540 kJ by 2 to get –270 kJ.

11. **B** For the temperature of the water to increase, the reaction must be exothermic. If this is the case, the magnitude of the hydration energy of the solution exceeds the magnitude of the lattice energy within the solute.

12. **C** The reaction must be exothermic, since the water heats up. In an exothermic reaction, heat is transferred from the system (the reaction itself) to the surroundings (the water).

13. **A** Hydration energy (like lattice energy) is based on Coulomb's Law, which has factors of both charge and size. The charges are the same in both substances, but fluoride is smaller than chloride. As size is in the denominator of Coulomb's Law, a smaller size means there will be more hydration energy.

Free-Response

1. (a) The heat of fusion is represented by the lower horizontal line, and the heat of vaporization by the higher one. As the higher line is longer, that means more heat is required to vaporize the substance than to melt it.

   (b) The substance is a gas at its highest temperature, and it would not change phase again.

   (c) The student is incorrect. The greater the specific heat of a substance, the more heat that is required to change the temperature of that substance. On the graph, that is represented by a shallower slope. As the slope of the solid (leftmost) line is greater than the liquid (central) or gas (rightmost) lines, the solid would actually have a lower specific heat than either the liquid or the gas.

2. (a) $q = mc\Delta T$
   $q = (25.0 \text{ g})(4.18 \text{ J/g·°C})(3.6°C)$
   $q = 380 \text{ J}$

   (b) The heat gained by the water is the same as the heat lost by the aluminum.

   $q = mc\Delta T$

   $-380 \text{ J} = (5.86 \text{ g})(c)(-82.3°C)$

   $c = 0.79 \text{ J/g·°C}$

   (c) $5.86 \text{ g Al} \times \dfrac{1 \text{mol Al}}{26.98 \text{ g Al}} = 0.217 \text{ mol Al}$

   $380 \text{ J}/0.217 \text{ mol} = 1,800 \text{ J/mol} = 1.8 \text{ kJ/mol}$

   (d) $\% \text{ error} = \dfrac{|\text{experimental } - \text{ accepted}|}{\text{accepted}} \times 100\%$

   $\dfrac{|0.79 - 0.900|}{0.900} \times 100\% = 12\% \text{ error}$

(e)   Error 1: If some of the heat that was lost by the aluminum was not absorbed by the water, that would cause the calculated heat gained by the water in part (a) to be artificially low. This, in turn, would reduce the value of the specific heat of aluminum as calculated in part (b).

Error 2: If there was more than 25.0 mL of water in the calorimeter, that would mean the mass in part (a) was artificially low, which would make the calculation for the heat gained by the water also artificially low. This, in turn, would reduce the value of the specific heat of aluminum calculated in part (b).

There are many potential errors here, but as long as you can quantitatively follow them to the conclusion that the experimental value would be too low, any error (within reason) can be acceptable.

Chapter 9
Unit 7: Equilibrium

THE EQUILIBRIUM CONSTANT, K_{eq}

Most chemical processes are reversible. That is, reactants react to form products, but those products can also react to form reactants.

A good way to examine this topic is to look at a graph which shows the concentrations of the various species in a reaction as the reaction progresses. Let's use the **Haber process**, which is used in the industrial preparation of ammonia, as an example.

$$3 \, H_2(g) + N_2(g) \rightarrow 2 \, NH_3(g)$$

If we started with both reactants present at the same concentration and no products, a graph of concentration versus time might look like this.

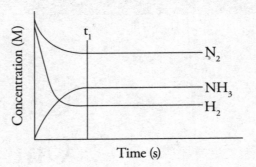

There are a few things to note here. First, at time t_1, the concentration of all three species stops changing. At this point, the reaction is said to have reached equilibrium. It's important to understand that equilibrium does NOT mean the reaction has stopped. Instead, it simply means that the rates of the forward and reverse reactions are equal, which means the concentration of all the species has stopped changing.

Also note that the coefficients in the balanced equation tell you how much the concentration of each species will change prior to reaching equilibrium. Three H_2 molecules reacted for every N_2 molecule, and so the concentration of the H_2 changed three times as much as the N_2. Likewise, the concentration of the NH_3 increased by twice as much as the N_2 decreased.

The relationship between the concentrations of reactants and products in a reaction at equilibrium is given by the equilibrium expression, also called the **Law of Mass Action.**

The Equilibrium Expression

For the reaction

$$aA + bB \rightleftharpoons cC + dD$$

$$K_{eq} = \frac{[C]^c [D]^d}{[A]^a [B]^b}$$

1. [A], [B], [C], and [D] are molar concentrations or partial pressures at equilibrium.
2. Products are in the numerator, and reactants are in the denominator.
3. Coefficients in the balanced equation become exponents in the equilibrium expression.
4. Solids and pure liquids are not included in the equilibrium expression because they cannot change their concentration. Only gaseous and aqueous species are included in the expression.
5. Units are not given for K_{eq}.

Let's look at a few examples:

1. $HC_2H_3O_2(aq) \rightleftharpoons H^+(aq) + C_2H_3O_2^-(aq)$

$$K_{eq} = K_a = \frac{[H^+][C_2H_3O_2^-]}{[HC_2H_3O_2]}$$

This reaction shows the dissociation of acetic acid in water. All of the reactants and products are aqueous particles, so they are all included in the equilibrium expression. None of the reactants or products have coefficients, so there are no exponents in the equilibrium expression. This is the standard form of K_a, the acid dissociation constant.

2. $2 H_2S(g) + 3 O_2(g) \rightleftharpoons 2 H_2O(g) + 2 SO_2(g)$

$$K_{eq} = K_c = \frac{[H_2O]^2 [SO_2]^2}{[H_2S]^2 [O_2]^3}$$

$$K_{eq} = K_p = \frac{(P_{H_2O})^2 (P_{SO_2})^2}{(P_{H_2S})^2 (P_{O_2})^3}$$

All of the reactants and products in this reaction are gases, so K_{eq} can be expressed in terms of concentration (K_c, moles/liter or molarity) or in terms of partial pressure (K_p, atmospheres). All of the reactants and products are included here, and the coefficients in the reaction become exponents in the equilibrium expression.

3. $CaF_2(s) \rightleftharpoons Ca^{2+}(aq) + 2\,F^-(aq)$
       $K_{eq} = K_{sp} = [Ca^{2+}][F^-]^2$

This reaction shows the dissociation of a slightly soluble salt. There is no denominator in this equilibrium expression because the reactant is a solid. Solids are left out of the equilibrium expression because the concentration of a solid is constant. There must be some solid present for equilibrium to exist, but you do not need to include it in your calculations. This form of K_{eq} is called the solubility product, K_{sp}.

4. $NH_3(aq) + H_2O(l) \rightleftharpoons NH_4^+(aq) + OH^-(aq)$

$$K_{eq} = K_b = \frac{\left[NH_4^+\right]\left[OH^-\right]}{\left[NH_3\right]}$$

This is the acid-base reaction between ammonia and water. We can leave water out of the equilibrium expression because it is a pure liquid. This is the standard form for K_b, the base dissociation constant.

Here is a roundup of the equilibrium constants you need to be familiar with for the test.

> A large value for K_{eq} means that products are favored over reactants at equilibrium, while a small value for K_{eq} means that reactants are favored over products at equilibrium.

- K_c is the constant for molar concentrations.
- K_p is the constant for partial pressures.
- K_{sp} is the solubility product, which has no denominator because the reactants are solids.
- K_a is the acid dissociation constant for weak acids.
- K_b is the base dissociation constant for weak bases.
- K_w describes the ionization of water ($K_w = 1 \times 10^{-14}$).

The equilibrium constant has a lot of aliases, but they all take the same form and tell you the same thing. The **equilibrium constant** tells you the relative amounts of products and reactants at equilibrium.

Manipulating K_{eq}

Much as Hess's Law allows for the determination of the enthalpy for a reaction given the enthalpy values for similar reactions, you can determine the equilibrium constant of a reaction by manipulating similar reactions with known equilibrium constants. However, the rules for doing so are different from the rules for shifting enthalpy values:

1.   If you flip a reaction, you take the reciprocal of the equilibrium constant to get the new equilibrium constant.

2.   If you multiply a reaction by a coefficient, you take the equilibrium constant to that power to get the new constant.

3.   If you add two reactions together, you multiply the equilibrium constants of those reactions to get the new constant.

Example:

$$2\,SO_2(g) + O_2(g) \rightleftharpoons 2\,SO_3(g) \qquad K = 4.0 \times 10^{24}$$

$$2\,NO(g) + O_2(g) \rightleftharpoons 2\,NO_2(g) \qquad K = 0.064$$

According to the information above, what is the equilibrium constant for each of the following reactions?

(a) $2\,SO_3(g) \rightleftharpoons 2\,SO_2(g) + O_2(g)$

This reaction is the opposite of the first reaction, so $K = \dfrac{1}{4.0 \times 10^{24}} = 2.5 \times 10^{-25}$.

(b) $SO_2(g) + \dfrac{1}{2}\,O_2(g) \rightleftharpoons SO_3(g)$

This reaction is the first reaction multiplied by one-half. Thus, the new equilibrium constant is:

$$K = (4.0 \times 10^{24})^{1/2} = 2.0 \times 10^{12}$$

(c) $SO_2(g) + NO_2(g) \rightleftharpoons SO_3(g) + NO(g)$

To get this reaction, we first have to flip the second reaction:

$$2\,NO_2 \rightleftharpoons 2\,NO + O_2$$

$$K = \dfrac{1}{0.065} = 15$$

We can then add the flipped second reaction to the original first reaction. The O_2 will cancel out, yielding the following:

$$2\,SO_2(g) + 2\,NO_2(g) \rightleftharpoons 2\,SO_3(g) + 2\,NO(g)$$

$$K = 4.0 \times 10^{24}\,(15) = 6.0 \times 10^{25}$$

Finally, we have to multiply that equation by half to get our desired reaction.

$$SO_2(g) + NO_2(g) \rightleftharpoons SO_3(g) + NO(g)$$

$$K = (6.0 \times 10^{25})^{1/2} = 7.7 \times 10^{12}$$

LE CHÂTELIER'S PRINCIPLE

At equilibrium, the rates of the forward and reverse reactions are equal. A "shift" in a certain direction means the rate of the forward or reverse reaction increases. Le Châtelier's Principle states that whenever a stress is placed on a system at equilibrium, the system will shift in response to that stress. If the forward rate increases, we say the reaction has shifted right, which will create more products. If the reverse rate increases, we say the reaction has shifted left, which will create more reactants.

We'll use the Haber process here.

$$N_2(g) + 3\ H_2(g) \rightleftharpoons 2\ NH_3(g) \qquad\qquad \Delta H° = -92.6\ kJ/mol_{rxn}$$

Concentration

When the concentration of a reactant or product is increased, the reaction will shift in the direction that allows it to use up the added substance. If N_2 or H_2 is added, the reaction shifts right. If NH_3 is added, the reaction shifts left.

When the concentration of a species is decreased, the reaction will shift in the direction that allows it to create the substance that has been removed. If N_2 or H_2 is removed, the reaction shifts left. If NH_3 is removed, the reaction shifts right.

Pressure

When the external pressure on a system is increased, that will increase the partial pressures of all of the gases inside the container and cause a shift to the side with fewer gas molecules. In the Haber process, there are four gas molecules on the left and two on the right, so an increase in partial pressures would cause a shift to the right. Decreasing the partial pressure of all gases would have the opposite effect and cause a shift to the left.

The most common way to cause a change in pressure is by changing the size of the container in which a reaction is occurring. If the Haber process were occurring inside a balloon, and we were to decrease the total volume of that balloon by squeezing it, that would increase the partial pressures of all gases involved (assuming the temperature stayed constant). Increasing the volume of the container would have the opposite effect.

In addition to altering partial pressures by changing the volume of a container, another way to change the partial pressure of gaseous species in a reaction is to add a non-reacting gas, usually a noble gas, while maintaining the same total pressure in the container. If you are dealing with a non-rigid container, adding a noble gas will simply expand the volume of the container while not changing the total pressure inside of it.

Look at it this way. Let's say we have a non-rigid container filled with 1.0 atm of N_2, 1.0 atm of H_2, and 1.0 atm of NH_3. The total pressure is thus 3.0 atm. If some helium is added but the container expands and the total pressure remains at 3.0 atm, then you might end up with the N_2, H_2, NH_3, and He all at a pressure of 0.75 atm. If the partial pressure of the reacting gases decreases, the reaction would shift to the side with more gaseous species—in this case, to the left.

On the other hand, if you add a noble gas to a rigid container with a fixed volume, the partial pressure of the noble gas would simply increase the total pressure in the container. In the above example where all three reacting species started at 1.0 atm, adding the He at 0.75 atm would simply increase the total pressure in the container to 3.75 atm and not affect the partial pressure of the other species. Thus, in this situation, no reaction shift would occur.

The above is undeniably somewhat confusing, but the best way to remember it is this. If a noble gas is added at a constant **pressure**, the equilibrium shifts to the side with more gaseous molecules. If a noble gas is added at constant **volume**, no shift occurs.

Note that for any equilibrium, there has to be a different number of gas molecules on each side of the equilibrium for a pressure change to cause any kind of shift. In equilibria where the number of gas molecules is the same on both sides, any of the above changes would cause no shift at all.

Temperature

There's a trick to figure out what happens when the temperature changes. First, rewrite the equation to include the heat energy on the side that it would be present on. The Haber process is exothermic, so heat is generated:

$$N_2(g) + 3\,H_2(g) \rightleftharpoons 2\,NH_3(g) + \text{energy}$$

Then, if the temperature goes up, the reaction will proceed in the reverse direction (shifting away from the added energy). If the temperature goes down, the reaction will proceed in the forward direction (creating more energy). The reverse would be true in an endothermic reaction, as the energy would be part of the reactants.

Dilution

One last type of shift is caused by dilution. For an aqueous equilibrium, diluting that equilibrium with extra water can cause a shift. The Haber process won't work for this one, so let's examine another equilibrium:

$$Fe^{3+}(aq) + SCN^-(aq) \rightleftharpoons FeSCN^{2+}(aq)$$

If you were to dilute the above equilibrium with extra water, it would shift to the side that has more aqueous species. In the above example, this causes a shift to the left. If some water were removed (say, by evaporation), the reaction would shift to the side with less aqueous species (right).

CHANGES IN THE EQUILIBRIUM CONSTANT

It is important to understand that shifts caused by concentration or pressure changes are temporary shifts, and do not change the value of the equilibrium constant itself. Eventually, the concentrations of the products and reactants will reestablish the same ratio as they originally had at equilibrium.

However, shifts caused by temperature changes are different. Because changing temperature also affects reaction kinetics by adding (or removing) energy from the equilibrium system, a change in temperature will also affect the equilibrium constant for the reaction itself, in addition to causing a shift. This means the ratio of the products to reactants at equilibrium will change as the temperature changes.

In the Haber process, an increase in temperature causes a shift to the left, and it would permanently affect the value of the equilibrium constant in a way that is consistent with the shift. A shift to the left causes an increase in the concentrations of the reactants (the denominator in the mass action expression) while simultaneously causing a decrease in the concentration of the product (the numerator in the mass action expression). Thus, increasing the temperature decreases the value of the equilibrium constant—that is, it causes there to be a larger amount of reactants present at equilibrium compared to the amount of products present. The reverse would be true for a temperature decrease; the shift to the right would cause an increase in the equilibrium constant.

THE REACTION QUOTIENT, Q

The reaction quotient is essentially the quantitative application of Le Châtelier's Principle. It is determined using the Law of Mass Action. However, you can use the concentration or pressure values at any point in the reaction to calculate Q. (Remember, you can only use concentrations or pressures of reactions at equilibrium to calculate K.) The value for the reaction quotient can be compared to the value for the equilibrium constant to predict in which direction a reaction will shift from the given set of initial conditions.

The Reaction Quotient

For the reaction

$$a\text{A} + b\text{B} \rightleftharpoons c\text{C} + d\text{D}$$

$$Q = \frac{[\text{C}]^c [\text{D}]^d}{[\text{A}]^a [\text{B}]^b}$$

[A], [B], [C], and [D] are initial molar concentrations or partial pressures.

- If Q is less than K, the reaction shifts right.
- If Q is greater than K, the reaction shifts left.
- If $Q = K$, the reaction is already at equilibrium.

Let's take a look at an example. The following reaction takes place in a sealed flask:

$$2\ CH_4(g) \rightleftharpoons C_2H_2(g) + 3\ H_2(g)$$

(a) Determine the equilibrium constant if the following concentrations are found at equilibrium.

$[CH_4] = 0.0032\ M$ $\qquad$ $[C_2H_2] = 0.025\ M$ $\qquad$ $[H_2] = 0.040\ M$

To solve this, we first need to create the equilibrium constant expression:

$$K_c = \frac{[C_2H_2][H_2]^3}{[CH_4]^2}$$

By plugging in the numbers, we get:

$$K_c = \frac{(0.025)(0.040)^3}{(0.0032)^2} \qquad K_c = \frac{1.6 \times 10^{-6}}{1.0 \times 10^{-5}} \qquad K_c = 0.16$$

(b) Upon testing this reaction at another point, the following concentrations are found:

$[CH_4] = 0.0055\ M$ $\qquad$ $[C_2H_2] = 0.026\ M$ $\qquad$ $[H_2] = 0.029\ M$

Use the reaction quotient to determine which way the reaction needs to shift to reach equilibrium.

Q uses the exact same ratios as the equilibrium constant expression, so:

$$Q = \frac{[C_2H_2][H_2]^3}{[CH_4]^2} = \frac{(0.026)(0.029)^3}{(0.0055)^2} = \frac{6.3 \times 10^{-7}}{3.0 \times 10^{-5}} = 0.021$$

0.021 is less than the equilibrium constant value of 0.16, so the reaction must proceed to the right, creating more products and reducing the amount of reactant in order to come to equilibrium.

SOLUBILITY

Roughly speaking, a salt can be considered "soluble" if more than 1 gram of the salt can be dissolved in 100 milliliters of water. Soluble salts are usually assumed to dissociate completely in aqueous solution. Most, but not all, solids become more soluble in a liquid as the temperature is increased.

Solubility Product (K_{sp})

Salts that are "slightly soluble" and "insoluble" still dissociate in solution to some extent. The solubility product (K_{sp}) is a measure of the extent of a salt's dissociation in solution. The K_{sp} is one of the forms of the equilibrium expression. The greater the value of the solubility product for a salt, the more soluble the salt.

Solubility Product

For the reaction

$$A_aB_b(s) \rightleftharpoons a\,A^{b+}(aq) + b\,B^{a-}(aq)$$

The solubility expression is

$$K_{sp} = [A^{b+}]^a[B^{a-}]^b$$

Here are some examples:

$$CaF_2(s) \rightleftharpoons Ca^{2+}(aq) + 2\,F^-(aq) \qquad K_{sp} = [Ca^{2+}][F^-]^2$$

$$Ag_2CrO_4(s) \rightleftharpoons 2\,Ag^+(aq) + CrO_4^{2-}(aq) \qquad K_{sp} = [Ag^+]^2[CrO_4^{2-}]$$

$$CuI(s) \rightleftharpoons Cu^+(aq) + I^-(aq) \qquad K_{sp} = [Cu^+][I^-]$$

The solubility of salts can be described by the K_{sp} or by the molar solubility. The molar solubility of the salt describes the number of moles of salt that can be dissolved per liter of solution. The K_{sp} value for Ag_2CrO_4 is 8.0×10^{-12}. We can determine the molar solubility using the following calculations:

$$K_{sp} = [Ag^+]^2[CrO_4^{2-}]$$

$$8.0 \times 10^{-12} = (2x)^2(x)$$

$$8.0 \times 10^{-12} = 4x^3$$

$$x = [CrO_4^{2-}] = 1.3 \times 10^{-4}\ M$$

$$2x = [Ag^+] = 2.6 \times 10^{-4}\ M$$

The molar solubility of the salt will also be equal to the concentration of any ion that occurs in a 1:1 ratio with the salt. In the example above, the molar solubility of the salt is 1.3×10^{-4} M, the same as the concentration of the CrO_4^{2-} ions.

Note that the concentration of the silver ions is both doubled (because twice as many of them will be in solution) AND squared in the K_{sp} expression. Thus, the 2 coefficient on the silver is represented twice in the solubility product calculations. Forgetting to account for that coefficient in the ion concentration (like using just x instead of $2x$) is a very common mistake made by students; take care to avoid it!

Typically, the molar solubility of most salts will increase with rising temperatures. This is because a higher temperature has more energy available to force the water molecules apart and make room for the solute ions. There are some salts that see their solubility decrease with increasing temperature, but there is no easy way to predict when that will occur.

The Common Ion Effect

Let's look at the solubility expression for AgCl.

$$K_{sp} = [Ag^+][Cl^-] = 1.6 \times 10^{-10}$$

If we throw a block of solid AgCl into a beaker of water, we can tell from the K_{sp} what the concentrations of Ag^+ and Cl^- will be at equilibrium. For every unit of AgCl that dissociates, we get one Ag^+ and one Cl^-, so we can solve the equation above as follows:

$$[Ag^+][Cl^-] = 1.6 \times 10^{-10}$$

$$(x)(x) = 1.6 \times 10^{-10}$$

$$x^2 = 1.6 \times 10^{-10}$$

$$x = [Ag^+] = [Cl^-] = 1.3 \times 10^{-5} \ M$$

So there are very small amounts of Ag^+ and Cl^- in the solution.

Let's say we add 0.10 mole of NaCl to 1 liter of the AgCl solution. NaCl dissociates completely, so that's the same thing as adding 0.1 mole of Na^+ ions and 0.1 mole of Cl^- ions to the solution. The Na^+ ions will not affect the AgCl equilibrium, so we can ignore them; but the Cl^- ions must be taken into account. That's because of the **common ion effect.**

The common ion effect says that the newly added Cl^- ions will affect the AgCl equilibrium, although the newly added Cl^- ions did not come from AgCl.

Let's look at the solubility expression again. Now we have 0.10 mole of Cl^- ions in 1 liter of the solution, so $[Cl^-] = 0.10\ M$.

$$[Ag^+][Cl^-] = 1.6 \times 10^{-10}$$

$$[Ag^+](0.10\ M) = 1.6 \times 10^{-10}$$

$$[Ag^+] = \frac{\left(1.6 \times 10^{-10}\right)}{(0.10)} M$$

$$[Ag^+] = 1.6 \times 10^{-9}\ M$$

Now the number of Ag^+ ions in the solution has decreased drastically because of the Cl^- ions introduced to the solution by NaCl. So when solutions of AgCl and NaCl, which share a common Cl^- ion, are mixed, the more soluble salt (NaCl) can cause the less soluble salt (AgCl) to precipitate. In general, when two salt solutions that share a common ion are mixed, the salt with the lower value for K_{sp} will precipitate first.

UNIT 7 QUESTIONS

Multiple-Choice Questions

Use the following information to answer questions 1–4.

The following reaction is found to be at equilibrium at 25°C:

$$2\,SO_3(g) \rightleftharpoons O_2(g) + 2\,SO_2(g) \qquad\qquad \Delta H = -198 \text{ kJ/mol}$$

1. What is the expression for the equilibrium constant, K_c?

   (A) $\dfrac{[SO_3]^2}{[O_2][SO_2]^2}$

   (B) $\dfrac{2[SO_3]}{[O_2]2[SO_2]}$

   (C) $\dfrac{[O_2][SO_2]^2}{[SO_3]^2}$

   (D) $\dfrac{[O_2]2[SO_2]}{2[SO_3]}$

2. Which of the following would cause the reverse reaction to speed up?

   (A) Adding more SO_3
   (B) Raising the pressure
   (C) Lowering the temperature
   (D) Removing some SO_2

3. The value for K_c at 25°C is 8.1. What must happen in order for the reaction to reach equilibrium if the initial concentrations of all three species was 2.0 M?

   (A) The rate of the forward reaction would increase, and $[SO_3]$ would decrease.
   (B) The rate of the reverse reaction would increase, and $[SO_2]$ would decrease.
   (C) Both the rate of the forward and reverse reactions would increase, and the value for the equilibrium constant would also increase.
   (D) No change would occur in either the rate of reaction or the concentrations of any of the species.

4. Which of the following would cause a reduction in the value for the equilibrium constant?

   (A) Increasing the amount of SO_3
   (B) Reducing the amount of O_2
   (C) Raising the temperature
   (D) Lowering the temperature

5. The solubility product, K_{sp}, of AgCl is 1.8×10^{-10}. Which of the following expressions is equal to the solubility of AgCl?

   (A) $\left(1.8 \times 10^{-10}\right)^2$ molar

   (B) $\dfrac{1.8 \times 10^{-10}}{2}$ molar

   (C) 1.8×10^{-10} molar

   (D) $\sqrt{1.8 \times 10^{-10}}$ molar

Use the following information to answer questions 6–8:

150 mL of saturated SrF_2 solution is present in a 250 mL beaker at room temperature. The molar solubility of SrF_2 at 298 K is 1.0×10^{-3} M.

6. What are the concentrations of Sr^{2+} and F^- in the beaker?

   (A) $[Sr^{2+}] = 1.0 \times 10^{-3}\ M$ $[F^-] = 1.0 \times 10^{-3}\ M$
   (B) $[Sr^{2+}] = 1.0 \times 10^{-3}\ M$ $[F^-] = 2.0 \times 10^{-3}\ M$
   (C) $[Sr^{2+}] = 2.0 \times 10^{-3}\ M$ $[F^-] = 1.0 \times 10^{-3}\ M$
   (D) $[Sr^{2+}] = 2.0 \times 10^{-3}\ M$ $[F^-] = 2.0 \times 10^{-3}\ M$

7. If some of the solution evaporates overnight, which of the following will occur?

   (A) The mass of the solid and the concentration of the ions will stay the same.
   (B) The mass of the solid and the concentration of the ions will increase.
   (C) The mass of the solid will decrease, and the concentration of the ions will stay the same.
   (D) The mass of the solid will increase, and the concentration of the ions will stay the same.

8. How could the concentration of Sr^{2+} ions in solution be decreased?

   (A) Adding some $NaF(s)$ to the beaker
   (B) Adding some $Sr(NO_3)_2(s)$ to the beaker
   (C) Heating the solution in the beaker
   (D) Adding a small amount of water to the beaker, but not dissolving all the solid

9. For a reaction involving nitrogen monoxide inside a sealed flask, the value for the reaction quotient (Q) was found to be 1.1×10^2 at a given point. If, after this point, the amount of NO gas in the flask increased, which reaction is most likely taking place in the flask?

   (A) $NOBr(g) \rightleftharpoons NO(g) + \frac{1}{2} Br_2(g)$ $\qquad K_c = 3.4 \times 10^{-2}$
   (B) $2\ NOCl(g) \rightleftharpoons 2\ NO(g) + Cl_2(g)$ $\qquad K_c = 1.6 \times 10^{-5}$
   (C) $2\ NO(g) + 2\ H_2(g) \rightleftharpoons N_2(g) + 2\ H_2O(g)$ $\qquad K_c = 4.0 \times 10^6$
   (D) $N_2(g) + O_2(g) \rightleftharpoons 2\ NO(g)$ $\qquad K_c = 4.2 \times 10^2$

10.
$$2 \, NOBr(g) \rightleftharpoons 2 \, NO(g) + Br_2(g)$$

The reaction above came to equilibrium at a temperature of 100°C. At equilibrium, the partial pressure due to NOBr was 4 atmospheres, the partial pressure due to NO was 4 atmospheres, and the partial pressure due to Br_2 was 2 atmospheres. What is the equilibrium constant, K_p, for this reaction at 100°C?

(A) $\dfrac{1}{4}$

(B) $\dfrac{1}{2}$

(C) 1

(D) 2

11.
$$Br_2(g) + I_2(g) \rightleftharpoons 2 \, IBr(g)$$

At 150°C, the equilibrium constant, K_c, for the reaction shown above has a value of 300. This reaction was allowed to reach equilibrium in a sealed container and the partial pressure due to $IBr(g)$ was found to be 3 atm. Which of the following could be the partial pressures due to $Br_2(g)$ and $I_2(g)$ in the container?

	$Br_2(g)$	$I_2(g)$
(A)	0.1 atm	0.3 atm
(B)	0.3 atm	1 atm
(C)	1 atm	1 atm
(D)	1 atm	3 atm

Use the information below to answer questions 12–14.

Silver sulfate, Ag_2SO_4, has a solubility product constant of 1.0×10^{-5}. The below diagram shows the products of a precipitation reaction in which some silver sulfate was formed.

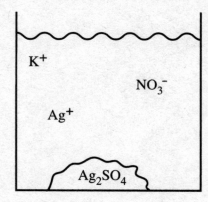

12. What is the identity of the excess reactant?

(A) $AgNO_3$
(B) Ag_2SO_4
(C) KNO_3
(D) K_2SO_4

13. If the beaker above were left uncovered for several hours

    (A) some of the Ag_2SO_4 would dissolve
    (B) some of the spectator ions would evaporate into the atmosphere
    (C) the solution would become electrically imbalanced
    (D) additional Ag_2SO_4 would precipitate

14. Which ion concentrations below would have led the precipitate to form?

    (A) $[Ag^+] = 0.01\ M\ \ [SO_4^{2-}] = 0.01\ M$
    (B) $[Ag^+] = 0.10\ M\ \ [SO_4^{2-}] = 0.01\ M$
    (C) $[Ag^+] = 0.01\ M\ \ [SO_4^{2-}] = 0.10\ M$
    (D) This is impossible to determine without knowing the total volume of the solution.

---

15. $$PCl_3(g) + Cl_2(g) \rightleftharpoons PCl_5(g) \quad \Delta H = -92.5 \text{ kJ/mol}$$

In which of the following ways could the reaction above be manipulated to create more product?

    (A) Decreasing the concentration of PCl_3
    (B) Increasing the pressure
    (C) Increasing the temperature
    (D) None of the above

Use the following information to answer questions 16–20.

A sample of H_2S gas is placed in an evacuated, sealed container and heated until the following decomposition reaction occurs at 1,000 K:

$$2\ H_2S(g) \rightarrow 2\ H_2(g) + S_2(g) \qquad K_c = 1.0 \times 10^{-6}$$

16. Which of the following represents the equilibrium constant for this reaction?

    (A) $K_c = \dfrac{[H_2]^2[S_2]}{[H_2S]^2}$

    (B) $K_c = \dfrac{[H_2S]^2}{[H_2]^2[S_2]}$

    (C) $K_c = \dfrac{2[H_2][S_2]}{2[H_2S]}$

    (D) $K_c = \dfrac{2[H_2S]}{2[H_2][S_2]}$

17. Which of the following graphs would best represent the change in concentration of the various species involved in the reaction over time?

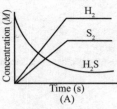
Time (s)
(A)

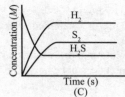
Time (s)
(C)

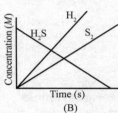
Time (s)
(B)

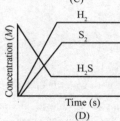
Time (s)
(D)

18. Which option best describes what will immediately occur to the reaction rates if the pressure on the system is increased after it has reached equilibrium?

(A) The rate of both the forward and reverse reactions will increase.
(B) The rate of the forward reaction will increase while the rate of the reverse reaction decreases.
(C) The rate of the forward reaction will decrease while the rate of the reverse reaction increases.
(D) Neither the rate of the forward reaction nor the rate of the reverse reaction will change.

19. If, at a given point in the reaction, the value for the reaction quotient Q is determined to be 2.5×10^{-8}, which of the following is occurring?

(A) The concentration of the reactant is decreasing while the concentration of the products is increasing.
(B) The concentration of the reactant is increasing while the concentration of the products is decreasing.
(C) The system has passed the equilibrium point, and the concentration of all species involved in the reaction will remain constant.
(D) The concentrations of all species involved are changing at the same rate.

20. As the reaction progresses at a constant temperature of 1,000 K, how does the value for the Gibbs free energy constant for the reaction change?

(A) It stays constant.
(B) It increases exponentially.
(C) It increases linearly.
(D) It decreases exponentially.

Free-Response Questions

1. The value of the solubility product, K_{sp}, for calcium hydroxide, $Ca(OH)_2$, is 5.5×10^{-6}, at 25°C.

   (a) Write the K_{sp} expression for calcium hydroxide.
   (b) What is the mass of $Ca(OH)_2$ in 500 mL of a saturated solution at 25°C?
   (c) What is the pH of the solution in (b)?
   (d) If 1.0 mole of OH^- is added to the solution in (b), what will be the resulting Ca^{2+} concentration? Assume that the volume of the solution does not change.

2. For sodium chloride, the solution process with water is endothermic.

   (a) Describe the change in entropy when sodium chloride dissociates into aqueous particles.
   (b) Two separate saturated aqueous NaCl solutions, one exposed to 1 atm of air pressure and one exposed to 2 atm of air pressure, are compared. Assuming all other conditions are the same, which will have the higher concentration? Justify your answer.
   (c) Which way will the solubility reaction shift if the temperature is increased?
   (d) If a saturated solution of NaCl is left out overnight and some of the solution evaporates, how will that affect the amount of solid NaCl present?

3. $N_2(g) + 3\ H_2(g) \rightleftharpoons 2\ NH_3(g)$ $\qquad \Delta H = -92.4$ kJ

   When the reaction above took place at a temperature of 570 K, the following equilibrium concentrations were measured:

   $[NH_3] = 0.20$ mol/L
   $[N_2] = 0.50$ mol/L
   $[H_2] = 0.20$ mol/L

   (a) Write the expression for K_c and calculate its value.
   (b) Calculate ΔG for this reaction.
   (c) Describe how the concentration of H_2 at equilibrium will be affected by each of the following changes to the system at equilibrium:
      (i)   The temperature is increased.
      (ii)   The reaction is run in the presence of an iron catalyst.
      (iii) N_2 gas is added to the reaction chamber.
      (iv)   Helium gas is added to the reaction chamber.

4. In an acidic medium, iron (III) ions will react with thiocyanate (SCN^-) ions to create the following complex ion:

$$Fe^{3+}(aq) + SCN^-(aq) \rightleftharpoons FeSCN^{2+}(aq)$$

Initially, the solution is a light yellow color due to the presence of the Fe^{3+} ions. As the $FeSCN^{2+}$ forms, the solution will gradually darken to a golden yellow. The reaction is not a fast one, and generally after mixing the ions the maximum concentration of $FeSCN^{2+}$ will occur between 2–4 minutes after mixing the solution.

A student creates four solutions with varying concentration of $FeSCN^{2+}$ and gathers the following data at 298 K using a spectrophotometer calibrated to 460 nm:

[FeSCN²⁺]	Absorbance
1.1×10^{-4} M	0.076
1.6×10^{-4} M	0.112
2.2×10^{-4} M	0.167
2.5×10^{-4} M	0.199

(a) (i) On the axes below, create a Beer's Law calibration plot for [$FeSCN^{2+}$]. Draw a best-fit line through your data points.

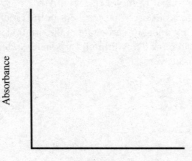

Concentration ($\times 10^{-4}$ M)

(ii) The slope of the best-fit line for the above set of data points is 879 and the y-intercept is -0.024. Write out the equation for this line.

To determine the equilibrium constant for the reaction, a solution is made up in which 5.00 mL of 0.0025 M $Fe(NO_3)_3$ and 5.00 mL of 0.0025 M KSCN are mixed. After 3 minutes, the absorbance of the solution is found to be 0.134.

(b) (i) Using your Beer's Law best-fit line from (a), calculate [$FeSCN^{2+}$] once equilibrium has been established.
(ii) Calculate [Fe^{3+}] and [SCN^-] at equilibrium.
(iii) Calculate K_{eq} for the reaction.

After equilibrium is established, the student heats the solution and observes that it becomes noticeably lighter.

(c) (i) Did heating the mixture increase the equilibrium constant, decrease it, or have no effect on it? Why?
(ii) Is the equilibrium reaction exothermic or endothermic? Justify your answer.

5. A student is tasked with determining the identity of an unknown carbonate compound with a mass of 1.89 g. The compound is first placed in water, where it dissolves completely. The K_{sp} value for several carbonate-containing compounds are given below.

Compound	K_{sp}
Lithium carbonate	8.15×10^{-4}
Nickel (II) carbonate	1.42×10^{-7}
Strontium carbonate	5.60×10^{-10}

(a) In order to precipitate the maximum amount of the carbonate ions from solution, which of the following should be added to the carbonate solution: $LiNO_3$, $Ni(NO_3)_2$, or $Sr(NO_3)_2$? Justify your answer.

(b) For the carbonate compound that contains the cation chosen in part (a), determine the concentration of each ion of that compound in solution at equilibrium.

(c) When mixing the solution, should the student ensure the carbonate solution or the nitrate solution is in excess? Justify your answer.

(d) After titrating sufficient solution to precipitate out all of the carbonate ions, the student filters the solution before placing it in a crucible and heating it to drive off the water. After several heatings, the final mass of the precipitate remains constant and is determined to be 2.02 g.
   (i) Determine the number of moles of precipitate.
   (ii) Determine the mass of carbonate present in the precipitate.

(e) Determine the percent, by mass, of carbonate in the original sample.

(f) Is the original compound most likely lithium carbonate, sodium carbonate, or potassium carbonate? Justify your answer.

UNIT 7 ANSWERS AND EXPLANATIONS

Multiple-Choice

1. **C** The equilibrium expression is always products over reactants, so that eliminates (A) and (B). The coefficients in the balanced equation turn into exponents in the expression, leading you to the correct answer.

2. **B** To speed up the reverse reaction, you are looking to cause a shift to the left via Le Châtelier's principle. If the pressure were to increase, there would be a shift to the side with fewer gas molecules, which in this case means a shift to the left. All other options cause a shift to the right.

3. **A** To determine which way the reaction would shift, the reaction quotient, Q, would need to be calculated. This can be done by plugging the concentrations into the Law of Mass Action. With all of the initial values being 2.0 M:

$$Q = \frac{(2.0)(2.0)^2}{(2.0)^2} = 2.0$$

2.0 < 8.1, so the reaction must shift right in order for the reaction to proceed to equilibrium. This would cause an increase in the rate of the forward reaction, along with a decrease in the $[SO_3]$.

4. **C** Changing the amounts of either reactants or products present will not cause a change in the equilibrium constant (nor would changing the pressure, if that were an option). The only way to actually change the value of the constant is by changing the temperature. As this is an exothermic reaction, adding more heat would cause a shift to the left, increasing the amount of reactants, and thus the denominator in the equilibrium constant expression. This, in turn, reduces the value for K_c.

5. **D** The solubility of a substance is equal to its maximum concentration in solution.

For every AgCl in solution, you get one Ag^+ and one Cl^-, so the solubility of AgCl—let's call it x—will be the same as $[Ag^+]$, which is the same as $[Cl^-]$.

So for AgCl, $K_{sp} = [Ag^+][Cl^-] = 1.8 \times 10^{-10} = x^2$.

$x = \sqrt{1.8 \times 10^{-10}}$

6. **B** When SrF_2 dissociates, it creates one Sr^{2+} ion and two F^- ions. That means the concentration of fluoride ions will be twice that of strontium ions in a saturated solution.

7. **D** If the solution is saturated, that means the concentrations of Sr^{2+} and F^- are at their maximum values and would remain unchanged even if some water evaporated. If that happens, thus decreasing the volume of the solution, there will not be as much room for the solute ions in the solution. These ions would fall out of solution, increasing the mass of the solid, and maintaining the same concentration of ions in solution.

8. **A** If extra F^- ions are added to the solution via the addition of NaF (which contains an alkali metal cation and is thus fully soluble), they will bond with some of the Sr^{2+} ions that were present to create more $SrF_2(s)$ and reduce the number of Sr^{2+} ions present in solution. This is called the common ion effect. Choice (D) is incorrect because as long as there is still solid on the bottom of the beaker, the equilibrium concentration of the ions will remain unchanged.

9. **D** When $Q > K_c$, the numerator of the equilibrium expression (the product concentration) is too big, and the equation shifts to the left. This is true for both (A) and (B), meaning [NO] would decrease. When $Q < K_c$, the numerator/product concentrations need to increase. This is the case in (C) and (D), but NO(g) is only a product in (D).

10. **D** $K_p = \dfrac{P^2_{NO} P_{Br_2}}{P^2_{NOBr}} = \dfrac{(4)^2(2)}{(4)^2} = 2$

11. **A** The equilibrium expression for the reaction is as follows:

$$\frac{P_{IBr}{}^2}{P_{Br_2} P_{I_2}} = 300$$

When all of the values are plugged into the expression, (A) is the only choice that works.

$$\frac{(3)^2}{(0.1)(0.3)} = \frac{9}{0.03} = 300$$

12. **A** There are no sulfate ions in solution, which means that whatever the sulfate was attached to was the limiting reactant. Thus, silver must have been attached to the excess reactant, which is $AgNO_3$. (Remember, Ag_2SO_4 was the product, not a reactant.)

13. **D** Leaving the container uncovered will cause some of the water molecules to evaporate. This allows some Ag^+ and SO_4^{2-} ions to "fall" out of solution and combine to create more Ag_2SO_4.

14. **B** For a precipitate to form, $Q > K_{sp}$. In this case, $Q = [Ag^+]^2[SO_4^{2-}]$. The concentrations in (B) lead to a Q of 1.0×10^{-4}, which is greater than the given K_{sp} of 1.0×10^{-5}. The other options lead to a Q value that is equal to or less than K_{sp}.

15. **B** Increasing the pressure in an equilibrium reaction with any gas molecules causes a shift to the side with fewer gas molecules—in this case, the product.

16. **A** Equilibrium is always products over reactants, and coefficients in a balanced equilibrium reaction become exponents in the equilibrium expression.

17. **C** The concentration of the H_2S will decrease exponentially until it reaches a constant value. The concentrations of the two products will increase exponentially (the H_2 twice as quickly as the S_2) until reaching equilibrium.

18. **C** Increasing the pressure causes a shift toward the side with fewer gas molecules—in this case, a shift to the left. This means the reverse reaction rate increases while the forward reaction rate decreases.

19. **A** The reaction will progress until $Q = K_c$. If $Q < K_c$, the numerator of the expression (the products) will continue to increase while the denominator (the reactant) decreases until equilibrium is established.

20. **A** If the temperature is constant, then the equilibrium constant K is unchanged. Via $\Delta G = -RT \ln K$, if K and T are both constant, then so is the value for ΔG.

Free-Response

1. (a) The solubility product is the same as the equilibrium expression, but because the reactant is a solid, there is no denominator.

   $K_{sp} = [Ca^{2+}][OH^-]^2$

   (b) Use the solubility product.

   $K_{sp} = [Ca^{2+}][OH^-]^2$

   $5.5 \times 10^{-6} = (x)(2x)^2 = 4x^3$

   $x = 0.011\ M$ for Ca^{2+}

   One mole of calcium hydroxide produces 1 mole of Ca^{2+}, so the concentration of $Ca(OH)_2$ must be 0.011 M.

   Moles = (molarity)(volume)

   Moles of $Ca(OH)_2$ = (0.011 M)(0.500 L) = 0.0055 mole

   Grams = (moles)(MW)

   Grams of $Ca(OH)_2$ = (0.0055 mol)(74 g/mol) = 0.41 g

(c) You can find $[OH^-]$ from (b).

If $[Ca^{2+}] = 0.011$ M, then $[OH^-]$ must be twice that, so $[OH^-] = 0.022$ M.

$pOH = -\log[OH^-] = 1.7$

$pH = 14 - pOH = 14 - 1.7 = 12.3$

(d) Find the new $[OH^-]$. The hydroxide already present is small enough to ignore, so use only the hydroxide just added.

$$\text{Molarity} = \frac{\text{moles}}{\text{liters}}$$

$$[OH^-] = \frac{(1.0 \text{ mol})}{(0.500 \text{ L})} = 2.0 \ M$$

Now use the K_{sp} expression.

$$K_{sp} = [Ca^{2+}][OH^-]^2$$

$$5.5 \times 10^{-6} = [Ca^{2+}](2.0 \ M)^2$$

$$[Ca^{2+}] = 1.4 \times 10^{-6} \ M$$

2. (a) Entropy increases when a salt dissociates because aqueous particles are more dispersed than a solid.

(b) Pressure will only affect an equilibrium reaction if the number of moles of gas changes in the course of the reaction. In this, the reactant is solid NaCl and the products are aqueous ions, no gases are involved at all so pressure should not affect this equilibrium. The concentrations of those saturated solutions should be the same.

(c) In an endothermic reaction, the energy is part of the reactants. Increasing the temperature would thus shift the reaction to the right.

(d) The solution was already saturated, which means $[Na^+]$ and $[Cl^-]$ were already at their maximum possible values. If the volume of water decreases, the number of moles of Na^+ and Cl^- present in solution must also decrease in order for their concentrations to remain constant. In order for that to happen, some of the sodium and chloride ions present will form a precipitate, increasing the mass of NaCl(s) present.

3. (a) $$K_c = \frac{\left[NH_3\right]^2}{\left[N_2\right]\left[H_2\right]^3}$$

$$K_c = \frac{(0.20)^2}{(0.50)(0.20)^3} = 10$$

(b) $\Delta G = -RT \ln K$

$\Delta G = -(8.31 \text{ J/mol·K})(570 \text{ K}) \ln (10)$

$\Delta G = (-4,740)(2.3)$

$\Delta G = -11,000 \text{ J/mol}$

(c) (i) An increase in temperature will shift the reaction to the left, causing an increase in the H_2 concentration in the short term. The temperature increase also causes a decrease in the value for the equilibrium constant. A smaller equilibrium constant means the species in the denominator of the mass action expression will be present in increased concentrations at equilibrium, so $[H_2]$ will remain increased.

(ii) Since a catalyst will lower the activation energies for the forward and reverse reactions equally, both forward and reverse reaction rates will increase by the same amount. Addition of a catalyst has no effect on equilibrium concentrations, only how long it takes to reach equilibrium. The $[H_2]$ will not change.

(iii) Adding N_2 will initially shift the reaction right, and the concentration of H_2 will decrease. However, by the time equilibrium is reestablished, there will be more moles of all three substances in the container, so ultimately $[H_2]$ will be increased.

(iv) The addition of He gas at a constant volume will increase the total pressure in the container, but the partial pressure of each species will remain unchanged. Thus, $[H_2]$ will be the same. Note that if the He were added at constant pressure, that would lead to a different result, as described on page 242.

4. (a) (i)

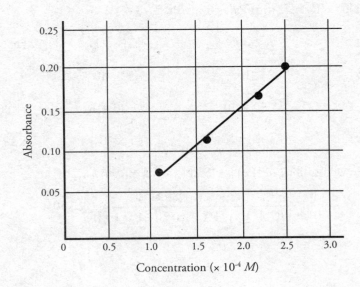

(ii) Slope-intercept form is $y = mx + b$. So:

$$y = 879x - 0.024$$

It is worth noting that in a perfect world, the y-intercept for this graph would be zero; that is, when there is no $FeSCN^{2+}$ present, the solution will not absorb light. Due to the nature of experimental data in the real world being not perfect, though, it is best in this case to include the approximate y-intercept in your best-fit line, as it may account for variations in your equipment.

(b) (i) Using the equation from part (a):

$$y = 879x - 0.024$$

$$0.134 = 879x - 0.024$$

$$x = 1.80 \times 10^{-4}\ M$$

(ii) An ICE chart will be very helpful here. To do that, you first need to calculate the initial concentration of the Fe^{3+} and the SCN^-. They both start with the same volumes and concentrations, so one calculation will suffice. Start with calculating the moles present:

$$0.0025\ M = \frac{n}{0.0050\ L} \qquad\qquad n = 1.3 \times 10^{-5}\ mol$$

Then, you have to divide by the total volume of both solutions combined (10.0 mL) in order to find out the concentration of each ion.

$$\frac{1.3 \times 10^{5}\ mol}{0.010\ L} = 1.3 \times 10^{-3}\ M$$

The ICE chart thus looks like this:

	$Fe^{3+}(aq)$	+	$SCN^-(aq)$	$\rightleftharpoons$	$FeSCN^{2+}(aq)$
I	1.3×10^{-3}		1.3×10^{-3}		0
C	-1.80×10^{-4}		-1.80×10^{-4}		$+1.80 \times 10^{-4}$
E	1.1×10^{-3}		1.1×10^{-3}		1.80×10^{-4}

So at equilibrium $[Fe^{3+}] = [SCN^-] = 1.1 \times 10^{-3}\ M$.

(iii) $K_{eq} = \dfrac{[FeSCN^{2+}]}{[Fe^{3+}][SCN^-]} = \dfrac{1.80 \times 10^{-4}}{(1.1 \times 10^{-3})(1.1 \times 10^{-3})} = 150$

(c) (i) Changing the temperature of a reaction is the only way to change the equilibrium constant, which is what is happening here. As the solution becomes lighter, the amount of $FeSCN^{2+}$ decreases, which indicates a shift to the left. Doing so causes there to be more products and fewer reactants at equilibrium, which lowers the value of the equilibrium constant.

(ii) Using the trick for temperature-caused shifts, you can see that a shift to the left means that there must have been heat on the products side of the equilibrium. This means the reaction is exothermic.

5. (a) The student should use the strontium nitrate. Using it would create strontium carbonate, which has the lowest K_{sp} value. That means it is the least soluble carbonate compound of the three and will precipitate the most possible carbonate ions out of solution.

(b) $SrCO_3(s) \rightleftharpoons Sr^{2+}(aq) + CO_3^{2-}(aq)$

$K_{sp} = [Sr^{2+}][CO_3^{2-}]$

$5.60 \times 10^{-10} = (x)(x)$

$x = 2.37 \times 10^{-5} \, M = [Sr^{2+}] = [CO_3^{2-}]$

(c) The nitrate solution should be in excess. In order to create the maximum amount of precipitate, enough strontium ions need to be added to react with all of the carbonate ions originally in solution. Having excess strontium ions in solution after the precipitate forms will not affect the calculated mass of the carbonate in the original sample.

(d) (i) The number of moles of precipitate:

$$2.02 \text{ g } SrCO_3 \times \frac{1 \text{ mol } SrCO_3}{147.63 \text{ g } SrCO_3} = 1.37 \times 10^{-2} \text{ mol } SrCO_3$$

(ii) The mass of carbonate present in the precipitate:

$$1.37 \times 10^{-2} \text{ mol } SrCO_3 \times \frac{1 \text{ mol } CO_3^{2-}}{1 \text{ mol } SrCO_3} \times \frac{60.01 \text{ g } CO_3^{2-}}{1 \text{ mol } CO_3^{2-}} = 0.822 \text{g } CO_3^{2-}$$

(e) The percent, by mass, of carbonate in the original sample:

$$\frac{0.822 \text{ g}}{1.89 \text{ g}} \times 100 = 43.5\% \, CO_3^{2-}$$

(f) The mass percent of carbonate in each compound must be compared to the experimentally determined mass percent of carbonate in the sample.

Li_2CO_3: $\dfrac{60.01}{73.89} \times 100 = 81.2\%$

Na_2CO_3: $\dfrac{60.01}{105.99} \times 100 = 56.6\%$

K_2CO_3: $\dfrac{60.01}{138.21} \times 100 = 43.4\%$

The compound is most likely potassium carbonate.

Chapter 10
Unit 8: Acids and Bases

pH

Many of the concentration measurements in acid-base problems are given in terms of pH and pOH.

$$p \text{ (anything)} = -\log \text{ (anything)}$$

$$pH = -\log [H^+]$$
$$pOH = -\log [OH^-]$$
$$pK_a = -\log K_a$$
$$pK_b = -\log K_b$$
$$pK_w = -\log K_w$$

In a solution

- when $[H^+] = [OH^-]$, the solution is neutral, and pH = 7
- when $[H^+]$ is greater than $[OH^-]$, the solution is acidic, and pH is less than 7
- when $[H^+]$ is less than $[OH^-]$, the solution is basic, and pH is greater than 7

It is important to remember that *increasing* pH means *decreasing* $[H^+]$, which means that there are fewer H^+ ions floating around and the solution is *less acidic*. Alternatively, *decreasing* pH means *increasing* $[H^+]$, which means that there are more H^+ ions floating around and the solution is *more acidic*.

As a solution's pH can tell us about both $[H^+]$ and $[OH^-]$, we can also make some determinations about the solubility of salts containing either of those ions. Specifically, many hydroxide salts will not dissolve as well in a solution with a high pH. This is due to the common ion effect discussed in the previous unit. Take the dissociation of magnesium hydroxide, shown below:

$$Mg(OH)_2(s) \rightleftharpoons Mg^{2+}(aq) + 2\,OH^-(aq)$$

If there are already a significant number of hydroxide ions in solution, that causes the above equilibrium to shift to the left, decreasing the amount of $Mg(OH)_2$ that would dissolve. Conversely, in a solution with a low pH, where there is an abundance of H^+ ions, the solubility of $Mg(OH)_2$ would increase. This is because the H^+ ions in solution will react with any dissociated OH^- ions, decreasing their concentration and shifting the equilibrium to the right.

ACID STRENGTHS

Strong Acids

Strong acids dissociate completely in water, so the reaction goes to completion and they never reach equilibrium with their conjugate bases. Because there is no equilibrium, there is no equilibrium constant, so there is no dissociation constant for strong acids or bases.

> **Important Strong Acids**
> HCl, HBr, HI, HNO_3, $HClO_4$, H_2SO_4
>
> **Important Strong Bases**
> $LiOH$, $NaOH$, KOH, $Ba(OH)_2$, $Sr(OH)_2$

Because the dissociation of a strong acid goes to completion, there is no tendency for the reverse reaction to occur, which means that the conjugate base of a strong acid must be extremely weak.

It's much easier to find the pH of a strong acid solution than it is to find the pH of a weak acid solution. That's because strong acids dissociate completely, so the final concentration of H^+ ions will be the same as the initial concentration of the strong acid.

Let's look at a 0.010-molar solution of HCl.

HCl dissociates completely, so $[H^+] = 0.010 \ M$
$pH = -\log [H^+] = -\log (0.010) = -\log (10^{-2}) = 2$

So you can always find the pH of a strong acid solution directly from its concentration.

Weak Acids

When a weak acid is placed in water, a small fraction of its molecules will dissociate into hydrogen ions (H^+) and conjugate base ions (A^-). Most of the acid molecules will remain in solution as undissociated aqueous particles.

The dissociation constants, K_a and K_b, are measures of the strengths of weak acids and bases. K_a and K_b are just the equilibrium constants specific to acids and bases.

Acid Dissociation Constant

$$K_a = \frac{[\text{H}^+][\text{A}^-]}{[\text{HA}]}$$

$[\text{H}^+]$ = concentration of hydrogen ions (M)

$[\text{A}^-]$ = concentration of conjugate base ions (M)

$[\text{HA}]$ = concentration of undissociated acid molecules (M)

Base Dissociation Constant

$$K_b = \frac{[\text{HB}^+][\text{OH}^-]}{[\text{B}]}$$

$[\text{HB}^+]$ = concentration of conjugate acid ions (M)

$[\text{OH}^-]$ = concentration of hydroxide ions (M)

$[\text{B}]$ = concentration of unprotonated base molecules (M)

The greater the value of K_a, the greater the extent of the dissociation of the acid and the stronger the acid. The same thing goes for K_b, but in the case of K_b, the base is not dissociating. Instead, it is accepting a hydrogen ion (a proton) from an acid. So, a base does not dissociate—it protonates (or ionizes).

If you know the K_a for an acid and the concentration of the acid, you can find the pH. For instance, let's look at a 0.20-molar solution of $HC_2H_3O_2$, with $K_a = 1.8 \times 10^{-5}$.

First, we set up the K_a equation, plugging in values.

$$HC_2H_3O_2 \rightleftharpoons H^+ + C_2H_3O_2^-$$

$$K_a = \frac{[\text{H}^+][\text{C}_2\text{H}_3\text{O}_2^-]}{[\text{HC}_2\text{H}_3\text{O}_2]}$$

Therefore, the ICE (Initial, Change, Equilibrium) table for the above problem is as follows:

	$[HC_2H_3O_2]$	$[H^+]$	$[C_2H_3O_2^-]$
Initial	0.20	0.0	0.0
Change	$-x$	$+x$	$+x$
Equilibrium	$0.20 - x$	x	x

Because every acid molecule that dissociates produces one H^+ and one $C_2H_3O_2^-$,

$$[H^+] = [C_2H_3O_2^-] = x$$

and because, strictly speaking, the molecules that dissociate should be subtracted from the initial concentration of $HC_2H_3O_2$, $[HC_2H_3O_2]$ should be ($0.20\ M - x$). In practice, however, x is almost always insignificant compared with the initial concentration of acid, so we just use the initial concentration in the calculation.

$$[HC_2H_3O_2] = 0.20\ M$$

Now we can plug our values and variable into the K_a expression.

$$1.8 \times 10^{-5} = \frac{x^2}{0.20}$$

Solve for x.

$$x = [H^+] = 1.9 \times 10^{-3}$$

Now that we know $[H^+]$, we can calculate the pH.

$$pH = -\log [H^+] = -\log (1.9 \times 10^{-3}) = 2.7$$

This is the basic approach to solving many of the weak acid/base problems that will be on the test.

Another way to write out the dissociation of a weak acid is:

$$HA(aq) + H_2O(l) \rightleftharpoons A^- + H_3O^+$$

The above includes water molecules in the reaction, and it is technically more accurate overall, as an acid will not dissociate unless it has a base to give protons to. The H_3O^+ ion is called the hydronium ion, and you can substitute it for H^+ in any acid/base reaction. So, $-\log [H_3O^+] = pH$.

In solving the above problem, you were told that the "$-x$" term in "$0.20 - x$" (henceforth called the dissociation value) is almost always insignificant. In that problem, "x" turned out to be 0.0019, and if you plug that into the original term you get $0.20 - 0.0019 = 0.20$ (when using significant figures correctly). Those values are just approximations, of course, but the point is that the very small amount of acid that dissociates does not change the effective concentration of the acid.

That being said, if the weak acid has a high enough K_a value, the dissociation value can become significant. It is true that you will never be asked to solve a problem like the one above in which the dissociation value term is significant, because that would require using the quadratic equation, which is not required on the exam. However, if instead of being given the K_a and concentration values for a weak acid, you are given the pH and asked to solve for either the acid's concentration or K_a, you should not neglect the dissociation value. The best way to show this is through another example.

A 0.50 M solution of HClO has a pH of 1.13. Determine the K_a value for HClO.

First, let's look at the K_a expression.

$$K_a = \frac{[H^+][ClO^-]}{[HClO]}$$

The ICE chart would then look like this:

	[HClO]	[H$^+$]	[ClO$^-$]
Initial	0.50	0	0
Change	$-x$	$+x$	$+x$
Equilibrium	$0.50 - x$	x	x

As we are given the pH, we can easily calculate [H$^+$], which is equal to x.

$1.13 = -\log[H^+]$

$[H^+] = 0.074\ M$

Plugging those numbers into the K_a expression yields:

$$K_a = \frac{(0.074)(0.074)}{(0.50 - 0.074)}$$

$$K_a = \frac{(0.074)(0.074)}{(0.43)}$$

$$K_a = 0.013 = 1.3 \times 10^{-2}$$

Solving without accounting for the dissociation value in the denominator here would have yielded a similar, but not identical, result. Even though the result would be similar, the points awarded on problems like these often focus on the process just as much as the actual result, which is why it is important to make sure the dissociation value is accounted for.

The exam is written in a way that is kind of tricky, in that it is safe to ignore the dissociation value when attempting to solve for pH because not doing so would lead to a quadratic equation, which is a skill not required on the exam. However, it is often **not** safe to ignore the dissociation value when attempting to solve for K_a or acid concentration. The best advice, therefore, is to only ignore the dissociation value if using it would lead to a quadratic equation.

Percent Dissociation

The primary factor when it comes to determining acid strength is that the more H^+ ions that an acid can donate, the stronger that acid will be. How easily an acid dissociates is often determined in part by its structure.

Consider the binary acids composed of hydrogen and a halogen. Of those, HI, HBr, and HCl are all strong acids, meaning they dissociate completely. HF, however, is not a strong acid. The reason for this is that fluorine is extremely electronegative, and thus fluorine will "hold on" to the hydrogen more effectively.

If you consider oxoacids, though, the reverse trend is true. Consider the Lewis diagrams for HOF and HOBr:

$$H - \ddot{\underset{\cdot\cdot}{O}} - \ddot{\underset{\cdot\cdot}{F}}: \qquad H - \ddot{\underset{\cdot\cdot}{O}} - \ddot{\underset{\cdot\cdot}{B}}r:$$

In this case, fluorine's very high electronegativity affects the O–F bond, drawing the shared electrons toward the fluorine. However, fluorine is so electronegative that it also attracts the shared electrons in the H–O bond, which weakens the overall H–O bond and makes the H more likely to dissociate. Thus, HOF is a stronger acid than HOBr.

Additionally, if you attach additional oxygens to an oxoacid, that affects the strength as well. If you compare $HBrO_2$ to $HBrO_3$, the stronger acid will be $HBrO_3$. The reason for this involves the conjugate bases of both acids. In BrO_3^-, the negative charge on the ion is spread out over more oxygen atoms, meaning the charge on each individual oxygen atom is weaker. Consequently, they are less likely to attract hydrogen ions and thus are more likely to stay dissociated as BrO_3^-. The more BrO_3^- there are in solution, the more dissociated H^+ there are as well, leading to a stronger acid. The overall trend is that the more oxygens that are attached to an oxoacid, the stronger it will be.

The strength of an acid is very case-specific, but the one rule that is true no matter what the situation is that the easier it is for the H^+ ion to break free, the stronger the acid will be.

> HAt is also considered to be a strong acid; however, given the extremely high reactivity as well as the radioactive nature of astatine, HAt is extremely unstable and of no practical laboratory use.

The other factor that affects how easily an acid can dissociate is the concentration of the acid. The lower the concentration is for an acid, the higher the percent dissociation will be. This is because the forward reaction involves the acid donating a proton to a water molecule. In an acid, there is an overabundance of water molecules, so it is very easy for the acid to find a water molecule to donate to. However, the reverse reaction is the hydronium ion donating a proton to the conjugate base. There are a much smaller number of both hydronium and conjugate base ions in a dilute solution, so the reverse reaction is kinetically hindered from happening. This means that more H_3O^+ ions will stay dissociated.

The greater the concentration of the acid, the more conjugate base there will be, and the easier it will be for that reverse reaction to take place. The easier it is for the reverse reaction to take place, the more HA there will be present in solution, and the less H_3O^+, leading to a lower overall percent dissociation.

Acid/Base Structure

You may have noticed that not all hydrogens in an acid are dissociable. You may wonder why acetic acid is often written as $HC_2H_3O_2$. Why not $C_2H_4O_2$? To answer that, let's take a look at the structure of acetic acid:

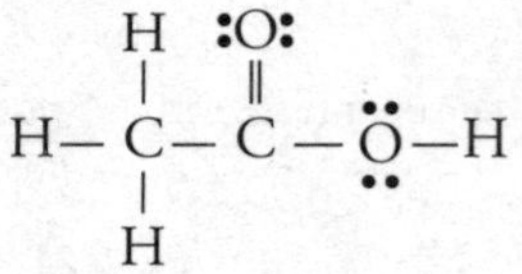

The dissociable hydrogen is the one attached to the oxygen atom at the end. That group, the –OH group, is called a hydroxyl group. When it comes to substances acting as acids, it is almost always hydrogen atoms at the end of hydroxyl groups that are dissociable. This is due to the high electronegativity difference between oxygen and hydrogen—the electrons in the bond spend so much time around the oxygen atom to begin with, that it's pretty easy for the proton (H^+) to just "fall" off.

On the other hand, hydrogen atoms that are bonded to carbon atoms are almost never dissociable, as carbon and hydrogen have a pretty similar electronegativity and thus share their electrons more equally. Generally, to help avoid confusion when acid formulas are written, any dissociable H^+ are listed at the beginning of the acid, while any non-dissociable H atoms are written in the middle.

The flip side of this is the fact that many bases have $–NH_2$ groups, which are called "amine" groups. Take methylamine below:

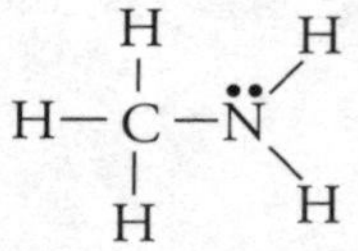

Any proton accepted by methylamine would attach itself to the nitrogen at the end. Weak bases that accept protons onto a nitrogen atom are called nitrogenous bases.

pH and Solubility

The pH of a solution can affect the solubility of compounds that contain either hydrogen ions or (more commonly) hydroxide ions. This is due to the common ion effect we discussed last chapter. Let's look at magnesium hydroxide as an example.

$$Mg(OH)_2(s) \rightleftharpoons Mg^{2+}(aq) + 2\ OH^-(aq)$$

$Mg(OH)_2$ would be less soluble in a solution with a high pH than it would in a solution with a lower pH. This is because a higher pH means more OH^- ions in solution, which would shift the equilibrium to the left. This would cause less solid to dissolve, reducing solubility. Conversely, a lower pH means more H^+ ions, which will react with the OH^- in solution, decreasing their concentration and shifting the equilibrium to the right. This would cause more solid to dissolve, increasing solubility.

Note this is no different than the normal common ion effect; however, it's kind of a round-about way to test it, as the exam can simply discuss the pH of various solutions without directly discussing the H^+ or OH^- concentrations.

Polyprotic Acids

Some acids, such as H_2SO_4 and H_3PO_4, can give up more than one hydrogen ion in solution. These are called **polyprotic** acids.

Polyprotic acids are always willing to give up their first proton more easily than future protons. Looking at the three K_a values for phosphoric acid:

$$H_3PO_4 \rightleftharpoons H_2PO^- + H^+ \qquad K_{a1} = 7.1 \times 10^{-3}$$

$$H_2PO_4^- \rightleftharpoons HPO_4^{2-} + H^+ \qquad K_{a2} = 6.3 \times 10^{-8}$$

$$HPO_4^{2-} \rightleftharpoons PO_4^{3-} + H^+ \qquad K_{a3} = 4.5 \times 10^{-13}$$

So, H_3PO_4 is a stronger acid than H_2PO^{4-}, which in turn is a stronger acid than HPO_4^{2-}. The reason behind this is that after the first proton dissociates, the resultant negative charge on the conjugate base attracts the remaining protons more strongly. That makes it more difficult for remaining protons to dissociate, weakening the acid.

This also means that the amount of each succeeding acid would decrease. In a solution of H_3PO_4, $[H_3PO_4] > [H_2PO_4^-] > [HPO_4^{2-}] > [PO_4^{3-}]$.

The Equilibrium Constant of Water (K_w)

Water comes to equilibrium with its ions according to the following reaction:

$$H_2O(l) \rightleftharpoons H^+(aq) + OH^-(aq) \qquad K_w = 1 \times 10^{-14} \text{ at } 25°C$$

$$K_w = 1 \times 10^{-14} = [H^+][OH^-]$$

$$pH + pOH = 14$$

The common ion effect tells us that the hydrogen ion and hydroxide ion concentrations for any acid or base solution must be consistent with the equilibrium for the ionization of water. That is, no matter where the H^+ and OH^- ions came from, when you multiply $[H^+]$ and $[OH^-]$, you must get 1×10^{-14}. So for any aqueous solution, if you know the value of $[H^+]$, you can find out the value of $[OH^-]$, and vice versa.

The acid and base dissociation constants for conjugates must also be consistent with the equilibrium for the ionization of water.

$$K_w = 1 \times 10^{-14} = K_a K_b$$
$$pK_a + pK_b = 14$$

So if you know K_a as a weak acid, you can find K_b for its conjugate base, and vice versa.

It is worth noting that pH is not limited to a 0 to 14 range. While most substances will fall into that range, substances that are very acidic can have negative pHs, and substances that are very basic can have pHs that are greater than 14. While you won't find these extreme pHs in most everyday substances, they do exist.

Most acids that you would find in an acids cabinet of a chemistry storeroom are very concentrated and may have negative pH values. For instance, nitric acid is typically manufactured at a concentration of 16 M. Doing the math:

$$-\log(16) = -1.2$$

Needless to say, you should exercise extreme caution when working with concentrated acids or bases, and it should be done only while wearing the proper personal protective equipment.

Something that is important to note is that the K_w for water is 1.0×10^{-14} at a temperature of 25°C only. Like any equilibrium constant, it will change if the temperature does. In this case, the dissociation of water is an endothermic process, as bonds are being broken without any new bonds being formed. So, as temperature increases, the reaction shifts to the right, which increases the value for K_w.

This temperature increase will have an effect on the pH of water. For instance, at 50°C, $K_w = 5.48 \times 10^{-14}$. Thus, $[H^+] = [OH^-] = 2.34 \times 10^{-7}$ M and pH = pOH = 6.63. So, the pH of pure water at 50°C is NOT 7. The K_w of water is often measured by looking at the pK_w value ($-\log K_w$). As temperature and K_w increase, pK_w decreases, as illustrated in the table below.

Temperature (°C)	K_w	pK_w	pH
0	1.14×10^{-15}	14.94	7.47
10	6.81×10^{-15}	14.17	7.27
25	1.00×10^{-14}	14.00	7.00
50	5.48×10^{-14}	13.26	6.63
100	5.13×10^{-13}	12.29	6.14

NEUTRALIZATION REACTIONS

When an acid and a base mix, the acid will donate protons to the base in what is called a neutralization reaction. There are four different mechanisms for this, depending on the strengths of the acids and bases.

1. Strong acid + strong base

When a strong acid mixes with a strong base, both substances are dissociated completely. The only important ions in this type of reaction are the hydrogen and hydroxide ions. (Even though not all bases have hydroxides, all strong bases do!)

Ex: $HCl + NaOH$

Net ionic: $H^+(aq) + OH^-(aq) \rightleftharpoons H_2O(l)$

The net ionic equation for all strong acid/strong base reactions is identical—it is always the creation of water. The other ions involved in the reaction (in the example above, Cl^- and Na^+) act as spectator ions and do not take part in the reaction.

2. Strong acid + weak base

In this reaction, the strong acid (which dissociates completely), will donate a proton to the weak base. The product will be the conjugate acid of the weak base.

Ex: $HCl + NH_3$

Net ionic: $H^+(aq) + NH_3(aq) \rightleftharpoons NH_4^+(aq)$

3. Weak acid + strong base

In this reaction, the strong base will accept protons from the weak acid. The products are the conjugate base of the weak acid and water.

Ex: $HC_2H_3O_2 + NaOH$

Net ionic: $HC_2H_3O_2(aq) + OH^-(aq) \rightleftharpoons C_2H_3O_2^-(aq) + H_2O(l)$

4. Weak acid + weak base

This is a simple proton transfer reaction, in which the acid gives protons to the base.

Ex: $HC_2H_3O_2 + NH_3$

Net ionic: $HC_2H_3O_2(aq) + NH_3(aq) \rightleftharpoons C_2H_3O_2^-(aq) + NH_4^+(aq)$

BUFFERS

A **buffer** is a solution with a very stable pH. You can add acid or base to a buffer solution without greatly affecting the pH of the solution. The pH of a buffer will also remain unchanged if the solution is diluted with water or if water is lost through evaporation.

A buffer is created by placing a large amount of a weak acid or base into a solution along with its conjugate, in the form of salt. A weak acid and its conjugate base can remain in solution together without neutralizing each other.

When both the acid and the conjugate base are together in the solution, any hydrogen ions that are added will be neutralized by the base, while any hydroxide ions that are added will be neutralized by the acid, without this having much of an effect on the solution's pH.

If some sodium acetate, $NaC_2H_3O_2$, is added to a solution of acetic acid, $HC_2H_3O_2$, a buffer solution is created, as the acetate ion is the conjugate base of acetic acid. If such a buffer were created and some strong acid were added to it, the protons from the strong acid would react with the acetate ion to create more acetic acid:

$$C_2H_3O_2^-(aq) + H^+(aq) \rightleftharpoons HC_2H_3O_2(aq)$$

Likewise, if a strong base were added to the buffer, the hydroxide ions from the strong base would react with the acetic acid to create more acetate ions:

$$HC_2H_3O_2(aq) + OH^-(aq) \rightleftharpoons C_2H_3O_2^-(aq) + H_2O(l)$$

Through this, we can see that the presence of the conjugate pair is what makes the buffer effective. It's not infinitely effective, of course—if enough strong acid or base is added that ALL of either the acid or its conjugate base is reacted, the buffer will "break." Thus, the higher the concentration of the conjugate pair, the more effective the buffer will be. A solution of 1.0 M $HC_2H_3O_2$ and 1.0 M $NaC_2H_3O_2$ will resist pH change (aka "be a better buffer") than a solution of 0.10 M $HC_2H_3O_2$ and 0.10 M $NaC_2H_3O_2$.

From a calculation standpoint, the Henderson-Hasselbalch equation allows us to calculate the exact pH of a buffer.

The Henderson-Hasselbalch Equation

$$pH = pK_a + \log\frac{[A^-]}{[HA]}$$

[HA] = molar concentration of undissociated weak acid (M)

[A$^-$] = molar concentration of conjugate base (M)

$$pOH = pK_b + \log\frac{[HB^+]}{[B]}$$

[B] = molar concentration of weak base (M)

[HB$^+$] = molar concentration of conjugate acid (M)

If in our acetic acid buffer, $[HC_2H_3O_2] = 0.20\ M$ and $[C_2H_3O_2^-] = 0.50\ M$, the pH of the buffer can be calculated using the below calculations:

$$pH = pK_a + \log \frac{\left[C_2H_3O_2^-\right]}{\left[HC_2H_3O\right]}$$

$$pH = -\log\left(1.8 \times 10^{-5}\right) + \log \frac{(0.50\,M)}{(0.20\,M)}$$

$$pH = -\log\left(1.8 \times 10^{-5}\right) + \log\left(2.5\right)$$

$$pH = (4.7) + (0.40) = 5.1$$

Now let's see what happens when $[HC_2H_3O_2]$ and $[C_2H_3O_2^-]$ are both equal to 0.20 M.

$$pH = pK_a + \log \frac{\left[C_2H_3O_2^-\right]}{\left[HC_2H_3O\right]}$$

$$pH = -\log\left(1.8 \times 10^{-5}\right) + \log \frac{(0.20M)}{(0.20M)}$$

$$pH = -\log\left(1.8 \times 10^{-5}\right) + \log\left(1\right)$$

$$pH = (4.7) + (0) = 4.7$$

Notice that when the concentrations of acid and conjugate base in a solution are the same, $pH = pK_a$ (and $pOH = pK_b$). When you choose an acid for a buffer solution, it is best to pick an acid with a pK_a that is close to the desired pH. That way you can have almost equal amounts of acid and conjugate base in the solution, which will make the buffer as flexible as possible in neutralizing both added H^+ and OH^-.

That being said, any buffer in which the concentration of the weak acid is greater than the concentration of the conjugate base will logically have a pH lower than the pK_a of the acid. Any buffer in which the concentration of the conjugate base exceeds the concentration of the weak acid will have a pH that is greater than the pK_a of the acid. Note that buffers CAN be created from a weak base (such as NH_3) and its conjugate acid (NH_4^+), and although these are far less common, the same sort of logic applies as to the weak acid/conjugate base buffers discussed above.

However, you cannot create a buffer solution from a strong acid and its conjugate, because the conjugate base of a strong acid will be very weak. Take HCl as an example; the Cl^- ion that is left after the acid dissociates completely is a very weak base and will not readily accept protons. The same is true for strong bases; you cannot form a buffer from a strong base and its conjugate for similar reasons.

INDICATORS

Indicators are weak acids which change colors in certain pH ranges due to Le Châtelier's principle. All indicators are weak acids in their own right, but often have very long, very complicated chemical formulas. For simplicity's sake, we will use the generic HIn to indicate a protonated indicator.

$$HIn \rightleftharpoons H^+ + In^-$$

$$K_a = \frac{[H^+][In^-]}{[HIn]}$$

For an indicator to be effective, the protonated (HIn) state has to be a different color than the deprotonated (In⁻) state. When an indicator is present in a titration and the environment is acidic, the excess H^+ ions drive the equilibrium to the left, causing the solution to consist primarily of HIn and take on that color. In a basic environment, the excess OH^- ions react with the H^+ from the indicator and drive the reaction right, causing the solution to consist of primarily In⁻, changing color.

The indicator will experience a color shift right around the point that the two species (HIn and In⁻) are present in equal amounts. When that is the case, those two values cancel out of the equilibrium expression, leaving $K_a = [H^+]$. Doing a quick logarithm, we can see that at the point of color change, $pK_a = pH$. This means that the indicator will change colors at a pH that matches the pK_a of the indicator.

When doing a titration, you want to add only a few drops of any indicator in order to avoid changing any kind of calculations involving the acid you are titrating against (or with!). It is always best to choose an indicator whose pK_a matches the pH at the equivalence point of the titration you are doing in order to be able to find the most accurate endpoint.

TITRATION

Neutralization reactions are generally performed by titration, where a base of known concentration is slowly added to an acid (or vice versa). The progress of a neutralization reaction can be shown in a titration curve. The diagram below shows the titration of a strong acid by a strong base.

For this titration, the pH at the equivalence point is exactly 7 because the titration of a strong acid by a strong base produces a neutral salt solution.

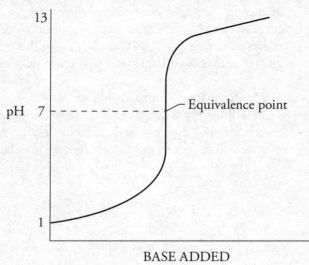

In the diagram on the previous page, the pH increases slowly but steadily from the beginning of the titration until just before the equivalence point. The **equivalence point** is the point in the titration when exactly enough base has been added to neutralize all the acid that was initially present. Just before the equivalence point, the pH increases sharply as the last of the acid is neutralized. The equivalence point of a titration can be recognized through the use of an indicator.

The following diagram shows the titration of a weak acid by a strong base:

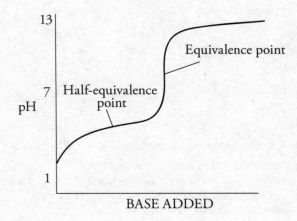

In this diagram, the pH increases more quickly at first and then levels out into a buffer region. At the center of the buffer region is the **half-equivalence point.**

At this point, enough base has been added to convert exactly half of the acid into conjugate base; here the concentration of acid is equal to the concentration of conjugate base ([HA] = [A⁻]). Putting that into our weak acid dissociation expression:

$$K_a = \frac{[H^+][A^-]}{[HA]}$$

If [A⁻] and [HA] are equal and cancel out, that leaves us with K_a = [H⁺], which is often represented by pK_a = pH. This is a good way to determine the K_a of a weak acid from a titration curve such as the one above.

The curve remains fairly flat until just before the equivalence point, when the pH increases sharply. For this titration, the pH at the equivalence point is greater than 7 because the only ion present in significant amounts at equilibrium is the conjugate base of the weak acid. Likewise, if you titrate a weak base with a strong acid, the solution will be acidic at equilibrium because the only ion present in significant amounts will be the conjugate acid of the weak base.

UNIT 8 QUESTIONS

Multiple-Choice Questions

Use the following information to answer questions 1–5.

A student titrates 20.0 mL of 1.0 M NaOH with 2.0 M formic acid, HCO_2H
($K_a = 1.8 \times 10^{-4}$). Formic acid is a monoprotic acid.

1. How much formic acid is necessary to reach the equivalence point?

   (A) 10.0 mL
   (B) 20.0 mL
   (C) 30.0 mL
   (D) 40.0 mL

2. At the equivalence point, is the solution acidic, basic, or neutral? Why?

   (A) Acidic; the strong acid dissociates more than the weak base
   (B) Basic; the only acidic or basic ion present at equilibrium is the conjugate base
   (C) Basic; the higher concentration of the base is the determining factor
   (D) Neutral; equal moles of both acid and base are present

3. If the formic acid were replaced with a strong acid such as HCl at the same concentration (2.0 M), how would that change the volume needed to reach the equivalence point?

   (A) The change would reduce the amount, as the acid now fully dissociates.
   (B) The change would reduce the amount, because the base will be more strongly attracted to the acid.
   (C) The change would increase the amount, because the reaction will now go to completion instead of equilibrium.
   (D) Changing the strength of the acid will not change the volume needed to reach equivalence.

4. Which of the following would create a good buffer when dissolved in formic acid?

   (A) $NaCO_2H$
   (B) $HC_2H_3O_2$
   (C) NH_3
   (D) H_2O

5. $CH_3NH_2(aq) + H_2O(l) \rightleftharpoons OH^-(aq) + CH_3NH_3^+(aq)$

   The above equation represents the reaction between the base methylamine ($K_b = 4.38 \times 10^{-4}$) and water. Which of the following best represents the concentrations of the various species at equilibrium?

   (A) $[OH^-] > [CH_3NH_2] = [CH_3NH_3^+]$
   (B) $[OH^-] = [CH_3NH_2] = [CH_3NH_3^+]$
   (C) $[CH_3NH_2] > [OH^-] > [CH_3NH_3^+]$
   (D) $[CH_3NH_2] > [OH^-] = [CH_3NH_3^+]$

6. A 0.1-molar solution of which of the following acids will be the best conductor of electricity?

    (A) H_2CO_3
    (B) H_2S
    (C) HF
    (D) HNO_3

7. A laboratory technician wishes to create a buffered solution with a pH of 5. Which of the following acids would be the best choice for the buffer?

    (A) $H_2C_2O_4$ $K_a = 5.9 \times 10^{-2}$

    (B) H_3AsO_4 $K_a = 5.6 \times 10^{-3}$

    (C) $H_2C_2H_3O_2$ $K_a = 1.8 \times 10^{-5}$

    (D) HOCl $K_a = 3.0 \times 10^{-8}$

8. A solution of sulfurous acid, H_2SO_3, is present in an aqueous solution. Which of the following represents the concentrations of three different ions in solution?

    (A) $[SO_3^{2-}] > [HSO_3^-] > [H_2SO_3]$

    (B) $[H_2SO_3] > [HSO_3^-] > [SO_3^{2-}]$

    (C) $[H_2SO_3] > [HSO_3^-] = [SO_3^{2-}]$

    (D) $[SO_3^{2-}] = [HSO_3^-] > [H_2SO_3]$

9. Nitrous acid, HNO_2, has a pK_a value of 3.3. If a solution of nitrous acid is found to have a pH of 4.2, what can be said about the concentration of the conjugate pair found in solution?

    (A) $[HNO_2] > [NO_2^-]$
    (B) $[NO_2^-] > [HNO_2]$
    (C) $[H_2NO_2^+] > [HNO_2]$
    (D) $[HNO_2] > [H_2NO_2^+]$

10. A 1-molar solution of a very weak monoprotic acid has a pH of 5. What is the value of K_a for the acid?

    (A) $K_a = 1 \times 10^{-10}$
    (B) $K_a = 1 \times 10^{-7}$
    (C) $K_a = 1 \times 10^{-5}$
    (D) $K_a = 1 \times 10^{-2}$

11. The value of K_a for HSO_4^- is 1×10^{-2}. What is the value of K_b for SO_4^{2-}?

    (A) $K_b = 1 \times 10^{-12}$
    (B) $K_b = 1 \times 10^{-8}$
    (C) $K_b = 1 \times 10^{-2}$
    (D) $K_b = 1 \times 10^2$

Use the following information to answer questions 12–14.

The following curve is obtained during the titration of 30.0 mL of 1.0 M NH_3, a weak base, with a strong acid:

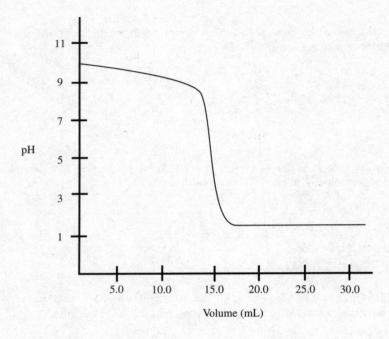

Volume (mL)

12. Why is the solution acidic at the equivalence point?

   (A) The strong acid dissociates fully, leaving excess $[H^+]$ in solution.
   (B) The conjugate acid of NH_3 is the only acidic or basic ion present at the equivalence point.
   (C) The water which is being created during the titration acts as an acid.
   (D) The acid is diprotic, donating two protons for every unit dissociated.

13. What is the concentration of the acid?

   (A) 0.5 M
   (B) 1.0 M
   (C) 1.5 M
   (D) 2.0 M

14. What ions are present in significant amounts during the first buffer region?

   (A) NH_3 and NH_4^+
   (B) NH_3 and H^+
   (C) NH_4^+ and OH^-
   (D) H_3O^+ and NH_3

15. Which of the following could be added to an aqueous solution of weak acid HF to increase the percent dissociation?

    (A) $NaF(s)$
    (B) $H_2O(l)$
    (C) $NaCl(s)$
    (D) $HCl(g)$

16. A bottle of water is left outside early in the morning. The bottle warms gradually over the course of the day. What will happen to the pH of the water as the bottle warms?

    (A) Nothing; pure water always has a pH of 7.00.
    (B) Nothing; the volume would have to change in order for any ion concentration to change.
    (C) It will increase because the concentration of $[H^+]$ is increasing.
    (D) It will decrease because the auto-ionization of water is an endothermic process.

17. The structure of two oxoacids is shown below:

$$H - \ddot{O} - \ddot{\underset{..}{Cl}}: \qquad H - \ddot{O} - \ddot{\underset{..}{F}}:$$

    Which would be a stronger acid, and why?

    (A) HOCl, because the H–O bond is weaker than in HOF as chlorine is larger than fluorine
    (B) HOCl, because the H–O bond is stronger than in HOF as chlorine has a higher electronegativity than fluorine
    (C) HOF, because the H–O bond is stronger than in HOCl as fluorine has a higher electronegativity than chlorine
    (D) HOF, because the H–O bond is weaker than in HOCl as fluorine is smaller than chlorine

18. Which of the following pairs of substances would make a good buffer solution when combined in equal molar amounts?

    (A) $HC_2H_3O_2(aq)$ and $NaC_2H_3O_2(aq)$
    (B) $H_2SO_4(aq)$ and $LiOH(aq)$
    (C) $HCl(aq)$ and $KCl(aq)$
    (D) $HF(aq)$ and $NH_3(aq)$

Free-Response Questions

1. A student tests the conductivity of three different acid samples, each with a concentration of 0.10 M and a volume of 20.0 mL. The conductivity was recorded in microsiemens per centimeter in the table below:

Sample	Conductivity (μS/cm)
1	26,820
2	8,655
3	35,120

(a)   The three acids are known to be HCl, H_2SO_4, and H_3PO_4. Identify which sample is which acid. Justify your answer.

(b)   The HCl solution is then titrated with a 0.150 M solution of the weak base methylamine, CH_3NH_2.
$(K_b = 4.38 \times 10^{-4})$
(i)   Write out the net ionic equation for this reaction.
(ii)   Determine the pH of the solution after 20.0 mL of methylamine has been added.

2.
$$HA + OH^- \rightleftharpoons A^- + H_2O(l)$$

A student titrates a weak acid, HA, with some 1.0 M NaOH, yielding the following titration curve:

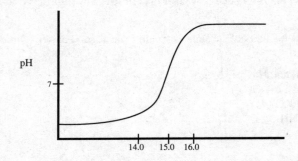

(a)   Which chemical species are present in solution dictates the pH of the solution in each of the volume ranges listed below?
(i)   1.0 mL–14.0 mL
(ii)   15.0 mL
(iii) 16.0 mL–30.0 mL

(b)   At which volumes is
(i)   [HA] > [A$^-$]?
(ii)   [HA] = [A$^-$]?
(iii) [HA] < [A$^-$]?

(c)   At which point in the titration (if any) would the concentration of the following species be equal to zero? Justify your answers.
(i)   HA
(ii)   A$^-$

(d)   If the titration were performed again, but this time with 2.0 M NaOH, name two things that would change about the titration curve, and explain the reasoning behind your identified changes.

3. A student performs an experiment to determine the concentration of a solution of hypochlorous acid, HOCl ($K_a = 3.5 \times 10^{-8}$). The student starts with 25.00 mL of the acid in a flask and titrates it against a standardized solution of sodium hydroxide with a concentration of 1.47 M. The equivalence point is reached after the addition of 34.23 mL of NaOH.

   (a) Write the net ionic equation for the reaction that occurs in the flask.

   (b) What is the concentration of the HOCl?

   (c) What would the pH of the solution in the flask be after the addition of 28.55 mL of NaOH?

   (d) The actual concentration of the HOCl is found to be 2.25 M. Quantitatively discuss whether or not each of the following errors could have caused the error in the student's results.

      (i) The student added additional NaOH past the equivalence point.

      (ii) The student rinsed the buret with distilled water but not with the NaOH solution before filling it with NaOH.

      (iii) The student measured the volume of acid incorrectly; instead of adding 25.00 mL of HOCl, only 24.00 mL was present in the flask prior to titration.

4. Aniline, $C_6H_5NH_2$, is a weak base with $K_b = 3.8 \times 10^{-10}$.

   (a) Write out the reaction that occurs when aniline reacts with water.

   (b)   (i) What is the concentration of each species at equilibrium in a solution of 0.25 M $C_6H_5NH_2$?

      (ii) What is the pH value for the solution in (i)?

UNIT 8 ANSWERS AND EXPLANATIONS

Multiple-Choice

1. **A** The equivalence point is defined as the point at which the moles of acid are equal to the moles of base. The amount of NaOH is equal to $(1.0\ M)(0.0200\ L) = 0.0200$ mol. To calculate the volume of acid:

$$2.0\ M = 0.0200\ mol/V$$

$$V = 0.0100\ L = 10.0\ mL$$

2. **B** Using the ICE chart:

	HCO$_2$H	OH$^-$	CO$_2$H$^-$	H$_2$O
Initial	0.020	0.020	0	X
Change	−0.020	−0.020	+ 0.020	X
Equilibrium	0	0	0.020	X

Both the hydroxide ion and the weak acid are not present at equilibrium. The only ionic species left, the O$_2$CH$^-$, is basic, and so the solution will be as well.

3. **D** The number of moles of base is staying the same, and so is the concentration of the acid. Therefore, the same volume of that acid will be needed to get to the equivalence point.

4. **A** A buffer is made up of an acid and its conjugate base. The conjugate base of HCO$_2$H is CO$_2$H$^-$, which is present as the anion in the NaCO$_2$H salt.

5. **D** Weak bases do not ionize fully in solution, and most of the methylamine molecules will not deprotonate. The hydroxide and conjugate acid ions are created in a 1:1 ratio and therefore will be equal.

6. **D** The best conductor of electricity (also called the strongest electrolyte) will be the solution that contains the most charged particles. HNO$_3$ is the only strong acid listed in the answer choices, so it is the only choice where the acid has dissociated completely in solution into H$^+$ and NO$_3^-$ ions. So a 0.1-molar HNO$_3$ solution will contain the most charged particles and therefore be the best conductor of electricity.

7. **C** The best buffered solution occurs when pH = pK_a. That happens when the solution contains equal amounts of acid and conjugate base. If you want to create a buffer with a pH of 5, the best choice would be an acid with a pK_a that is as close to 5 as possible. Because the exponent in the K_a of (C) is −5, the pK_a will be closer to 5 than any of the other answer choices.

8. **B**   Weak acids have low dissociation values, meaning the vast majority of the H_2SO_3 will not dissociate.

$$H_2SO_3\ (aq) \rightleftharpoons HSO_3^-\ (aq) + H^+(aq)$$

Thus, only a small amount of HSO_3^- will be produced. The second dissociation of the HSO_3^-, in turn, will be even less than the original dissociation.

$$HSO_3^-\ (aq) \rightleftharpoons SO_3^{2-}\ (aq) + H^+(aq)$$

This means very little SO_3^{2-} will be created.

9. **B**   In this case, the pH is greater than the pK_a. This means that there will be more conjugate base present in solution than the original acid. The conjugate base of HNO_2 is NO_2^-.

10. **A**   A pH of 5 means that $[H^+] = 1 \times 10^{-5}$.

$$K_a = \frac{[H^+][A^-]}{[HA]}$$

For every HA that dissociates, you get one H^+ and one A^-, so $[H^+] = [A^-] = 1 \times 10^{-5}$.

The acid is weak, so you can assume that very little HA dissociates and that the concentration of HA remains 1-molar.

$$K_a = \frac{[H^+][A^-]}{[HA]} = \frac{(1 \times 10^{-5})(1 \times 10^{-5})}{(1)} = 1 \times 10^{-10}.$$

11. **A**   For conjugates, $(K_a)(K_b) = K_w = 1 \times 10^{-14}$

$$K_b = \frac{K_w}{K_a} = \frac{(1 \times 10^{-14})}{(1 \times 10^{-2})} = 1 \times 10^{-12}$$

12. **B**   The reaction here is $NH_3 + H^+ \rightleftharpoons NH_4^+$. At the equivalence point, the moles of NH_3 and H^+ would be equal, leaving behind NH_4^+ ions, which will then donate ions to water, creating an acidic medium.

13. **D**   At the equivalence point, the moles of acid are equal to the moles of base.

Moles base = $(1.0\ M)(0.030\ L) = 0.030$ mol base = 0.030 moles acid

It requires 15.0 mL of acid to reach equivalence, so:

$$\frac{0.030\ \text{mol}}{0.015\ \text{L}} = 2.0\ M$$

14. **A** The reaction occurring is $NH_3 + H^+ \rightleftharpoons NH_4^+$. During the first buffer region, all added hydrogen ions immediately react with NH_3 to create NH_4^+. NH_3 remains in excess until the half-equivalence point.

15. **B** Percent dissociation is inversely proportional to concentration. Adding some water to the solution will cause the concentration of the HF to decrease, leading to a greater percent dissociation.

16. **D** $H_2O \rightleftharpoons H^+ + OH^-$ is an endothermic process, requiring energy in order for the bonds within the water molecules to break. Thus, heat is a reactant, and as temperature increases, the reaction will shift right. Because this is a temperature shift, this causes a permanent increase in the value of K_w. This, in turn, causes the concentrations of H^+ and OH^- to increase, and the $-\log$ of an increased concentration leads to a decreased value for pH.

17. **D** The weaker the O–H bond is in an oxoacid, the stronger the acid will be, because the H^+ ions are more likely to dissociate. The O–F bond in HOF is stronger than the O–Cl bond in HOCl because fluorine is smaller (and thus more electronegative) than chlorine. If the O–F bond is stronger, the O–H bond is correspondingly weaker, making HOF the stronger acid.

18. **A** A buffer is made up of either a weak acid and its salt or a weak base and its salt. Choice (B) has a strong acid and strong base, (C) has a strong acid and its salt, and (D) has a weak acid and a weak base.

Free-Response

1. (a) H_3PO_4 is the only weak acid, meaning it will not dissociate completely in solution. Therefore, it is acid 2. H_2SO_4 and HCl are both strong acids that will dissociate completely, but H_2SO_4 is diprotic, meaning there will be more H^+ ions in solution and thus a higher conductivity. Therefore, H_2SO_4 is acid 3, and HCl is acid 1.

(b) (i) The net ionic equation for this reaction:

$$H^+(aq) + CH_3NH_2(aq) \rightleftharpoons CH_3NH_3^+(aq)$$

(ii) First, the number of moles of hydrogen ions and methylamine need to be determined.

H^+: $0.10\ M = \dfrac{n}{0.020\ L}$ $\qquad n = 0.0020$ mol

CH_3NH_2: $0.150\ M = \dfrac{n}{0.020\ L}$ $\qquad n = 0.0030$ mol

	H^+ +	CH_3NH_2	$\rightleftharpoons$	$CH_3NH_3^+$
I	0.0020	0.0030		0
C	−0.0020	−0.0020		+0.0020
E	0	0.0010		0.0020

The new volume of the solution is 40.0 mL, which was used to calculate the concentrations of the ions in solution at equilibrium:

$[CH_3NH_2]\ = \dfrac{0.0010\ \text{mol}}{0.040\ L} = 0.025\ M$

$[CH_3NH_3^+]\ = \dfrac{0.0020\ \text{mol}}{0.040\ L} = 0.050\ M$

Finally, use the Henderson-Hasselbalch equation.

$pOH = pK_b + \log\dfrac{[CH_3NH_3^+]}{[CH_3NH_2]}$

$pOH = -\log(4.38 \times 10^{-4}) + \log\dfrac{0.050}{0.025}$

$pOH = 3.36 + 0.30 = 3.66$

$pH + pOH = 14$

$pH + 3.66 = 14$

$pH = 10.34$

2. (a) (i) Between these volumes lies the buffer region of the titration, in which both [HA] and [A⁻] contribute significantly to the pH of the solution.

   (ii) At equivalence, the only species present which affects the pH of the solution is the conjugate base, A⁻.

   (iii) In this region, the strong base is in excess, and [OH⁻] determines the pH of the solution.

   (b) (i) As the reaction progress, OH⁻ will take protons from HA to create A⁻. There will be more HA in solution until half of it has been deprotonated. The point at which that occurs is the half-equivalence point at 7.50 mL. So, [HA] > [A⁻] from 0 to 7.5 mL.

(ii) The only time these values are equal is when exactly half of the HA has been converted to A⁻, which occurs at 7.50 mL.

(iii) For the remainder of the titration (greater than 7.50 mL), there will be more conjugate base present than there will be of original weak acid.

(c) (i) and (ii) Neither of these values will ever be equal to zero. Prior to the titration beginning, there is some A⁻ in solution via the reaction of HA with water, as follows:

$$HA + H_2O \rightleftharpoons A^- + H_3O^+$$

Adding OH⁻ through titrating the strong base will cause the HA molecules present to deprotonate fully, but the conjugate base that is created will, in turn, also react with water to create additional HA molecules:

$$A^- + H_2O \rightleftharpoons HA + OH^-$$

As both the reaction of the weak acid with water and that of its conjugate base with water will always be occurring, there will always be some of each species in solution.

(d) One: Twice as much OH⁻ is being added per drop, so the pH changes charted will occur over only 15.0 mL instead of 30.0 mL (for instance, equivalence will occur at 7.50 mL instead of 15.0 mL).

Two: The final pH of the solution will be slightly higher, as a higher concentration of NaOH will lead to a higher pH in the region where [OH⁻] dictates the pH of the solution.

3. (a) The hypochlorous acid will donate its proton to the hydroxide. The sodium ions are spectators and would not appear in the net ionic equation.

$$HOCl(aq) + OH^-(aq) \rightleftharpoons OCl^-(aq) + H_2O(l)$$

(b) At the equivalence point, the moles of acid are equal to the moles of base.

Moles of base = $(1.47\ M)(0.03423\ L) = 0.0503$ mol

Concentration of acid = $(0.0503\ \text{mol}/0.02500\ L) = 2.01\ M$

(c) This calls for an ICE chart. You will first need the moles of both the acid and base.

Moles HOCl = (2.01 *M*)(0.02500 L) = 0.0503 mol HOCl

Moles OH⁻ = (1.47 *M*)(0.02855 L) = 0.0420 mol OH⁻

Putting those numbers into the ICE chart:

	HOCl	**OH⁻**	**OCl⁻**	**H₂O**
Initial	0.0503	0.0420	0	X
Change	−0.0420	−0.0420	0.0420	X
Equilibrium	0.0083	0	0.0420	X

Now that you know the number of moles at equilibrium of each species, you need to determine their new concentrations. The total volume of the solution at equilibrium is 25.00 mL + 28.55 mL = 53.55 mL. So the concentrations at equilibrium will be:

[HOCl] = 0.0083 mol/0.05355 L = 0.15 *M*
[OCl⁻] = 0.0420 mol/0.05355 L = 0.784 *M*

To finish off, use Henderson-Hasselbalch:

$$\text{pH} = \text{p}K_a + \log \frac{\left[\text{OCl}^- \right]}{\left[\text{HOCl} \right]}$$

$$\text{pH} = -\log \left(3.5 \times 10^{-8} \right) + \log \frac{(0.784)}{(0.15)}$$

$$\text{pH} = 7.5 + \log (5.2)$$

$$\text{pH} = 7.5 + 0.72$$

$$\text{pH} = 8.2$$

(d) (i) If additional NaOH is added, that would mean a larger number of moles of NaOH added to the flask, and thus more apparent moles of HOCl would be present at the equivalence point. More apparent moles of HOCl would lead to a larger numerator in the acid molarity calculation and thus a larger apparent molarity. As the student's calculated molarity was too low, this could not have caused the error.

(ii) If the buret was not rinsed with the NaOH prior to filling it, the concentration of the NaOH would be diluted by the water inside the buret. The student would then have to add a greater volume of NaOH to reach equivalence. However, in the calculations, using the original concentration of NaOH multiplied by the higher volume would lead to an artificially high number of moles of NaOH and thus, more apparent moles of HOCl at equivalence. More apparent moles of HOCl would lead to a larger numerator in the acid molarity calculation and thus a larger apparent molarity. As the student's calculated molarity was too low, this could not have caused the error.

(iii) If the student did not add enough acid to the flask, that would cause the denominator in the molarity of the acid calculation to be artificially high. This, in turn, would make the calculated acid molarity be artificially low. This matches up with the student's results and could be an acceptable source of error.

4. (a) The aniline will act as a proton acceptor and take a proton from the water.

$$C_6H_5NH_2(aq) + H_2O(l) \rightleftharpoons C_6H_5NH_3^+(aq) + OH^-(aq)$$

(b) (i) For the above equilibrium, $K_b = [C_6H_5NH_3^+][OH^-]/[C_6H_5NH_2]$. The concentrations of both the conjugate acid and the hydroxide ion will be equal, and the concentration of the aniline itself will be approximately the same, as it is a weak base which has a very low protonation rate. You can do an ICE chart to confirm this, but it is not really necessary if you understand the concepts underlying weak acids and bases.

$3.8 \times 10^{-10} = (x)(x)/(0.25)$

$x^2 = 9.5 \times 10^{-11}$

$x = 9.7 \times 10^{-6}$

$[C_6H_5NH_3^+] = [OH^-] = 9.7 \times 10^{-6}\,M$ and $[C_6H_5NH_2] = 0.25\,M$

(ii) $pOH = -\log[OH^-]$

$pOH = -\log(9.7 \times 10^{-6}\,M)$

$pOH = 5.0$

$pOH + pH = 14$

$5.0 + pH = 14$

$pH = 9.0$

Chapter 11
Unit 9: Applications of Thermodynamics

ENTROPY

The entropy, S, of a system is a measure of the randomness or dispersion of the system; the greater the dispersion of a system, the greater its entropy. Because zero entropy is defined as a solid crystal at 0 K, and because 0 K has never been reached experimentally, all substances that we encounter will have some positive value for entropy. Standard entropies, $S°$, are calculated at 25°C (298 K).

The standard entropy change, $\Delta S°$, that has taken place at the completion of a reaction is the difference between the standard entropies of the products and the standard entropies of the reactants.

$$\Delta S° = \Sigma S°_{products} - \Sigma S°_{reactants}$$

You can often predict the sign of ΔS for a reaction by studying the reaction itself. The first way to do this is by examining phase changes. Solids are less dispersed than liquids, which in turn are less dispersed than gases. Look at the following reaction for an example.

$$CaCO_3(s) \rightarrow CaO(s) + CO_2(g)$$

You should guess that ΔS is positive because some of the matter is turned into a gas in the products, meaning the overall dispersion of the system increased. Additionally, an aqueous solution is more dispersed than an organized precipitate.

$$Pb^{2+}(aq) + CO_3^{2-}(aq) \rightarrow PbCO_3(s)$$

The above reaction would have a negative value for ΔS, as the solid product is less dispersed than the aqueous ions. One more way you can predict the sign for ΔS is for reactions that consist of solely gaseous molecules, such as the one below.

$$C_2H_4(g) + 2\ O_2(g) \rightarrow 2\ CO_2(g) + 2\ H_2O(g)$$

There are three gaseous species in the reactants, and four in the products. That means the matter is more dispersed in the products, as gases have no set volume and will spread out all over the place. Thus, the sign for ΔS should be positive. This would also hold true for any fully aqueous reaction.

$$Fe^{3+}(aq) + SCN^-(aq) \rightarrow FeSCN^{2+}(aq)$$

The above reaction would have a negative ΔS, as there are more ions (and thus more dispersion) in the reactants. Also, if you dilute any aqueous solution, that would yield a positive value for ΔS, as the ions in, say, a 0.10 M solution of NaCl are more spread out/dispersed than those in a 1.0 M solution of NaCl.

Note you can only count ions/molecules when you are looking at substances in the gaseous or aqueous phase. The coefficients in front of any solids or liquids would not help predict dispersion, as both solids and liquids have set volumes.

GIBBS FREE ENERGY

The change in **Gibbs free energy**, or simply free energy, G, of a process determines whether that process is thermodynamically favored or unfavored. Prior to the 2014 exam, the terms *spontaneous* and *nonspontaneous* were used to described thermodynamically favored and unfavored reactions, respectively. It would be good for you to be familiar with both sets of terms, although this text will use the updated terms.

Free Energy Change, ΔG

The standard free energy change, $\Delta G°$, for a reaction can be calculated from the standard free energies of formation, $G_f°$, of its products and reactants in the same way that $\Delta S°$ was calculated.

$$\Delta G° = \Sigma G_{f \text{ products}}° - \Sigma G_{f \text{ reactants}}°$$

For a given reaction:

- if ΔG is negative, the reaction is thermodynamically favored
- if ΔG is positive, the reaction is thermodynamically unfavored
- if $\Delta G = 0$, the reaction is at equilibrium

ΔG, ΔH, and ΔS

In general, nature likes to move toward two different and seemingly contradictory states—low energy and high dispersion, so thermodynamically favored processes must result in decreasing enthalpy or increasing entropy or both.

There is an important equation that relates favorability (ΔG), enthalpy (ΔH), and entropy (ΔS) to one another.

$$\Delta G° = \Delta H° - T\Delta S°$$
$$T = \text{absolute temperature (K)}$$

Note that you should make sure your units always match up here. Frequently, ΔS is given in J/mol·K and ΔH is given in kJ/mol. The convention is to convert the $T\Delta S$ term to kilojoules (kJ), as that is what Gibbs free energy is usually measured in. However, you can also convert the ΔH term to joules instead, so long as you are keeping track and labeling all units appropriately.

Looking at an example:

$$CO(g) + H_2O(g) \rightarrow CO_2(g) + H_2(g) \quad \Delta H° = -41.2 \text{ kJ/mol} \quad \Delta S° = -135 \text{ J/mol·K}$$

(a)   Is the above reaction thermodynamically favorable at 25°C?

To calculate $\Delta G°$, the units need to match. If we change the entropy units to kJ, $\Delta S° = -0.135$ kJ/mol · K.

From there, it's just a matter of plugging numbers in:

$$\Delta G° = (-41.2 \text{ kJ/mol}) - (298 \text{ K})(-0.135 \text{ kJ/mol·K})$$

$$\Delta G° = -0.97 \text{ kJ/mol}$$

As the Gibbs value is negative, the above reaction is favored at 25°C.

(b)   At what temperature would the reaction no longer be thermodynamically favored? Assume the enthalpy and entropy values are independent of temperature.

The favorability of the reaction would change if the Gibbs value became positive. To figure out the required temperature for that, we can set $\Delta G° = 0$.

$$0 = (-41.2 \text{ kJ/mol}) - (T)(-0.135 \text{ kJ/mol·K})$$

$$T = 305 \text{ K}$$

Above 305 K, this reaction would not be thermodynamically favorable.

The chart below shows how different values of enthalpy and entropy affect favorability.

ΔH	ΔS	T	ΔG	
–	+	Low	–	Always favored
		High	–	
+	–	Low	+	Never favored
		High	+	
+	+	Low	+	Not favored at low temperature
		High	–	Favored at high temperature
–	–	Low	–	Favored at low temperature
		High	+	Not favored at high temperature

You should note that at low temperature, enthalpy is dominant, while at high temperature, entropy is dominant.

$\Delta G°$ and Phase Changes

A substance at a normal phase transition temperature is equally stable in either of those two phases. As a result, $\Delta G = 0$ at that temperature. With that in mind, if we have data regarding a substance's thermodynamic properties at the point of fusion or vaporization, we can determine the melting or boiling point of that substance.

Example:

Acetone, CH_3COCH_3, is a liquid at room temperature with a $\Delta H_{vap} = 29.1$ kJ/mol and a $\Delta S_{vap} = 88.4$ J/mol·K. What is the boiling point of acetone?

$0 = 29.1$ kJ/mol $- T(0.0884$ kJ/mol·K$)$
$T = 329$ K

Standard Free Energy Change and the Equilibrium Constant

The amount of Gibbs free energy in any given reaction can also be calculated if you know the equilibrium constant expression for that reaction.

$$\Delta G° = -RT \ln K$$

R = the gas constant, 8.31 J/mol·K
T = absolute temperature (K)
K = the equilibrium constant

While this is the same "R" as the ideal gas law R, it has different units and thus a different magnitude. Be very careful you are using the correct value for R when doing calculations. If $\Delta G°$ is negative, K must be greater than 1, and products will be favored at equilibrium. Alternatively, if $\Delta G°$ is positive, K must be less than 1, and reactants will be favored at equilibrium.

Reduction Potentials

Every half-reaction has an electric potential, or voltage, associated with it. You will be given the necessary values for the standard reduction potential of half-reactions for any question in which they are required. Potentials are always given as reduction half-reactions, but you can read them in reverse and flip the sign on the voltage to get oxidation potentials.

Look at the reduction potential for Zn^{2+}.

$$Zn^{2+}(aq) + 2\ e^- \rightarrow Zn(s) \qquad E° = -0.76 \text{ V}$$

Read the reduction half-reaction in reverse and change the sign on the voltage to get the oxidation potential for Zn.

$$Zn(s) \rightarrow Zn^{2+}(aq) + 2\ e^- \qquad\qquad E^\circ = +0.76\ V$$

The larger the potential for a half-reaction, the more likely it is to occur.

$$F_2(g) + 2\ e^- \rightarrow 2\ F^-(aq) \qquad\qquad E^\circ = +2.87\ V$$

$F_2(g)$ has a very large reduction potential, so it is likely to gain electrons and be reduced.

$$Li(s) \rightarrow Li^+(aq) + e^- \qquad\qquad E^\circ = +3.05\ V$$

Li(s) has a very large oxidation potential, making it very likely to lose electrons and be oxidized.

You can calculate the potential of a redox reaction if you know the potentials for the two half-reactions that constitute it. There are two important things to remember when calculating the potential of a redox reaction.

- Add the potential for the oxidation half-reaction to the potential for the reduction half-reaction.
- Never multiply the potential for a half-reaction by a coefficient.

Let's look at the following reaction:

$$Zn(s) + 2\ Ag^+(aq) \rightarrow Zn^{2+}(aq) + 2\ Ag(s)$$

The two half-reactions are:

Oxidation: $\quad Zn(s) \rightarrow Zn^{2+}(aq) + 2\ e^- \qquad\qquad E^\circ = +0.76\ V$
Reduction: $\quad Ag^+(aq) + e^- \rightarrow Ag(s) \qquad\qquad\quad E^\circ = +0.80\ V$

$$E = E_{oxidation} + E_{reduction}$$
$$E = 0.76\ V + 0.80\ V = 1.56\ V$$

Notice that we ignored that silver has a coefficient of 2 in the balanced equation.

The relative reduction strengths of two different metals can also be determined qualitatively. In the above reaction, if Zn(s) were placed in a solution containing Ag^+ ions, the silver ions have a high enough reduction potential that they would take electrons from the zinc and start forming solid silver, which would precipitate out on the surface of the zinc.

However, if Ag(s) were placed in a solution containing Zn^{2+} ions, Zn^{2+} does not have a high enough reduction potential to take electrons from silver and so no reaction would occur. So, when a solid metal is placed into a solution containing cations of another metal and a new solid starts to form, the reduction potential of the cations in solution is greater than the reduction potential of the original solid metal's cations. If no solid forms, the reduction potential of the original solid metal's cations is greater than that of the solution-phase cations.

GALVANIC CELLS

In a **galvanic cell** (also called a voltaic cell), a favored redox reaction is used to generate a flow of current.

Look at the following favored redox reaction:

$$Zn(s) + Cu^{2+}(aq) \rightarrow Zn^{2+}(aq) + Cu(s) \qquad E° = 1.10 \text{ V}$$

Oxidation: $\qquad Zn(s) \rightarrow Zn^{2+}(aq) + 2\ e^- \qquad E° = 0.76 \text{ V}$

Reduction: $\qquad Cu^{2+}(aq) + 2\ e^- \rightarrow Cu(s) \qquad E° = 0.34 \text{ V}$

> **Going Platinum**
> Because it is unreactive with most solutions, platinum is often used as a conductor in an electrolytic cell.

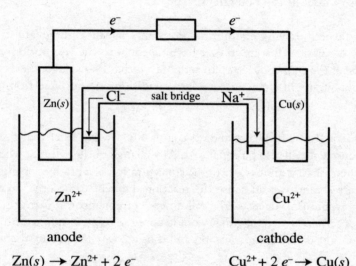

anode $\qquad\qquad$ cathode

$$Zn(s) \rightarrow Zn^{2+} + 2\ e^- \qquad\qquad Cu^{2+} + 2\ e^- \rightarrow Cu(s)$$

In a galvanic cell, the two half-reactions take place in separate chambers, and the electrons that are released by the oxidation reaction pass through a wire to the chamber where they are consumed in the reduction reaction. That's how the current is created. **Current** is defined as the flow of electrons from one place to another.

In any electric cell (either a galvanic cell or an electrolytic cell, which we'll discuss a little later), oxidation takes place at the electrode called the **anode.** Reduction takes place at the electrode called the **cathode.**

The salt bridge maintains the electrical neutrality in the cell. At the cathode, where reduction occurs and the solution is becoming less positively charged, the positive cations from the salt bridge solution flow into the half-cell. At the anode, where oxidation occurs and the solution is becoming more positively charged, the negative anions from the salt bridge solution flow into the half-cell. If the salt bridge were to be removed, the solutions in the half-cells would become electrically imbalanced, and the voltage of the cell would drop to 0. In the previous diagram, the salt bridge is filled with aqueous NaCl. As the cell functions, the Na^+ cations flow to the cathode and the Cl^- anions flow to the anode.

> There's a mnemonic device to remember oxidation and reduction:
>
> AN OX
> RED CAT

Under standard conditions (when all concentrations are 1 M), the voltage of the cell is the same as the total voltage of the redox reaction. To determine what would happen to the cell potential under nonstandard conditions, you must first learn more about equilibrium, which is covered in Unit 7.

Non-Standard Conditions

When examining voltaic cells, the reduction potentials are always given at standard conditions; that is 25°C, 1.0 atm, and with all species having a concentration of 1.0 M. If any of those conditions deviate, it will also cause the cell potential to deviate.

Voltaic cells are all very favored, having equilibrium constants significantly greater than 1. If the reaction quotient for a voltaic cell were to ever become equal to the equilibrium constant (so, at $Q = K$), the voltage of the cell would drop to zero. Given that knowledge, the best way to determine how a cell's potential will change if standard conditions are deviated from is to use the reaction quotient.

As all concentrations are equal to 1.0 M at standard conditions, we can infer that at standard conditions, the reaction quotient will also be equal to 1. Any change to the cell that would cause the reaction quotient to increase (say, an increase in the concentration of a product or the decrease in the concentration of a reactant), would cause the reaction quotient to become closer to the equilibrium constant and would thus decrease the cell potential (remember, if Q ever reaches K, the cell potential becomes zero!). Any change that would cause the reaction quotient to decrease would move it further from the equilibrium constant, and thus increase the potential of the cell.

If a cell has a gas at either the cathode or the anode, a change in pressure can also affect the reaction quotient, and a change in temperature may as well. However, concentration changes, which usually occur constantly as the reaction in a cell is progressing, are the most common reason that a voltaic cell will deviate from its standard potential.

Example:

A voltaic cell using silver and zinc is connected, and the below reaction occurs.

$2\,Ag^+(aq) + Zn(s) \rightarrow Zn^{2+}(aq) + 2\,Ag(s)$

What would happen to the voltage of the above cell as the reaction progresses?

As the reaction progresses, $[Ag^+]$ decreases as it is reduced into $Ag(s)$, and $[Zn^{2+}]$ increases as it is oxidized from $Zn(s)$. Both of these changes would have the effect of increasing the reaction quotient, which would bring the cell closer to equilibrium and thus decrease the overall potential.

There is also an equation—known as the Nernst equation—that allows us to calculate the cell potential when there are deviations from the standard conditions.

$$E_{cell} = E_{cell}^{\circ} - \frac{RT}{nF}\ln Q$$

In the above equation, $E°_{cell}$ is the standard cell potential and E_{cell} is the potential under the non-standard conditions. You WILL NOT be required to use the Nernst equation quantitatively on the exam; however, you can use it to mathematically justify the change in cell potential under non-standard conditions. In our example on the previous page, the increase in the reaction quotient would cause the value of the ln Q term to increase, which in turn would cause E_{cell} to decrease.

ELECTROLYTIC CELLS

In an electrolytic cell, an outside source of voltage is used to force an unfavored redox reaction to take place. Most electrolytic cells occur in aqueous solutions, which are created when a chemical dissolves in water; either the ions or the water molecules can be reduced or oxidized.

Let's look at a solution of nickel (II) chloride as an example. To determine which substance is reduced, you must compare the reduction potential of the cation with that of water. The half-reaction with the more positive value is the one that will occur.

$$Ni^{2+}(aq) + 2\ e^- \rightarrow Ni(s) \qquad\qquad E° = -0.25\ V$$

$$2\ H_2O(l) + 2\ e^- \rightarrow H_2(g) + 2\ OH^-(aq) \qquad E° = -0.80\ V$$

In this case, the Ni^{2+} reduction will occur. For the oxidation, the oxidation potential of the anion versus that of water must be examined. As before, the half-reaction with the most positive value is the one that will occur. Remember that potentials are always given as reduction potentials, so you must flip the sign when you flip the equation to an oxidation.

$$2\ Cl^-(aq) \rightarrow Cl_2(g) + 2\ e^- \qquad\qquad E° = -1.36\ V$$

$$2\ H_2O(l) \rightarrow O_2(g) + 4\ H^+(aq) + 4\ e^- \qquad E° = -1.23\ V$$

So, in this case, the water itself would be oxidized. When the equations are balanced for electron transfer, the net ionic equation looks like this:

$$2\ Ni^{2+}(aq) + 2\ H_2O(l) \rightarrow 2\ Ni(s) + O_2(g) + 4\ e^-$$

$$E = -0.25\ V + -1.24V = -1.49\ V$$

The anode and cathode in an electrolytic reaction are usually just metal bars that conduct current and do not take part in the reaction. In the above reaction, solid nickel is being created at the cathode, and oxygen gas is being evolved at the anode. The sign of your total cell potential (E) for an electrolytic reaction is always negative.

Occasionally, a current will either be run through a molten compound or pure water. In this case, you do not need to determine which redox reactions are taking place, as you will only have one choice for each.

The AN OX/RED CAT rule applies to the electrolytic cell in the same way that it applies to the galvanic cell.

Electroplating

Electrolytic cells are used for electroplating. You may see a question on the test that gives you an electrical current and asks you how much metal "plates out."

There are roughly four steps for figuring out electrolysis problems.

1. If you know the current and the time, you can calculate the charge in coulombs.

Current

$$I = \frac{q}{t}$$

I = current (amperes, A)
q = charge (coulombs, C)
t = time (seconds, s)

---

2. Once you know the charge in coulombs, you know how many electrons were involved in the reaction.

$$\text{Moles of electrons} = \frac{\text{coulombs}}{96{,}500 \text{ coulombs/mol}}$$

3. When you know the number of moles of electrons and you know the half-reaction for the metal, you can find out how many moles of metal plated out. For example, from this half-reaction for gold,

$$Au^{3+}(aq) + 3\ e^- \rightarrow Au(s),$$

you know that for every 3 moles of electrons consumed, you get 1 mole of gold.

4. Once you know the number of moles of metal, you can use what you know from stoichiometry to calculate the number of grams of metal.

For instance, if a current of 2.50 A is run through a solution of iron (III) chloride for 15 minutes, it would cause the following mass of iron to plate out:

$$15 \text{ minutes} \times \frac{60 \text{ seconds}}{1 \text{ minute}} \times \frac{2.50 \text{ C}}{1 \text{ second}} \times \frac{1 \text{ mol } e^-}{96{,}500 \text{ C}} \times \frac{1 \text{ mol Fe}}{3 \text{ mol } e^-} \times \frac{55.85 \text{ g}}{1 \text{ mol Fe}} = 0.43 \text{ g Fe}$$

It is particularly important to keep track of your units in an electroplating problem; there are many different conversions before you come up with your final answer. In general, as long as all of your conversions are set up correctly your final answer will have the correct units.

VOLTAGE AND FAVORABILITY

A redox reaction will be favored if its potential has a positive value. We also know from thermodynamics that a reaction that is favored has a negative value for free energy change. The relationship between reaction potential and free energy for a redox reaction is given by the equation below, which serves as a bridge between thermodynamics and electrochemistry.

$$\Delta G° = -nFE°$$

$\Delta G°$ = Standard Gibbs free energy change (J/mol)

n = the number of moles of electrons exchanged in the reaction (mol)

F = Faraday's constant, 96,500 coulombs/mole (that is, 1 mole of electrons has a charge of 96,500 coulombs)

$E°$ = Standard reaction potential (V, which is equivalent to J/C)

From this equation we can see a few important things. If $E°$ is positive, $\Delta G°$ is negative and the reaction is thermodynamically favored, and if $E°$ is negative, $\Delta G°$ is positive and the reaction is thermodynamically unfavored.

Let's take a look at an example.

Calculate the ΔG value for the below reaction under standard conditions:

$$Zn(s) + 2\,Ag^+(aq) \rightarrow Zn^{2+}(aq) + 2\,Ag(s) \qquad E° = +1.56\ V$$

For this reaction, two moles of electrons are being transferred as the silver is reduced and the zinc is oxidized. So, $n = 2$. With that in mind:

$$\Delta G° = -(2)(96,500)(1.56)$$

$$\Delta G° = -301,000\ J/mol$$

As the $\Delta G°$ value is negative, we would expect this reaction to be favored under standard conditions, which is in line with the positive value for the cell.

Note that your units on the free energy are in J/mol and not kJ/mol; this will always be the case with this equation, as one volt is equivalent to one joule/coulomb, so the answer comes out in joules instead of kilojoules.

UNIT 9 QUESTIONS

Multiple-Choice Questions

Use the following information to answer questions 1–5.

Reaction 1: $N_2H_4(l) + H_2(g) \rightarrow 2\ NH_3(g)$ $\Delta H = ?$

Reaction 2: $N_2H_4(l) + CH_4O(l) \rightarrow CH_2O(g) + N_2(g) + 3\ H_2(g)$ $\Delta H = -37\ kJ/mol_{rxn}$

Reaction 3: $N_2(g) + 3\ H_2(g) \rightarrow 2\ NH_3(g)$ $\Delta H = -46\ kJ/mol_{rxn}$

Reaction 4: $CH_4O(l) \rightarrow CH_2O(g) + H_2(g)$ $\Delta H = -65\ kJ/mol_{rxn}$

1. What is the enthalpy change for reaction 1?

    (A) $-148\ kJ/mol_{rxn}$
    (B) $-56\ kJ/mol_{rxn}$
    (C) $-18\ kJ/mol_{rxn}$
    (D) $+148\ kJ/mol_{rxn}$

2. If reaction 2 were repeated at a higher temperature, how would the reaction's value for ΔG be affected?

    (A) It would become more negative because entropy is a driving force behind this reaction.
    (B) It would become more positive because the reactant molecules would collide more often.
    (C) It would become more negative because the gases will be at a higher pressure.
    (D) It will stay the same; temperature does not affect the value for ΔG.

3. Under what conditions would reaction 3 be thermodynamically favored?

    (A) It is always favored.
    (B) It is never favored.
    (C) It is only favored at low temperatures.
    (D) It is only favored at high temperatures.

4. If 64 g of CH_4O were to decompose via reaction 4, approximately how much energy would be released or absorbed?

    (A) 65 kJ of energy will be absorbed.
    (B) 65 kJ of energy will be released.
    (C) 130 kJ of energy will be absorbed.
    (D) 130 kJ of energy will be released.

5. A 2.0 L flask holds 0.40 g of helium gas. If the helium is evacuated into a larger container while the temperature is held constant, what will the effect on the entropy of the helium be?

    (A) It will remain constant because the number of helium molecules does not change.
    (B) It will decrease because the gas will be more ordered in the larger flask.
    (C) It will decrease because the molecules will collide with the sides of the larger flask less often than they did in the smaller flask.
    (D) It will increase because the gas molecules will be more dispersed in the larger flask.

6. $2 Al(s) + 3 Cl_2(g) \rightarrow 2 AlCl_3(s)$

The reaction above is not thermodynamically favored under standard conditions, but it becomes thermodynamically favored as the temperature decreases toward absolute zero. Which of the following is true at standard conditions?

(A) ΔS and ΔH are both negative.
(B) ΔS and ΔH are both positive.
(C) ΔS is negative, and ΔH is positive.
(D) ΔS is positive, and ΔH is negative.

7. $SF_4(g) + H_2O(l) \rightarrow SO_2(g) + 4 HF(g)$ $\Delta H = -828$ kJ/mol

Which of the following statements accurately describes the above reaction?

(A) The entropy of the reactants exceeds that of the products.
(B) $H_2O(l)$ will always be the limiting reagent.
(C) This reaction is never thermodynamically favored.
(D) The temperature of the surroundings will increase as this reaction progresses.

8. $H_2O(l) \rightarrow H_2O(s)$

Which of the following is true for this process at STP?

(A) The value for ΔS is positive.
(B) The value for ΔG is zero.
(C) The value for ΔH is positive.
(D) The reaction is favored.

9. In which of the following reactions is entropy increasing?

(A) $2 SO_2(g) + O_2(g) \rightarrow 2 SO_3(g)$
(B) $CO(g) + H_2O(g) \rightarrow H_2(g) + CO_2(g)$
(C) $H_2(g) + Cl_2(g) \rightarrow 2 HCl(g)$
(D) $2 NO_2(g) \rightarrow 2 NO(g) + O_2(g)$

10. The reaction shown in the diagram below is accompanied by a large increase in temperature. If all molecules shown are in their gaseous state, which statement accurately describes the reaction?

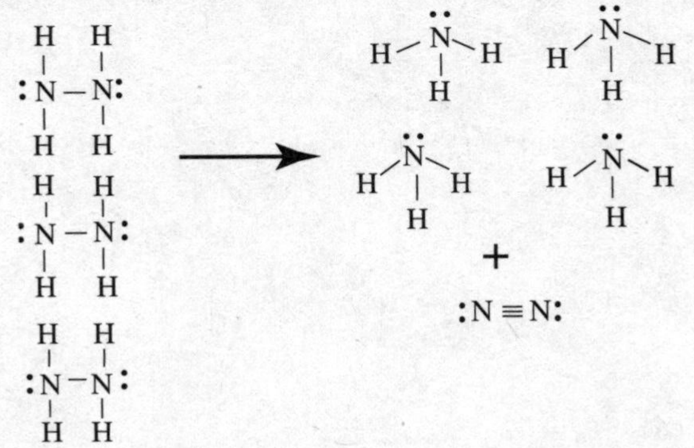

(A) It is an exothermic reaction in which entropy increases.
(B) It is an exothermic reaction in which entropy decreases.
(C) It is an endothermic reaction in which entropy increases.
(D) It is an endothermic reaction in which entropy decreases.

11. In which of the following circumstances is the value for K_{eq} always greater than 1?

	ΔH	ΔS
(A)	Positive	Positive
(B)	Positive	Negative
(C)	Negative	Negative
(D)	Negative	Positive

12. Which expression below should be used to calculate the mass of copper that can be plated out of a 1.0 M $Cu(NO_3)_2$ solution using a current of 0.75 A for 5.0 minutes?

(A) $\dfrac{(5.0)(60)(0.75)(63.55)}{(96,500)(2)}$

(B) $\dfrac{(5.0)(60)(63.55)(2)}{(0.75)(96,500)}$

(C) $\dfrac{(5.0)(60)(96,500)(0.75)}{(63.55)(2)}$

(D) $\dfrac{(5.0)(60)(96,500)(63.55)}{(0.75)(2)}$

13. What is the general relationship between temperature and entropy for diatomic gases?

(A) They are completely independent of each other; temperature has no effect on entropy.
(B) There is a direct relationship, because at higher temperatures there is an increase in energy dispersal.
(C) There is an inverse relationship, because at higher temperatures substances are more likely to be in a gaseous state.
(D) It depends on the specific gas and the strength of the intermolecular forces between individual molecules.

14. A strip of metal X is placed into a solution containing Y^{2+} ions and no reaction occurs. When metal X is placed in a separate solution containing Z^{2+} ions, metal Z starts to form on the strip. Which of the following choices organizes the reduction potentials for these cations from greatest to least?

(A) $X^{2+} > Y^{2+} > Z^{2+}$
(B) $Y^{2+} > Z^{2+} > X^{2+}$
(C) $Z^{2+} > X^{2+} > Y^{2+}$
(D) $Y^{2+} > X^{2+} > Z^{2+}$

Use the following information to answer questions 15–17.

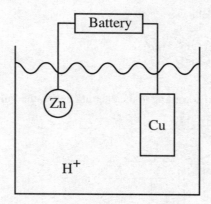

Pennies are made primarily of zinc, which is coated with a thin layer of copper through electroplating, using a setup like the one above. The solution in the beaker is a strong acid (which produces H^+ ions), and the cell is wired so that the copper electrode is the anode and the zinc penny is the cathode. Use the following reduction potentials to answer questions 15–17.

Half-Reaction	Standard Reduction Potential
$Cu^{2+} + 2\,e^- \rightarrow Cu(s)$	+0.34 V
$2\,H^+ + 2\,e^- \rightarrow H_2(g)$	0.00 V
$Ni^{2+} + 2\,e^- \rightarrow Ni(s)$	−0.25 V
$Zn^{2+} + 2\,e^- \rightarrow Zn(s)$	−0.76 V

15. When the cell is connected, which of the following reactions takes place at the anode?

    (A) $Cu^{2+} + 2\,e^- \rightarrow Cu(s)$
    (B) $Cu(s) \rightarrow Cu^{2+} + 2\,e^-$
    (C) $2\,H^+ + 2\,e^- \rightarrow H_2(g)$
    (D) $H_2(g) \rightarrow 2\,H^+ + 2\,e^-$

16. What is the required voltage to make this cell function?

    (A) 0.34 V
    (B) 0.42 V
    (C) 0.76 V
    (D) 1.10 V

17. If, instead of copper, a nickel bar were to be used, could nickel be plated onto the zinc penny effectively? Why or why not?

    (A) Yes, the SRP of Ni^{2+} is greater than that of Zn^{2+}, which is all that is required for Ni^{2+} to be reduced at the cathode.
    (B) Yes, Ni^{2+} is able to take electrons from the H^+ ions in solution, allowing it to be reduced.
    (C) No, the SRP of Ni^{2+} is lower than that of H^+ ions, which means the only product being produced at the cathode would be hydrogen gas.
    (D) No, the SRP of Ni^{2+} is negative, meaning it cannot be reduced in an electrolytic cell.

Use the following information to answer questions 18–22.

Two half-cells are set up as follows:

Half-Cell A: Strip of Cu(s) in $CuNO_3$(aq)
Half-Cell B: Strip of Zn(s) in $Zn(NO_3)_2$(aq)

When the cells are connected according to the diagram below, the following reaction occurs:

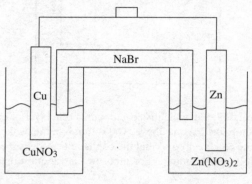

$$2\ Cu^+(aq) + Zn(s) \rightarrow 2\ Cu(s) + Zn^{2+}(aq) \qquad E° = +1.28\ V$$

18. Correctly identify the anode and cathode in this reaction as well as where oxidation and reduction are taking place.

    (A) Cu is the anode where oxidation occurs, and Zn is the cathode where reduction occurs.
    (B) Cu is the anode where reduction occurs, and Zn is the cathode where oxidation occurs.
    (C) Zn is the anode where oxidation occurs, and Cu is the cathode where reduction occurs.
    (D) Zn is the anode where reduction occurs, and Cu is the cathode where oxidation occurs.

19. How many moles of electrons must be transferred to create 127 g of copper?

    (A) 1 mole of electrons
    (B) 2 moles of electrons
    (C) 3 moles of electrons
    (D) 4 moles of electrons

20. If the $Cu^+ + e^- \rightarrow Cu(s)$ half-reaction has a standard reduction potential of +0.52 V, what is the standard reduction potential for the $Zn^{2+} + 2\ e^- \rightarrow Zn(s)$ half-reaction?

    (A) +0.76 V
    (B) –0.76 V
    (C) +0.24 V
    (D) –0.24 V

21. As the reaction progresses, what will happen to the overall voltage of the cell?

    (A) It will increase as $[Zn^{2+}]$ increases.
    (B) It will increase as $[Cu^+]$ increases.
    (C) It will decrease as $[Zn^{2+}]$ increases.
    (D) The voltage will remain constant.

22. What will happen in the salt bridge as the reaction progresses?
    (A) The Na^+ ions will flow to the Cu/Cu^+ half-cell.
    (B) The Br^- ions will flow to the Cu/Cu^+ half-cell.
    (C) Electrons will transfer from the Cu/Cu^+ half-cell to the Zn/Zn^{2+} half-cell.
    (D) Electrons will transfer from the Zn/Zn^{2+} half-cell to the Cu/Cu^+ half-cell.

Free-Response Questions

1.

Substance	Absolute Entropy, S° (J/mol·K)	Molar Mass (g/mol)
$C_6H_{12}O_6(s)$	212.13	180
$O_2(g)$	205	32
$CO_2(g)$	213.6	44
$H_2O(l)$	69.9	18

Energy is released when glucose is oxidized in the following reaction, which is a metabolism reaction that takes place in the body.

$$C_6H_{12}O_6(s) + 6\,O_2(g) \rightarrow 6\,CO_2(g) + 6\,H_2O(l)$$

The standard enthalpy change, $\Delta H°$, for the reaction is $-2,801$ kJ/mol$_{rxn}$ at 298 K.

(a) Calculate the standard entropy change, $\Delta S°$, for the oxidation of glucose.
(b) Calculate the standard free energy change, $\Delta G°$, for the reaction at 298 K.
(c) Using the axis below, draw an energy profile for the reaction and indicate the magnitude of ΔH.

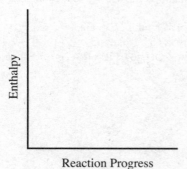

(d) How much enthalpy is given off by the oxidation of 1.00 gram of glucose?

2.

Bond	Average Bond Dissociation Energy (kJ/mol)
C–H	415
O=O	495
C=O	799
O–H	463

$$CH_4(g) + 2\ O_2(g) \rightarrow CO_2(g) + 2\ H_2O(g)$$

The standard free energy change, $\Delta G°$, for the reaction above is –801 kJ/mol$_{rxn}$ at 298 K.

(a) Use the table of bond dissociation energies to find $\Delta H°$ for the reaction above.

(b) How many grams of methane must react with excess oxygen in order to release 1,500 kJ of heat?

(c) What is the value of $\Delta S°$ for the reaction at 298 K?

(d) Give an explanation for the size of the entropy change found in (c).

3. $$CH_3OH(l) \rightarrow CH_3OH(g)$$

For the boiling of methanol, CH_3OH, $\Delta H° = +37.6$ kJ/mol and $\Delta S° = +111$ J/mol·K.

(a) (i) Why is the ΔH value positive for this process?

    (ii) Why is the ΔS value positive for this process?

(b) What is the boiling point of methanol in degrees Celsius?

(c) How much heat is required to boil 50.0 mL of methanol if the density of methanol is 0.789 g/mL?

(d) What will happen to the temperature of the methanol as it boils? Explain.

(e) Would methanol be soluble with water? Why or why not?

(f) Would you expect the boiling point of ethanol, CH_3CH_2OH, to be less than, greater than, or the same as methanol? Justify your answer.

4. Ammonia gas reacts with dinitrogen monoxide via the following reaction:

$$2\,NH_3(g) + 3\,N_2O(g) \rightarrow 4\,N_2(g) + 3\,H_2O(g)$$

The absolute entropy values for the varying substances are listed in the table below.

Substance	$S°$ (J/mol·K)
$NH_3(g)$	193
$N_2O(g)$	220
$N_2(g)$	192
$H_2O(g)$	189

(a) Calculate the entropy value for the overall reaction.

Several heats of formation are listed in the table below.

Substance	$\Delta H_f°$ (kJ/mol)
$NH_3(g)$	−46
$N_2O(g)$	85.5
$N_2(g)$	−242

(b) Calculate the enthalpy value for the overall reaction.
(c) Is this reaction thermodynamically favored at 25°C? Justify your answer.
(d) If 25.00 g of NH_3 reacts with 25.00 g of N_2O:
   (i) Will energy be released or absorbed?
   (ii) What is the magnitude of the energy change?
(e) On the reaction coordinates below, draw a line showing the progression of this reaction. Label both ΔH and E_a on the graph.

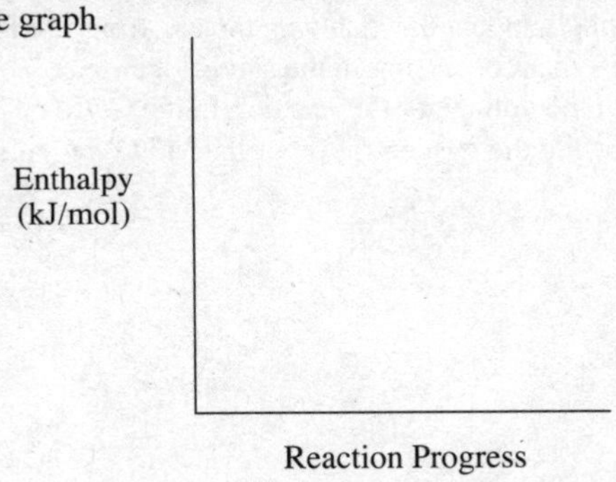

Enthalpy (kJ/mol)

Reaction Progress

5. A student performs an experiment in which a bar of unknown metal M is placed in a solution with the formula MNO_3. The metal is then hooked up to a copper bar in a solution of $CuSO_4$ as shown below. A salt bridge that contains aqueous KCl links the cell together.

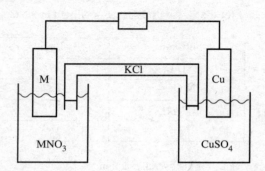

The cell potential is found to be +0.74 V. Separately, when a bar of metal M is placed in the copper sulfate solution, solid copper starts to form on the bar. When a bar of copper is placed in the MNO_3 solution, no visible reaction occurs.

The following gives some reduction potentials for copper:

Half-reaction	$E°$
$Cu^{2+} + 2\,e^- \rightarrow Cu(s)$	0.34 V
$Cu^{2+} + e^- \rightarrow Cu^+$	0.15 V
$Cu^+ + e^- \rightarrow Cu(s)$	0.52 V

(a) Write the net ionic equation that takes place in the Cu/M cell.
(b) What is the standard reduction potential for metal M?
(c) Which metal acted as the anode and which as the cathode? Justify your answer.
(d) On the diagram of the cell, indicate which way the electrons are flowing in the wire. Additionally, indicate any ionic movement occurring in the salt bridge.
(e) What would happen to the voltage of the reaction in the Cu/M cell if the concentration of the $CuSO_4$ increased while the concentration of the MNO_3 remained constant? Justify your answer.

6. Two electrodes are inserted into a solution of nickel (II) fluoride and a current of 2.20 A is run through them. A list of standard reduction potentials is as follows:

Half-reaction	$E°$
$O_2(g) + 4\,H^+(aq) + 4\,e^- \rightarrow H_2O(l)$	1.23 V
$F_2(g) + 2\,e^- \rightarrow 2\,F^-(aq)$	2.87 V
$2\,H_2O(l) + 2\,e^- \rightarrow H_2(g) + 2\,OH^-(aq)$	−0.83 V
$Ni^{2+}(aq) + 2\,e^- \rightarrow Ni(s)$	−0.25 V

(a) Write the net ionic equation that takes place during this reaction.
(b) Qualitatively describe what an observer would see taking place at each electrode.
(c) Will the solution become acidic, basic, or remain neutral as the reaction progresses?
(d) How long would it take to create 1.2 g of Ni(s) at the cathode?

UNIT 9 ANSWERS AND EXPLANATIONS

Multiple-Choice

1. **C** To get reaction 1, all that is needed is to flip reaction 4 and then add it to reactions 2 and 3.

   $$N_2H_4(l) + CH_4O(l) \rightarrow CH_2O(g) + N_2(g) + 3\ H_2(g)$$
   $$N_2(g) + 3\ H_2(g) \rightarrow 2\ NH_3(g)$$
   $$CH_2O(g) + H_2(g) \rightarrow CH_4O(l)$$

   $$(-37\ \text{kJ/mol}_{rxn}) + (-46\ \text{kJ/mol}_{rxn}) + (65\ \text{kJ/mol}_{rxn}) = -18\ \text{kJ/mol}_{rxn}$$

2. **A** Entropy is positive during reaction 2, as creating gas molecules out of liquid molecules demonstrates an increase in dispersion. Using $\Delta G = \Delta H - T\Delta S$, if the temperature increases, the $T\Delta S$ term will also increase. As the overall value for that term is negative, increasing $T\Delta S$ makes ΔG more negative.

3. **C** Use the $\Delta G = \Delta H - T\Delta S$ equation. To do that, determine the signs of both ΔH and the $T\Delta S$ term.

   You know that ΔH is negative. ΔS is negative as well, because the reaction is going from four molecules to two molecules, meaning it is becoming more ordered. That, in turn, means the $T\Delta S$ term is positive.

   Reactions are only favored when ΔG is negative. As temperature increases, the $T\Delta S$ term becomes larger and more likely to be greater in magnitude than the ΔH term. If the temperature remains low, the $T\Delta S$ value is much more likely to be smaller in magnitude than the ΔH value, meaning ΔG is more likely to be negative and the reaction will be favored.

4. **D** The molar mass of CH_4O is 32 g/mol, so 64 g of it represents two moles. For every one mole of CH_4O that decomposes, 65 kJ of energy is released—this is indicated by the negative sign on the H value. If two moles decompose, twice that amount—130 kJ—will be released.

5. **D** Entropy is a measure of a system's dispersion. In a larger flask, the gas molecules will spread farther apart and become more dispersed.

6. **A** Remember that $\Delta G = \Delta H - T\Delta S$.

   If the reaction is thermodynamically favored only when the temperature is very low, then ΔG is negative only when T is very small. This can happen only when ΔH is negative and ΔS is negative. A very small value for T will eliminate the influence of ΔS.

7. **D** A negative enthalpy change indicates that energy is lost to the surroundings as the reaction occurs.

8. **B** During a phase change, the products and reactants are at equilibrium with each other, meaning ΔG is zero. When a liquid turns into a solid, dispersion decreases (negative ΔS) and energy is released (negative ΔH). Choice (D) is incorrect because during a phase change the system is in equilibrium. The forward and reverse reaction rates are equal so neither direction is favored—solid ice melts and liquid water freezes at this temperature.

9. **D** Choice (D) is the only reaction where the number of moles of gas is increasing, going from 2 moles of gas on the reactant side to 3 moles of gas on the product side. In all the other choices, the number of moles of gas either decreases or remains constant.

10. **A** The temperature increase is indicative of energy being released, meaning the reaction is exothermic. The entropy (dispersion) of the system is increasing as it moves from three gas molecules to five.

11. **D** Via $K_{eq} = e^{-\frac{\Delta G}{RT}}$, if ΔG is negative the value for K will be greater than one. Via $\Delta G = \Delta H - T\Delta S$, ΔG is always negative when ΔH is negative and ΔS is positive.

12. **A** $5.0 \text{ min} \times \dfrac{60.0 \text{ s}}{1.0 \text{ min}} \times \dfrac{0.75 \text{ C}}{1.0 \text{ s}} \times \dfrac{1 \text{ mol } e^-}{96{,}500 \text{ C}} \times \dfrac{1 \text{ mol Cu}}{2 \text{ mol } e^-} \times \dfrac{63.55 \text{ g Cu}}{1 \text{ mol Cu}}$

13. **B** The higher the temperature of any substance, the larger the range of velocities the molecules of that substance can have, and thus the more dispersion the substance can have. A Maxwell-Boltzmann diagram represents this in a graphical form.

14. **C** Z^{2+} was able to reduce to solid Z by taking electrons from metal X, so Z^{2+} must have a higher reduction potential than X^{2+}. Y^{2+} was unable to take electrons from metal X, and therefore Y^{2+} must have a lower reduction potential than X^{2+}.

15. **B** Oxidation occurs at the anode, and that entails the loss of electrons. As there is no hydrogen gas present in the solution to start, the only substance that can be reduced at the anode is the solid copper itself.

16. **A** The half-reaction occurring at the cathode is $2 \text{ H}^+ + 2 \text{ } e^- \rightarrow \text{H}_2(g)$, as the only substance present at the beginning of the reaction that can be reduced is the hydrogen ion (initially, there are no Zn^{2+} or Cu^{2+} ions in solution). With copper being oxidized at the anode, the calculation then becomes $(0.00 \text{ V}) + (-0.34 \text{ V}) = -0.34 \text{ V}$. That means a voltage differential of 0.34 V will be required for this reaction to occur.

17. **C** In the initial Cu/Zn cell, as the electrolysis proceeds, Cu^{2+} ions are produced and become part of the solution. The SRP of those ions is greater than that of the SRP of H^+ ions, so later in the reaction, the Cu^{2+} is reduced into solid copper at the zinc cathode—this is what causes the copper to plate out on the zinc. If nickel were used as the anode, Ni^{2+} would be in solution; however, H^+ has a higher reduction potential than Ni^{2+}, which means the H^+ reduction would continue to occur at the cathode, and no nickel would be plated out of solution.

18. **C** The oxidation state of copper changes from +1 to 0, meaning it has gained electrons and is being reduced, and reduction occurs at the cathode. Zinc's oxidation state changes from 0 to +2, meaning it has lost electrons and is being oxidized, which occurs at the anode.

19. **B** 127 g is equal to 2 moles of copper, which is what appears on the balanced equation. To change one mole of copper from +1 to 0, 1 mole of electrons is required. Twice as many moles being created means twice as many electrons are needed.

20. **B** $E_{cell} = E_{red} + E_{ox}$

$1.28\ V = 0.52\ V + E_{ox}$

$E_{ox} = 0.76\ V$

$-E_{ox} = E_{red}$

$E_{red} = -0.76\ V$

21. **C** As the reaction progresses, $[Cu^+]$ will decrease and $[Zn^{2+}]$ will increase. Doing this will increase the value of Q ($Q = [Zn^{2+}]/[Cu^+]$), bringing the reaction closer to equilibrium and decreasing the overall potential of the reaction.

22. **A** The electron transfer does not happen across the salt bridge, eliminating (C) and (D). As the reaction progresses and $[Cu^+]$ decreases in the copper half-cell, positively charged sodium ions are transferred in to keep the charge balanced within the half-cell.

Free-Response

1. (a) Use the entropy values in the table.

$\Delta S° = \Sigma \Delta S°_{products} - \Sigma \Delta S°_{reactants}$

$\Delta S° = [(6)(213.6) + (6)(69.9)] - [(212.13) + (6)(205)]$ J/mol·K

$\Delta S° = 259$ J/mol·K

(b) Use the equation below. Remember that enthalpy values are given in kJ and entropy values are given in J.

$$\Delta G° = \Delta H° - T\Delta S°$$

$$\Delta G° = (-2{,}801 \text{ kJ/mol}_{rxn}) - (298 \text{ K})(0.259 \text{ kJ/mol}_{rxn}\cdot\text{K}) = -2{,}880 \text{ kJ/mol}_{rxn}$$

(c)

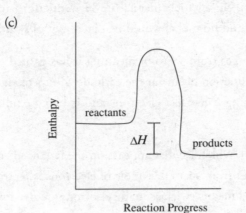

Reaction Progress

The reaction is exothermic; therefore the reactants must have more energy than the products, as indicated. The difference in energy is equal to the ΔH for this reaction.

(d) $1.00 \text{ g glucose} \times \dfrac{1 \text{ mol glucose}}{180 \text{ g glucose}} \times \dfrac{1 \text{ mol}_{rxn}}{1 \text{ mol glucose}} \times \dfrac{-2{,}801 \text{ kJ}}{1 \text{ mol}_{rxn}} = -15.6 \text{ kJ}$

2. (a) Use the relationship below.

$$\Delta H° = \Sigma \text{ Energies of the bonds broken } - \Sigma \text{ Energies of the bonds formed}$$

$$\Delta H° = [(4)(415) + (2)(495)] - [(2)(799) + (4)(463)] \text{ kJ/mol}$$

$$\Delta H° = -800 \text{ kJ/mol}$$

(b) $1{,}500 \text{ kJ} \times \dfrac{1 \text{ mol CH}_4}{800 \text{ kJ}} = 1.9 \text{ mol CH}_4 \times \dfrac{16.04 \text{ g CH}_4}{1 \text{ mol CH}_4} = 30 \text{ g CH}_4$

(c) Use $\Delta G° = \Delta H° - T\Delta S°$

Remember that enthalpy values are given in kJ and entropy values are given in J.

$$\Delta S° = \frac{\Delta H° - \Delta G°}{T} = \frac{(-800 \text{ kJ/mol}) - (-801 \text{ kJ/mol})}{(298 \text{ K})}$$

$$\Delta S° = 0.003 \text{ kJ/K} = 3 \text{ J/K}$$

(d) $\Delta S°$ is very small, which means that the entropy change for the process is very small. This makes sense because the number of moles remains constant, the number of moles of gas remains constant, and the complexity of the molecules remains about the same.

3. (a) (i)   In order to boil a liquid, the intermolecular forces between the various molecules of the liquid must be broken. This requires the input of energy; thus, ΔH is positive.

        (ii) A liquid turning into a gas leads to an increase in dispersion, which yields a positive ΔS.

    (b) When a phase change is occurring, the value for ΔG is always zero, as neither phase is favored at the phase transition temperature and the system is in equilibrium. Using that knowledge and making sure the units are matching (by converting the entropy to kJ/mol × K), you get:

    $\Delta G = \Delta H° - T\Delta S°$

    $0 = 37.6 \text{ kJ/mol} - T(0.111 \text{ kJ/mol·K})$

    $T = 339 \text{ K}$

    Converting to Celsius: 339 K – 273.15 = 66°C

    (c) First, figure out how many moles of methanol are present.

    $D = \dfrac{m}{V}$ $0.789 \text{ g/mL} = \dfrac{m}{50.0 \text{ mL}}$ $m = 39.5 \text{ g CH}_3\text{OH}$

    $39.5 \text{ g} \times \dfrac{1 \text{mol CH}_3\text{OH}}{32.04 \text{ g}} = 1.23 \text{ mol CH}_3\text{OH}$

    Then, use the $\Delta H°$ value to calculate the amount of heat needed.

    $1.23 \text{ mol} \times \dfrac{37.6 \text{ kJ}}{\text{mol}} = 46.2 \text{ kJ}$

    (d) As methanol (or any other liquid) boils, the temperature will remain constant, as the added heat is going into breaking intermolecular forces instead of adding extra kinetic energy to the liquid.

    (e) Yes, methanol would be soluble with water. Methanol is a polar molecule, as is water, and in terms of molecular solubility, like dissolves like.

    (f) It would be greater than that of methanol. Both molecules are polar with H-bonds, but ethanol has more electrons, and thus its electron cloud is more polarizable. This causes the London dispersion forces in ethanol to be stronger than those in methanol, leading to a higher boiling point.

4. (a) $\Delta S = \Delta S_{products} - \Delta S_{reactants}$

$\Delta S = [3(H_2O) + 4(N_2)] - [2(NH_3) + 3(N_2O)]$

$\Delta S = [3(189) + 4(192)] - [2(193) + 3(220)]$

$\Delta S = 1{,}335 - 1{,}046$

$\Delta S = 289 \ J/mol \cdot K$

(b) Use the balanced chemical reaction and values in the table.

$\Delta H° = \Sigma \Delta H°_{f \ (product)} - \Sigma \Delta H°_{f \ (reactant)}$

$\Delta H° = [4(N_2) + 3(H_2O)] - [2(NH_3) + 3(N_2O)]$

$\Delta H° = [4(0) + 3(-242)] - [2(-46) + 3(85.5)]$

$\Delta H° = -726 - 152.5 = -879 \ kJ/mol$

(c) $\Delta G = \Delta H - T\Delta S$

$\Delta G = -879 \ kJ/mol - (298 \ K)(0.289 \ kJ/mol \cdot K)$

$\Delta G = -879 \ kJ/mol - 86.1 \ kJ/mol$

$\Delta G = -965 \ kJ/mol$

The value for ΔG is negative; the reaction is thermodynamically favored at 25°C.

(d) (i) The value for ΔH is negative; therefore, it is an exothermic reaction and energy will be released.

(ii) To determine how much energy is released, the limiting reagent must be determined.

$$25.00 \ g \ NH_3 \times \frac{1 \ mol \ NH_3}{17.03 \ g \ NH_3} \times \frac{1 \ mol_{rxn}}{2 \ mol \ NH_3} \times \frac{-879 \ kJ}{1 \ mol_{rxn}} = -645 \ kJ$$

$$25.00 \ g \ N_2O \times \frac{1 \ mol \ NH_3}{44.02 \ g \ N_2O} \times \frac{1 \ mol_{rxn}}{3 \ mol \ N_2O} \times \frac{-879 \ kJ}{1 \ mol_{rxn}} = -166 \ kJ$$

As the N_2O would produce less energy, it would run out first and is thus limiting. The answer is thus −166 kJ.

(e) The activation energy describes the distance between the bond energy of the reactants and the energy of the activated state. In terms of enthalpy, the value for ΔH is negative, so that means that the bond energy of the products will be lower than that of the reactants.

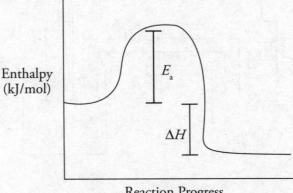

5. (a) When metal M was placed in the copper solution, a reaction occurred. Therefore, the copper cations must have the higher reduction potential and thus are reduced in the Cu/M cell. The half-reactions are:

$$\text{Reduction: } Cu^{2+}(aq) + 2\ e^- \rightarrow Cu(s)$$
$$\text{Oxidation: } M(s) \rightarrow M^+(aq) + e^-$$

The oxidation half-reaction must be multiplied by two to balance the electrons before combining the reactions to yield:

$$Cu^{2+}(aq) + 2\ M(s) \rightarrow Cu(s) + 2\ M^+(aq)$$

(b) $E_{red} + E_{ox} = +0.74\ V$

The reduction potential for $Cu^{2+} + 2\ e^- \rightarrow Cu(s)$ is known:

$$0.34\ V + E_{ox} = +0.74\ V \qquad E_{ox} = +0.40\ V$$

The reduction potential for metal M is the opposite of its oxidation potential.

$$E_{red} = -0.40\ V$$

(c) Reduction occurs at the cathode, so copper is the cathode. Oxidation occurs at the anode, so M is the anode.

(d)

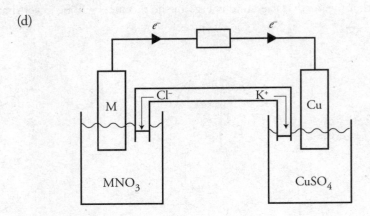

Electrons should be flowing from metal M to the copper bar. In the salt bridge, the K⁺ ions will flow toward the copper solution to replace the Cu^{2+} being reduced into Cu(s). The Cl⁻ will flow toward the solution for metal M in order to balance out the charge of the extra M⁺ ions being created via the oxidation of M(s).

(e) If the concentration of the Cu^{2+} increases, it will cause a shift to the right in accordance with Le Châtelier's principle. This will increase the cell's overall voltage.

6. (a) The two potential reduction reactions are the bottom two, as their reactants (H_2O and Ni^{2+}) are actually present at the start of the reaction. Of the two, the nickel reduction is more positive, and thus will take place. The top two reactions must be flipped to have the reactants (H_2O and F⁻) on the reactants side, which also flips the sign. As a result, the water oxidation has the more positive value (–1.23 as opposed to –2.87) and will therefore occur.

Reduction: $Ni^{2+}(aq) + 2\ e^- \rightarrow Ni(s)$
Oxidation: $2\ H_2O(l) \rightarrow O_2(g) + 4\ H^+(aq) + 4\ e^-$

After multiplying the reduction half-reaction by 2 to balance the electrons and combining both half-reactions, the net cell reaction is:

$2\ Ni^{2+}(aq) + 2\ H_2O(l) \rightarrow 2\ Ni(s) + O_2(g) + 4\ H^+(aq)$

(b) At the cathode, solid nickel would begin to plate out of solution. At the anode, oxygen gas would form and bubble up to the surface.

(c) The solution will become more acidic due to the creation of hydrogen ions at the anode.

(d) $1.20\text{g Ni} \times \dfrac{1\ \text{mol Ni}}{58.69\text{g Ni}} \times \dfrac{2\ \text{mol}\,e^-}{1\ \text{mol Ni}} \times \dfrac{96,500\ \text{C}}{1\ \text{mol}\,e^-} \times \dfrac{1\ \text{second}}{2.20\ \text{C}} = 1,790\ \text{seconds}$

Chapter 12
Laboratory Overview

INTRODUCTION

The AP Chemistry Exam will test your knowledge of basic lab techniques, as well as your understanding of accuracy and precision, and your ability to analyze potential sources of error in a lab. In this section, we will discuss safety and accuracy precautions, laboratory equipment, and laboratory procedures.

SAFETY

Here are some basic safety rules that might turn up in test questions.

- Don't put chemicals in your mouth. You were told this when you were four years old, and it still holds true for the AP Chemistry Exam.
- When diluting an acid, always add the acid to the water. This is to avoid the spattering of hot solution.
- Always work with good ventilation; many common chemicals are toxic.
- When heating substances, do it slowly. When you heat things too quickly, they can spatter, burn, or explode.

ACCURACY

Here are some rules for ensuring the accuracy of experimental results.

- When titrating, rinse the buret with the solution to be used in the titration instead of with water. If you rinse the buret with water, you might dilute the solution, which will cause the volume added from the buret to be too large.
- Allow hot objects to return to room temperature before weighing. Hot objects on a scale create convection currents that may make the object seem lighter than it is.
- Don't weigh reagents directly on a scale. Use a glass or porcelain container to prevent corrosion of the balance pan.
- When collecting a gas over water, remember to take into account the pressure and volume of the water vapor.
- Don't contaminate your chemicals. Never insert another piece of equipment into a bottle containing a chemical. Instead you should always pour the chemical into another clean container. Also, don't let the inside of the stopper for a bottle containing a chemical touch another surface.
- When mixing chemicals, stir slowly to ensure even distribution.
- Be conscious of significant figures when you record your results. The number of significant figures you use should indicate the precision of your results.
- Be aware of the difference between accuracy and precision. A measurement is accurate if it is close to the accepted value. A series of measurements is precise if the values of all of the measurements are close together.

SIGNIFICANT FIGURES

When taking measurements in lab, your measurement will always have a certain number of significant figures. For instance, if you are using a balance to measure the weight of an object to the hundredths place, you might get 23.15 g. That number has four significant figures. The balance is no more precise than that; you cannot say the mass of the object is 23.15224 g.

It's important to be able to identify the number of significant figures in any number given to you on the AP Exam. There's a plethora of information on ways to count significant figures out there, but we've reduced it to two simple rules.

1. For numbers without a decimal place, you count every number except for trailing zeroes (those which appear after all nonzeroes). So, 100 mL has only one significant figure, and 250 mL has two. On the other hand, 105 g has three significant figures—the zero in that measurement does not trail all other numbers.

2. For numbers with a decimal place, you count every number except for leading zeroes (those which appear before all nonzeroes). The number 0.052 has 2 significant figures, but 0.0520 g would have three (trailing zeroes DO count in numbers with decimals). Most (but not all) values you get on the AP Exam will have decimal places.

This leads to numbers that otherwise might look very strange. For instance, it is not unusual to see numbers with a decimal at the end but no numbers past it. That's because a number like 100. g has three significant figures, but 100 g only has one. In science, 100. g is not the same as 100 g, which is also not the same as 100.0 g. All of those values have different numbers of significant figures, and that implies different levels of accuracy. Significant figures only apply to measurements with units; pure unitless numbers (such as those you find in math) do not follow these rules. Nonetheless, many math books do have units on practice problems and then ignore significant figures. Yes, it's wrong to do that, but don't tell your math teacher.

When doing calculations, it is important that any calculated value cannot be more accurate than the measurements used in the calculation. Essentially, significant figures can tell you how to round your answers correctly. Again, this can be divided into two categories:

1. When multiplying and dividing, your answer cannot have more significant figures than your least precise measurement. For instance:

    2.50 g / 12 cm³ = 0.20833 g/cm³

    is wrong. Your two measurements have three significant figures and two significant figures, respectively. Your answer cannot have five. It has to have only two because that's how many figures your least-precise measurement had. The correct answer is 0.21 g/cm³.

More Great Books
If all this math talk is worrying you, check out these other books from The Princeton Review to help you in your math review:
High School Geometry Unlocked
High School Algebra I Unlocked
High School Algebra II Unlocked

2. When adding and subtracting, your answer cannot have more figures after the decimal place than your value with the least number of figures after its decimal place. For instance:

1.435 cm + 12.1 cm = 13.535 cm

is wrong. Your values have three figures after the decimal and one figure after the decimal, respectively. Your answer cannot have more than one figure past the decimal, because that's how many figures your least-precise measurement had. The correct answer is 13.5 cm.

On a side note, counting numbers are considered to have an infinite number of significant figures. If you are doing a molar mass conversion and you identify the conversion as 22.99 g of sodium = 1 mol of sodium, your eventual answer will have four significant figures. That 1 mol represents *exactly* 1 mol of sodium, which is essentially a 1 with an infinite number of zeroes past the decimal place. This will usually come up when dealing with stoichiometry—the coefficients in a balanced equation are counting numbers of moles and don't count when considering the number of significant figures your answer should have.

You should get into the habit of making sure your calculations always have the correct number of significant figures as you do them, both in labs and on any practice problems you do in class. The AP Exam may have questions on significant figures in a lab, and they are typically integrated into the free-response section. Additionally, you may lose points for the incorrect number of significant figures when you do calculations. Along with making sure you have the correct units on your number, making sure you do your calculations to the correct number of significant figures is a good laboratory practice.

EXPERIMENTAL DESIGN

The AP Chemistry Exam is heavily focused on students being able to understand the various types of experiments that can be undertaken in order to reinforce the concepts that are taught in the classroom. Some, if not most, of the free-response questions will be based on laboratory investigations. Here are the seven types that, based on the current framework for the exam, you should be familiar with. We've listed the test-makers' expectations (in their own words), a brief description of the experiment, and where in this book you can find additional explanations and practice problems.

Spectrophotometry

Use the absorption of light to determine the identity and/or concentration of an analyte in solution.

Spectrophotometry (aka colorimetry) is often used to measure the concentration of colored solutions. This type of experiment is often done in connection with reaction rates and/or equilibrium calculations.

Conceptual Explanation: Beer's Law on Page 130

Free-Response Examples: Unit 3 FRQ #9, Unit 5 FRQ #7

Gravimetric Analysis
Use gravimetric analysis to determine the amount of an analyte in a mixture.

Gravimetric analysis is intrinsically tied in with precipitation reactions and mass percent calculations.

Conceptual Explanation: Pages 162–163

Free-Response Examples: Unit 4 FRQ #1, Unit 7 FRQ #5

Acid-Base Titrations
Use titration to characterize an acid or base solution.

Acid-base titrations are by far the most common form of this learning objective that you will see on the exam. Acid-base titrations typically use indicators to cause the color to change at or near the equivalence point.

Conceptual Explanation: Pages 278–279

Free-Response Examples: Unit 8 FRQ #2 and FRQ #3

Component Separation
Use differences in intermolecular forces to separate a mixture into its components or to determine the identity of components of a mixture.

Understanding the best way to separate unique substances out of a mixture ties directly in with concepts of intermolecular forces and bonding. Out of the seven types of experiments, this one is generally the least mathematical and the most conceptual.

Conceptual Explanation: Pages 119–122

Free-Response Examples: Unit 3 FRQ #3 and #7

Redox Titrations
Use titration to determine the concentration of an analyte in a solution.

As the name indicates, these are titrations that are focused on oxidation-reduction reactions. These do not always use indicators (although they can), but instead some of the solutions involved are colored themselves.

Conceptual Explanation: Pages 165–166

Free-Response Examples: Unit 4 FRQ #2 and #3

Kinetics
Determine the rate law of a chemical reaction.

Determining the rate of reaction can usually only be done experimentally. As such, that makes kinetics an easy topic to study via experimental design. Graphical analysis is particularly common in these experiments.

Conceptual Explanation: Pages 184–195

Free-Response Examples: Unit 5 FRQ #1–3

Calorimetry
Use calorimetry to determine the change in enthalpy of a process.

The most common experimental data you will deal with involving any kind of thermodynamics calculations begins with calorimetry. Most thermodynamic data is gathered from existing data tables/research, but calorimetry is something that any high school student can experience during lab.

Conceptual Explanation: Pages 225–226

Free-Response Example: Unit 6 FRQ #1 and #2

There will certainly be other types of experiments tested on the AP Exam—there are countless ways to set up and design experiments to test various chemistry concepts. However, focusing on these is an excellent place to start.

LABORATORY EQUIPMENT
The following charts show some standard chemistry lab equipment.

 Beaker	Used to hold and pour liquids, only precise for measuring volume to the ones place—also your favorite Muppet.
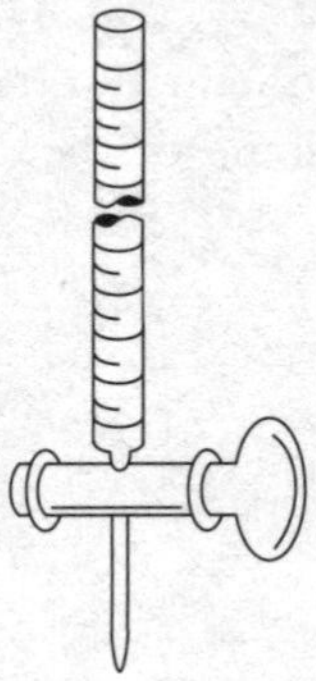 Buret	Used to add small but precisely measured volumes of liquid to a solution, usually estimated to the hundredths place by eye. Burets are used frequently in titration experiments.

Bunsen burner	Used to apply heat and to wake up sleeping AP Chemistry students.
Crucible tongs	Used to handle objects that are too hot to touch (careful though, they can break test tubes!).
Dropper pipette	Used to add small amounts of liquid to a solution, but only when a precise volume is not needed, because the drops themselves should be consistent only for one particular dropper (for example, adding an indicator, comparing the amount of drops of strong acid or base needed for a pH change in different solutions).
Erlenmeyer flask	A flask used for heating liquids, only precise for measuring volume to the ones place. The conic shape reduces evaporation.
Evaporating dish	Used to hold liquids for evaporation. The wide mouth allows vapor to escape.
Florence flask	Used for boiling of liquids. The small neck prevents excessive evaporation and splashing.

Forceps	A fancy name for tweezers.
Funnel	Used to get liquids into a smaller container. Doubles as a cute hat when inverted.
Graduated cylinder	Used for measuring a volume of liquid precisely, often to the tenths place, to be poured all at once (rather than dripped from a buret).
Graduated pipette	Used to transfer small and precise volumes of liquid, usually to the tenths place, from one container to another. The gradations indicate the volume.
Metal spatula	Used to scoop and transport powders.
Mortar and pestle	Used to grind solids into powders suitable for dissolving or mixing.

Pipette bulb	Rubber bulb used to draw liquid into pipette. In a pinch, could be used as a miniature clown nose.
Platform balance (triple beam)	A very precise scale operated by moving a set of three weights (typically corresponding to 100-, 10-, and 1-gram increments). A measurement will proceed like this: Rear weight is in the notch reading 30 g. Middle weight is in the notch reading 200 g. Front beam weight reads 3.86 g. The sample weighs 200 + 30 + 3.86 = 233.86 g.
Ring clamp	Used to hold funnels or other vessels in conjunction with a stand.
Rubber policeman	A hard tipped rubber scraper used to transfer precipitate.
Safety goggles	Used by chemists to protect their eyes during all laboratory experiments. Many students like to wear these on their foreheads or as a headband, but they really must be worn over your eyes to protect you.
Test tube	Used to contain samples, especially when heating.

Thermometer	Measures temperature of a solution (don't use these to measure your body temperature, you don't know where they've been).
Volumetric flask	Used to contain solutions of a fixed volume, indicated by the meniscus reaching a premarked line in the neck. Considered infinitely precise for the volume they're marked for.
Volumetric pipette	Used to transfer small amounts of liquid. The big difference between a graduated pipette and a volumetric pipette is that the volumetric type is marked for only one particular volume, and is considered infinitely precise to that volume.

Part VI

Practice Tests 2 and 3

Practice Test 2

The Exam

AP® Chemistry Exam

SECTION I: Multiple-Choice Questions

DO NOT OPEN THIS BOOKLET UNTIL YOU ARE TOLD TO DO SO.

At a Glance

Total Time
1 hour and 30 minutes
Number of Questions
60
Percent of Total Grade
50%
Writing Instrument
Pencil required

Instructions

Section I of this examination contains 60 multiple-choice questions. Fill in only the ovals for numbers 1 through 60 on your answer sheet.

Indicate all of your answers to the multiple-choice questions on the answer sheet. No credit will be given for anything written in this exam booklet, but you may use the booklet for notes or scratch work. After you have decided which of the suggested answers is best, completely fill in the corresponding oval on the answer sheet. Give only one answer to each question. If you change an answer, be sure that the previous mark is erased completely. Here is a sample question and answer.

Sample Question   Sample Answer

Chicago is a   Ⓐ ● Ⓒ Ⓓ
(A) state
(B) city
(C) country
(D) continent

Use your time effectively, working as quickly as you can without losing accuracy. Do not spend too much time on any one question. Go on to other questions and come back to the ones you have not answered if you have time. It is not expected that everyone will know the answers to all the multiple-choice questions.

About Guessing

Many candidates wonder whether or not to guess the answers to questions about which they are not certain. Multiple-choice scores are based on the number of questions answered correctly. Points are not deducted for incorrect answers, and no points are awarded for unanswered questions. Because points are not deducted for incorrect answers, you are encouraged to answer all multiple-choice questions. On any questions you do not know the answer to, you should eliminate as many choices as you can, and then select the best answer among the remaining choices.

CHEMISTRY
SECTION I
Time—1 hour and 30 minutes

INFORMATION IN THE TABLE BELOW AND ON THE FOLLOWING PAGES MAY BE USEFUL IN
ANSWERING THE QUESTIONS IN THIS SECTION OF THE EXAMINATION

DO NOT DETACH FROM BOOK.

PERIODIC TABLE OF THE ELEMENTS

1	2	3	4	5	6	7	8	9	10	11	12	13	14	15	16	17	18
1 H 1.008																	2 He 4.00
3 Li 6.94	4 Be 9.01											5 B 10.81	6 C 12.01	7 N 14.01	8 O 16.00	9 F 19.00	10 Ne 20.18
11 Na 22.99	12 Mg 24.30											13 Al 26.98	14 Si 28.09	15 P 30.97	16 S 32.06	17 Cl 35.45	18 Ar 39.95
19 K 39.10	20 Ca 40.08	21 Sc 44.69	22 Ti 47.87	23 V 50.94	24 Cr 52.00	25 Mn 54.94	26 Fe 55.85	27 Co 58.93	28 Ni 58.69	29 Cu 63.55	30 Zn 65.38	31 Ga 69.72	32 Ge 72.63	33 As 74.92	34 Se 78.97	35 Br 79.90	36 Kr 83.80
37 Rb 85.47	38 Sr 87.62	39 Y 88.91	40 Zr 91.22	41 Nb 92.91	42 Mo 95.95	43 Tc	44 Ru 101.07	45 Rh 102.91	46 Pd 106.42	47 Ag 107.87	48 Cd 112.41	49 In 114.82	50 Sn 118.71	51 Sb 121.76	52 Te 127.60	53 I 126.90	54 Xe 131.29
55 Cs 132.91	56 Ba 137.33	57 57-71 *	72 Hf 178.49	73 Ta 180.95	74 W 183.94	75 Re 186.21	76 Os 190.23	77 Ir 192.22	78 Pt 195.08	79 Au 196.97	80 Hg 200.59	81 Tl 204.38	82 Pb 207.2	83 Bi 208.98	84 Po	85 At	86 Rn
87 Fr	88 Ra	89-103 †	104 Rf	105 Db	106 Sg	107 Bh	108 Hs	109 Mt	110 Ds	111 Rg	112 Cn	113 Nh	114 Fl	115 Mc	116 Lv	117 Ts	118 Og

*Lanthanoids	57 La 138.91	58 Ce 140.12	59 Pr 140.91	60 Nd 144.24	61 Pm	62 Sm 150.36	63 Eu 151.97	64 Gd 157.25	65 Tb 158.93	66 Dy 162.50	67 Ho 164.93	68 Er 167.26	69 Tm 168.93	70 Yb 173.05	71 Lu 174.97
†Actinoids	89 Ac	90 Th 232.04	91 Pa 231.04	92 U 238.03	93 Np	94 Pu	95 Am	96 Cm	97 Bk	98 Cf	99 Es	100 Fm	101 Md	102 No	103 Lr

GO ON TO THE NEXT PAGE.

AP® CHEMISTRY EQUATIONS & CONSTANTS

Throughout the exam the following symbols have the definitions specified unless otherwise noted.

L, mL	= liter(s), milliliter(s)		mm Hg	= millimeters of mercury
g	= gram(s)		J, kJ	= joule(s), kilojoule(s)
nm	= nanometer(s)		V	= volt(s)
atm	= atmosphere(s)		mol	= mole(s)

ATOMIC STRUCTURE

$$E = h\nu$$
$$c = \lambda\nu$$

E = energy
ν = frequency
λ = wavelength

Planck's constant, $h = 6.626 \times 10^{-34}$ J s

Speed of light, $c = 2.998 \times 10^8$ m s^{-1}

Avogadro's number $= 6.022 \times 10^{23}$ mol^{-1}

Electron charge, $e = -1.602 \times 10^{-19}$ coulomb

EQUILIBRIUM

$$K_c = \frac{[C]^c[D]^d}{[A]^a[B]^b}, \text{ where } a\,A + b\,B \rightleftarrows c\,C + d\,D$$

$$K_p = \frac{(P_C)^c(P_D)^d}{(P_A)^a(P_B)^b}$$

$$K_a = \frac{[H^+][A^-]}{[HA]}$$

$$K_b = \frac{[OH^-][HB^+]}{[B]}$$

$$K_w = [H^+][OH^-] = 1.0 \times 10^{-14} \text{ at } 25°C$$
$$= K_a \times K_b$$

$$pH = -\log[H^+], \ pOH = -\log[OH^-]$$

$$14 = pH + pOH$$

$$pH = pK_a + \log\frac{[A^-]}{[HA]}$$

$$pK_a = -\log K_a, \ pK_b = -\log K_b$$

Equilibrium Constants

K_c (molar concentrations)

K_p (gas pressures)

K_a (weak acid)

K_b (weak base)

K_w (water)

KINETICS

$$[A]_t - [A]_0 = -kt$$

$$\ln[A]_t - \ln[A]_0 = -kt$$

$$\frac{1}{[A]_t} - \frac{1}{[A]_0} = kt$$

$$t_{1/2} = \frac{0.693}{k}$$

k = rate constant
t = time
$t_{1/2}$ = half-life

GO ON TO THE NEXT PAGE.

GASES, LIQUIDS, AND SOLUTIONS

$$PV = nRT$$

$$P_A = P_{total} \times X_A, \text{ where } X_A = \frac{\text{moles A}}{\text{total moles}}$$

$$P_{total} = P_A + P_B + P_C + \dots$$

$$n = \frac{m}{M}$$

$$K = {}^{\circ}C + 273$$

$$D = \frac{m}{V}$$

$$KE_{molecule} = \frac{1}{2}mv^2$$

Molarity, M = moles of solute per liter of solution

$$A = \varepsilon bc$$

P = pressure
V = volume
T = temperature
n = number of moles
m = mass
M = molar mass
D = density
KE = kinetic energy
v = velocity
A = absorbance
ε = molar absorptivity
b = path length
c = concentration

Gas constant, $R = 8.314 \text{ J mol}^{-1}\text{K}^{-1}$

$= 0.08206 \text{ L atm mol}^{-1}\text{K}^{-1}$

$= 62.36 \text{ L torr mol}^{-1}\text{K}^{-1}$

$1 \text{ atm} = 760 \text{ mm Hg} = 760 \text{ torr}$

STP = 273.15 K and 1.0 atm

Ideal gas at STP = 22.4 L mol^{-1}

THERMOCHEMISTRY/ ELECTROCHEMISTRY

$$q = mc\Delta T$$

$$\Delta S^{\circ} = \sum S^{\circ} \text{ products} - \sum S^{\circ} \text{ reactants}$$

$$\Delta H^{\circ} = \sum \Delta H_f^{\circ} \text{ products} - \sum \Delta H_f^{\circ} \text{ reactants}$$

$$\Delta G^{\circ} = \sum \Delta G_f^{\circ} \text{ products} - \sum \Delta G_f^{\circ} \text{ reactants}$$

$$\Delta G^{\circ} = \Delta H^{\circ} - T\Delta S^{\circ}$$

$$= -RT \ln K$$

$$= -nFE^{\circ}$$

$$I = \frac{q}{t}$$

$$E_{cell} = E_{cell}^{\circ} - \frac{RT}{nF}\ln Q$$

q = heat
m = mass
c = specific heat capacity
T = temperature
S° = standard entropy
H° = standard enthalpy
G° = standard Gibbs free energy
n = number of moles
E° = standard reduction potential
I = current (amperes)
q = charge (coulombs)
t = time (seconds)
Q = reaction quotient

Faraday's constant, F = 96,485 coulombs per mole of electrons

$$1 \text{ volt} = \frac{1 \text{ joule}}{1 \text{ coulomb}}$$

GO ON TO THE NEXT PAGE.

1. In a saturated solution of Na_3PO_4, $[Na^+] = 0.30$ M. What is the molar solubility of Na_3PO_4?

    (A) 0.10 M
    (B) 0.30 M
    (C) 0.60 M
    (D) 0.90 M

2. When some LiCl is dissolved in water, the temperature of the water increases. This means that

    (A) the strength of the intermolecular forces between the water molecules is stronger than the bond energy within the LiCl lattice
    (B) the attraction of the lithium ions to the negative partial charges of the water molecules is weaker than the attraction of the chloride ions to the positive partial charges of the water molecules
    (C) breaking the bonds between the lithium and chloride ions is an exothermic process
    (D) the strength of the ion-dipole attractions that are formed exceeds the lattice energy in LiCl

Use the following information to answer questions 3–6.

A student titrates some 1.0 M HCl into 20.0 mL of methylamine (CH_3NH_2), a weak base that accepts only a single proton. The following titration curve results:

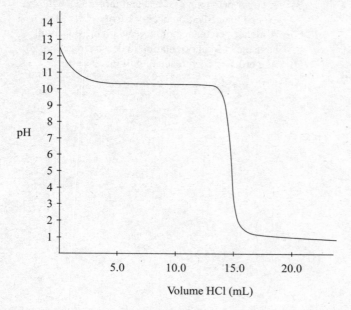

Volume HCl (mL)

3. What is the concentration of the methylamine?

    (A) 0.50 M
    (B) 0.75 M
    (C) 1.0 M
    (D) 1.25 M

4. What is the approximate pK_b for methylamine?

    (A) 3.5
    (B) 5.5
    (C) 10.5
    (D) 12.5

5. The buffer region of this titration is located

    (A) below 3.0 mL
    (B) between 3.0 mL and 14.0 mL
    (C) between 14.0 mL and 16.0 mL
    (D) above 16.0 mL

6. The methylamine is replaced by 20.0 mL of sodium hydroxide of an identical concentration. If the sodium hydroxide is titrated with the 1.0 M HCl, which of the following options accurately describes the pH levels at various points during the titration when compared to the pH levels at the same point in the HCl/methylamine titration?

	Initial pH	Equivalence pH	Ending pH
(A)	lower	same	higher
(B)	higher	higher	same
(C)	same	higher	same
(D)	higher	lower	lower

7. The formate ion, HCO_2^-, is best represented by the Lewis diagram below. Each bond is labeled with a different letter.

$$\left[\begin{array}{c} H \overset{X}{-} C \overset{Y}{\nearrow} \overset{:O:}{} \\ \underset{Z}{\searrow} \overset{..}{\underset{..}{O}}: \end{array} \right]^{-}$$

What is the bond order for each bond?

	X	Y	Z
(A)	1	1	2
(B)	2	2	1
(C)	1	1.5	1.5
(D)	1.33	1.33	1.33

GO ON TO THE NEXT PAGE.

$$Ag^+(aq) + 2\,NH_3(aq) \rightleftharpoons Ag(NH_3)_2^+(aq)$$

8. The reaction above is at equilibrium in a closed system. Which of the following will happen immediately when water is added?

(A) The rate of the reverse reaction will increase.
(B) Both ions will increase in concentration, while the NH_3 decreases in concentration.
(C) The reaction will shift to the right.
(D) Nothing will happen; adding water does not cause any changes to the equilibrium system.

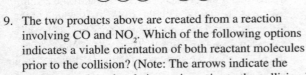

9. The two products above are created from a reaction involving CO and NO_2. Which of the following options indicates a viable orientation of both reactant molecules prior to the collision? (Note: The arrows indicate the direction each molecule is moving prior to the collision.)

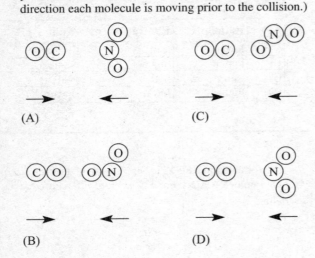

(A)

(B)

(C)

(D)

$$C_xH_Y(g) + O_2(g) \rightarrow CO_2(g) + H_2O(g)$$

10. When the above <u>unbalanced</u> reaction occurs at STP, 1.5 L of CO_2 and 1.0 L of H_2O are created. What is the empirical formula of the hydrocarbon?

(A) CH_2
(B) C_2H_3
(C) C_2H_5
(D) C_3H_4

$$2\,H_2O_2(aq) \rightarrow 2\,H_2O(l) + O_2(g)$$

11. For the decomposition of hydrogen peroxide, which element (if any) is being reduced, and which is being oxidized?

	Oxidized	Reduced
(A)	Hydrogen	Oxygen
(B)	Oxygen	None
(C)	None	Hydrogen
(D)	Oxygen	Oxygen

12. Identical amounts of the four gases listed below are present in four separate balloons. At STP, which balloon size experiences the greatest deviation from the volume calculated using the Ideal Gas Law?

(A) H_2
(B) O_2
(C) N_2
(D) F_2

13. Which of the following examples correctly explains one reason that increasing the temperature of a reaction increases its speed?

(A) All reactant molecules will have more kinetic energy.
(B) A larger percentage of reactant molecules will exceed the activation energy barrier.
(C) A higher percentage of molecular collisions will have the correct orientation to cause a reaction.
(D) The order of each reactant will increase.

GO ON TO THE NEXT PAGE.

Use the following information to answer questions 14–16.

The radius of atoms and ions is typically measured in Angstroms (Å), which is equivalent to 1×10^{-10} m. Below is a table of information for three different elements.

Element	Atomic Radius (Å)	Ionic Radius (Å)
Ne	0.38	N/A
P	0.98	1.00
Zn	1.42	1.35

14. The phosphorus ion is larger than a neutral phosphorus atom, yet a zinc ion is smaller than a neutral zinc atom. Which of the following statements best explains why?

   (A) The zinc atom has more protons than the phosphorus atom.
   (B) The phosphorus atom has fewer valence electrons than the zinc atom.
   (C) Phosphorus gains electrons when forming an ion, but zinc loses them.
   (D) The valence electrons in zinc are further from the nucleus than those in phosphorus.

15. Neon has a smaller atomic radius than phosphorus because

   (A) unlike neon, phosphorus has electrons present in its third energy level.
   (B) phosphorus has more protons than neon, which increases the repulsive forces in the atom.
   (C) the electrons in a neon atom are all found in a single energy level.
   (D) phosphorus can form anions, while neon is unable to form any ions.

16. Which of the following represents the correct electron configuration for the zinc ion, Zn^{2+}?

   (A) $[Ar]3d^{10}$
   (B) $[Ar]4s^23d^8$
   (C) $[Ar]4s^24d^8$
   (D) $[Kr]4s^23d^8$

17. The Lewis diagrams for $SiCl_4$ and PCl_3 are drawn above. What are the approximate bond angles between the terminal chlorine atoms in each structure?

	$SiCl_4$	PCl_3
(A)	90°	90°
(B)	109.5°	< 109.5°
(C)	90°	109.5°
(D)	< 109.5°	> 90°

$$2\,CrO_4^{2-}(aq) + 2\,H^+(aq) \rightleftharpoons Cr_2O_7^{2-}(aq) + H_2O(l)$$

18. The above reaction is present at equilibrium in a beaker. A student stirs the mixture. What effect will this have on the reaction rates?

   (A) It will increase both the forward and reverse reaction rates.
   (B) It will increase the forward rate, but decrease the reverse rate.
   (C) It will have no effect on the forward rate, but it will decrease the reverse rate.
   (D) It will have no effect on either rate.

19. A sample of water originally at 25°C is heated to 75°C. As the temperature increases, the vapor pressure of the water is also observed to increase. Why?

   (A) Water molecules are more likely to have enough energy to break free of the intermolecular forces holding them together.
   (B) The covalent bonds between the hydrogen and oxygen atoms within individual water molecules are more likely to be broken.
   (C) The strength of the hydrogen bonding between different water molecules will increase until it exceeds the covalent bond energy within individual water molecules.
   (D) The electron clouds surrounding each water molecule are becoming less polarizable, weakening the intermolecular forces between them.

GO ON TO THE NEXT PAGE.

20. The enthalpy change for which of the following reactions would be equal to the enthalpy of formation for ethanol (CH_3CH_2OH)?

    (A) $CH_3 + CH_2 + OH \rightarrow CH_3CH_2OH$
    (B) $2\,C + 5\,H + O \rightarrow CH_3CH_2OH$
    (C) $4\,C + 6\,H_2 + O_2 \rightarrow 2\,CH_3CH_2OH$
    (D) $2\,C + 3\,H_2 + \frac{1}{2}\,O_2 \rightarrow CH_3CH_2OH$

21. A chemist wants to plate out 1.00 g of solid iron from a solution containing aqueous Fe^{2+} ions. Which of the following expressions will equal the amount of time, in seconds, it takes if a current of 5.00 A is applied?

    (A) $\dfrac{(2)(55.85)(5.00)}{96,500}$

    (B) $\dfrac{(2)(96,500)}{(55.85)(5.00)}$

    (C) $\dfrac{(55.85)(96,500)}{(2)(5.00)}$

    (D) $\dfrac{(2)(55.85)(96,500)}{(5.00)}$

Use the following information to answer questions 22–24.

10.0 g each of three different gases are present in three glass containers of identical volume, as shown below. The temperature of all three flasks is held constant at 298 K.

22. The container with which gas would have the greatest pressure?

    (A) SO_2
    (B) CH_4
    (C) NCl_3
    (D) All three containers would have the same pressure.

23. Which of the gases would have the greatest density?

    (A) SO_2
    (B) CH_4
    (C) NCl_3
    (D) All three gases would have the same density.

24. If a small, pinhole-size leak were to be drilled into each container, the container with which gas would experience the fastest pressure decrease?

    (A) SO_2
    (B) CH_4
    (C) NCl_3
    (D) All three containers would decrease pressure at the same rate.

$$CO_2(g) + H_2(g) \rightleftharpoons CO(g) + H_2O(g) \qquad K_c = 1.4$$
$$CO(g) + 2\,H_2(g) \rightleftharpoons CH_3OH(g) \qquad K_c = 14.5$$

25. Given the above information, what would the equilibrium constant for the below reaction be?

$$3\,CO(g) + 2\,H_2O(g) \rightleftharpoons 2\,CO_2(g) + CH_3OH(g)$$

    (A) $(2)(1.4)(14.5)$

    (B) $\dfrac{(1.4)(14.5)}{2}$

    (C) $\dfrac{14.5}{(1.4)^2}$

    (D) $14.5 - 1.4^2$

$$2\,H_2(g) + O_2(g) \rightarrow 2\,H_2O(g)$$

26. When 1.0 mole of H_2 is combined with 1.0 mol of O_2 in a sealed flask, the reaction above occurs to completion at a constant temperature. After the reaction, the pressure in the container will have

    (A) increased by 25%
    (B) increased by 50%
    (C) decreased by 25%
    (D) decreased by 50%

27. A strong acid/strong base titration is completed using an indicator which changes color at the exact equivalence point of the titration. The protonated form of the indicator is HIn, and the deprotonated form is In⁻. At the equivalence point of the reaction

    (A) $[HIn] = [In^-]$
    (B) $[HIn] = 1/[In^-]$
    (C) $[HIn] = 2[In^-]$
    (D) $[HIn] = [In^-]^2$

GO ON TO THE NEXT PAGE.

H H
| | ..
H–C–C–Ö–H
| | ..
H H

Ethanol

H H H H H H H H
| | | | | | | |
H–C–C–C–C–C–C–C–C–H
| | | | | | | |
H H H H H H H H

Octane

28. The Lewis diagrams for both ethanol and octane are drawn above. Ethanol's boiling point is 78°C, while octane's is 125°C. This is best explained by the fact that

(A) octane has hydrogen bonding, while ethanol does not
(B) octane has a significantly lower molar mass than ethanol
(C) octane's temporary dipoles are stronger than those in ethanol
(D) octane is more symmetrical than ethanol

29. Which compound, $CaCl_2$ or CaO, would you expect to have a high melting point? Why?

(A) $CaCl_2$, because there are more ions per lattice unit
(B) $CaCl_2$, because a chlorine ion is smaller than an oxygen ion
(C) CaO, because the charge of an oxygen ion exceeds that of a chlorine ion
(D) CaO, because the common charges of calcium and oxygen ions are identical in magnitude

30. Even though it is a noble gas, xenon is known to form bonds with other elements. Which element from the options below would xenon most likely be able to bond with?

(A) Lithium
(B) Argon
(C) Fluorine
(D) Carbon

31. Stock nitric acid, HNO_3, has a concentration of 15.8 M. The pH of stock nitric acid would fall into which of the following pH ranges?

(A) Between –2 and –1
(B) Between –1 and 0
(C) Between 0 and 1
(D) Between 1 and 2

32. During gravimetric analysis experiments, collected precipitates are often rinsed with distilled water prior to being collected. What is the purpose of doing so?

(A) This ensures the precipitation reaction has gone to completion.
(B) It washes away any spectator ions stuck to the precipitate.
(C) The precipitate must be fully hydrated.
(D) Washing the precipitate ensures it has the correct density.

$$2 SO_3(g) \rightleftharpoons 2 SO_2(g) + O_2(g)$$

33. The above equilibrium is established in a sealed container. Some neon gas is injected into the container at constant pressure. If the neon does not react with any of the species in the container, what effect will this have on the reaction quotient and the reaction rates for this reaction?

(A) Q will increase, causing the forward reaction rate to increase.
(B) Q will increase, causing the reverse reaction rate to increase.
(C) Q will decrease, causing the forward reaction rate to decrease.
(D) Q will remain unchanged and the reaction rates will also remain unchanged.

GO ON TO THE NEXT PAGE.

Use the following information to answer questions 34–38.

A 10.0 g sample of NaNO₃ is dissolved in 200.0 mL of water with stirring, and the temperature of the water changes from 23.0°C to 18.0°C during the process. Assume the density and the specific heat of the final solution are identical to those of water.

34. Which of the following diagrams correctly shows a particulate representation of the species present in the beaker after the NaNO₃ has dissolved?

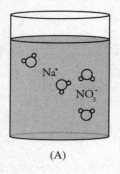

(A)

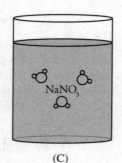

(C)

(B)

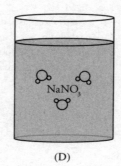

(D)

35. Which of the following statements best describes the enthalpy of solution for NaNO₃?

(A) It will be greater than zero because energy is absorbed from the water.

(B) It will be greater than zero because the magnitude of the ion-dipole attractive forces exceeds the lattice energy of NaNO₃.

(C) It will be less than zero because water is a highly polar solvent.

(D) It will be less than zero due to the presence of the polyatomic NO₃⁻ anion.

36. During the process, energy is released when

(A) the ionic bonds within the NaNO₃ lattice are broken

(B) the water molecules are spread apart to make room for the dissolved NaNO₃

(C) new attractive forces form between the dissociated ions and the water dipoles

(D) the covalent bonds within the water molecules are broken

37. Which of the following expressions will correctly calculate the heat change experienced by the water?

(A) (200.0 g)(4.18 J/g·°C))(–5.0°C)

(B) (210.0 g)(4.18 J/g·°C)(–5.0°C)

(C) (10.0 g)(4.18 J/g·°C)(–5.0 °C)

(D) (200.0 g)(4.18 J/g·°C)(5.0 °C)

38. After experimentally determining the enthalpy of solution for NaNO₃ using the gathered data, the magnitude of the experimental value is found to be lower than the accepted value. Which of the following correctly identifies a reason why?

(A) Some water evaporated while the NaNO₃ was dissolving.

(B) The specific heat of the final solution is actually lower than 4.18 J/g·°C.

(C) The stirring of the solution added significant energy to the system.

(D) Some energy from the dissolution process was lost to the surroundings.

---

39. Red light has a wavelength of 680 nm, and blue light has a wavelength of 470 nm. Which of the following options correctly identifies the relationships between the frequencies and energy levels of red and blue light?

	Frequency	Energy
(A)	Red > Blue	Red > Blue
(B)	Red > Blue	Red < Blue
(C)	Red < Blue	Red > Blue
(D)	Red < Blue	Red < Blue

GO ON TO THE NEXT PAGE.

Step 1: $N_2O_5(g) \rightarrow NO_2(g) + NO_3(g)$
Step 2: $NO_2(g) + NO_3(g) \rightarrow NO_2(g) + O_2(g) + NO(g)$
Step 3: $NO(g) + N_2O_5(g) \rightarrow 3\ NO_2(g)$

40. The rate law for the above reaction is determined to be rate = $k[N_2O_5]$. Which of the following options correctly identifies the overall order of the reaction as well as the molecularity of the rate-determining step?

(A) First order, unimolecular
(B) Second order, bimolecular
(C) Third order, bimolecular
(D) Fifth order, unimolecular

41. As temperature decreases, the pH of water increases. This means that

(A) the number of H^+ and OH^- ions is increasing
(B) water is more basic at lower temperatures
(C) $[H^+]$ begins to exceed the $[OH^-]$
(D) the auto-ionization of water is an endothermic process

42. The Lewis diagrams for four different compounds are drawn below. In which molecule would the dipole moment be closest to zero?

(A)

(C)

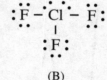
(B)

(D)

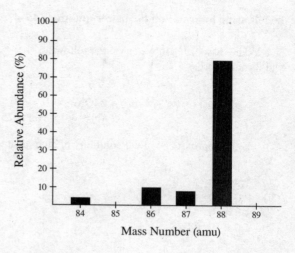

43. The mass spectrum above shows the distribution of various isotopes of strontium. Based on the data, which of the following conclusions can be drawn?

(A) Strontium most commonly forms ions with a charge of +2.
(B) Strontium isotopes with a mass of 86 or 87 are very unstable.
(C) The number of protons in a strontium atom nucleus can vary.
(D) The most common isotope of strontium has 50 neutrons.

$$\underline{\hphantom{xx}}\ MnO_4^-(aq) + \underline{\hphantom{xx}}\ H^+(aq) + \underline{\hphantom{xx}}\ C_2O_4^{2-}(aq)$$
$$\rightarrow \underline{\hphantom{xx}}\ Mn^{2+}(aq) + \underline{\hphantom{xx}}\ H_2O(l) + \underline{\hphantom{xx}}\ CO_2(g)$$

44. When the above oxidation-reduction reaction is completely balanced, what is the coefficient on the MnO_4^- ion?

(A) 1
(B) 2
(C) 3
(D) 4

GO ON TO THE NEXT PAGE.

Use the following information to answer questions 45–47.

Zinc iodate dissociates in water via the following equilibrium process

$$Zn(IO_3)_2(s) \rightleftharpoons Zn^{2+}(aq) + 2\ IO_3^-(aq)$$
$$K_{sp} = 4.0 \times 10^{-6} \text{ at } 25°C$$

45. What is the approximate molar solubility of zinc iodate at 25°C?

    (A) 1.0×10^{-2} M
    (B) 2.0×10^{-3} M
    (C) 4.0×10^{-6} M
    (D) 1.0×10^{-6} M

46. Which of the following solutions would zinc iodate be the LEAST soluble in?

    (A) $1.0\ M$ $BaCl_2$
    (B) $1.0\ M$ $NaIO_3$
    (C) $1.0\ M$ K_2CO_3
    (D) Pure water

47. A beaker of saturated zinc iodate is left out overnight. The following morning, some water is found to have evaporated. Assuming the temperature remained constant at 25°C, which of the following options correctly identifies the changes in the $[Zn^{2+}]$ and the mass of the zinc iodate present in the beaker compared to the previous night?

	$[Zn^{2+}]$	Mass $Zn(IO_3)_2$
(A)	Increase	Decrease
(B)	Increase	No Change
(C)	No Change	Increase
(D)	No Change	Decrease

48. Which of the following reactions would be favored at lower temperatures, but not favored at higher temperatures?

    (A) $2\ Fe_2O_3(s) + 3\ C(s) \rightarrow 4\ Fe(s) + 3\ CO_2(g)$
                      $\Delta H° = +468$ kJ/mol$_{rxn}$

    (B) $S(g) + ½\ O_2(g) \rightarrow SO_2(g)$
                      $\Delta H° = -297$ kJ/ mol$_{rxn}$

    (C) $N_2H_4(l) + CH_2O(l) \rightarrow CH_2O(g) + N_2(g) + 3\ H_2(g)$
                      $\Delta H° = -37$ kJ/mol$_{rxn}$

    (D) $2\ B(s) + 3\ H_2(g) \rightarrow B_2H_6(g)$
                      $\Delta H° = +36$ kJ/mol$_{rxn}$

49. Which of the following statements regarding covalent bonds is true?

    (A) Two atoms will form a covalent bond at the distance which minimizes the potential energy between them.
    (B) Covalent bonds always involve the equal sharing of valence electrons between two atoms.
    (C) The breaking of a covalent bond always occurs during a liquid-to-gas phase change for a covalent substance.
    (D) Covalent bonds typically form between atoms that demonstrate metallic properties.

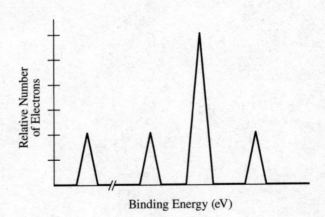

50. The photoelectron spectrum of an element is above. Based on the spectrum, what is the charge on the most common ion of the element?

    (A) -2
    (B) -1
    (C) $+1$
    (D) $+2$

GO ON TO THE NEXT PAGE.

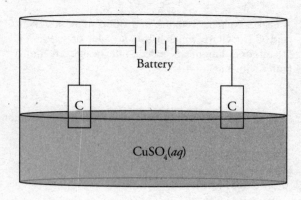

Bond	Energy (kJ/mol)
C-H	413
O=O	498
O-H	467

$$CH_4(g) + 2\,O_2(g) \rightarrow CO_2(g) + 2\,H_2O(g)$$
$$\Delta H° = -890 \text{ kJ/mol}$$

53. Determine the approximate bond energy of a C=O bond given the above data.

(A) 400 kJ/mol
(B) 600 kJ/mol
(C) 800 kJ/mol
(D) 1,000 kJ/mol

51. A beaker is filled with some 1.0 M $CuSO_4$, and two carbon electrodes are placed in the beaker with a battery wired between them, as shown above. As current is run through the system, solid copper plates out onto the carbon cathode. Which of the following changes would increase the amount of solid copper that is plated out of solution?

(A) Replacing the 1.0 M $CuSO_4$ solution with 1.0 M $CuCl_2$
(B) Using a 9.0 V battery instead of a 1.5 V battery
(C) Decreasing the pH of the solution
(D) Changing out the carbon electrodes with platinum electrodes

$$3\,NO_2(g) + H_2O(l) \rightarrow 2\,HNO_3(aq) + NO(g)$$
$$\Delta H° = -135 \text{ kJ/mol}_{rxn}$$

52. 1.0 mole of $NO_2(g)$ is bubbled through excess water, causing the above reaction to take place. Which of the following statements correctly describes the energy change that will occur during the reaction?

(A) 135 kJ of energy will be emitted.
(B) 45 kJ of energy will be emitted.
(C) 135 kJ of energy will be absorbed.
(D) 405 kJ of energy will be absorbed.

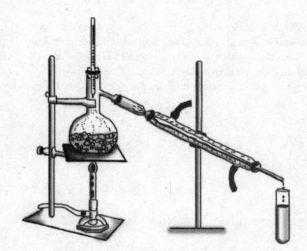

Substance	Vapor Pressure @ 25°C (torr)
H_2O	23.8
CH_3OH	99.0
C_6H_{12}	78.0

54. Three different liquids are mixed together in a flask, and that flask is then hooked up to a distillation apparatus, as shown above. The liquids are initially at 25°C, and the heat is turned up until the mixture starts to boil. Which liquid would be the first to separate out of the mixture?

(A) H_2O
(B) CH_3OH
(C) C_6H_{12}
(D) Distillation would be an ineffective method of separating the mixture.

GO ON TO THE NEXT PAGE.

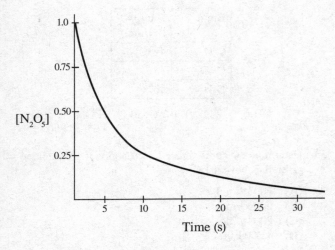

$$2 N_2O_5(g) \rightarrow 4 NO_2(g) + O_2(g)$$

55. A sample of some N_2O_5 gas is placed in a sealed container and allowed to decompose. The concentration of the N_2O_5 is tracked over time, and the results are plotted on the above graph. Which of the following represents the possible units on the rate constant for this reaction?

(A) s^{-1}
(B) sM^{-1}
(C) Ms^{-1}
(D) Ms^{-2}

$$Cu^+(aq) + 4 NH_3(aq) \rightleftharpoons Cu(NH_3)_4^{2+}(aq) \quad K = 5.0 \times 10^{13}$$

56. The above system is at equilibrium. If it were to be diluted, what would happen to the moles of reactants and the moles product after equilibrium is reestablished?

	Mole Reactants	Moles Product
(A)	Increase	Increase
(B)	Increase	Decrease
(C)	Decrease	Decrease
(D)	No Change	No Change

Use the following information to answer questions 57–60.

A beaker contains 100. mL of a 1.0 M solution of benzoic acid, C_6H_5COOH, which has a pK_a value of 4.19.
50. mL of Solution X is added to the beaker, creating a buffer with a pH of 4.19.

57. Which of the following could be the identity of Solution X?

(A) 1.0 M NaOH
(B) 1.0 M HCl
(C) 1.0 M NaCl
(D) 1.0 M NH_3

58. If the buffer solution was then diluted with 150. mL of distilled water, which of the following values would be closest to the pH of the diluted buffer solution?

(A) 2.08
(B) 4.19
(C) 5.19
(D) 6.27

GO ON TO THE NEXT PAGE.

59. In the diagrams below, benzoic acid is represented by HA, and the benzoate ion, $C_6H_5COO^-$, is represented by A^-. Which of the beakers shows the correct ratio of HA:A$^-$ in a buffer solution that has a pH of 3.19 ?

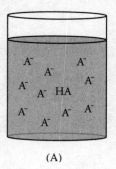

(A)

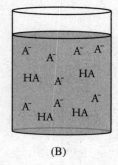

(B)

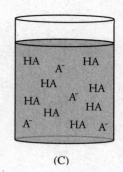

(C)

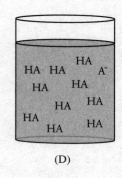

(D)

60. The benzoic acid solution can be made into an effective buffer. Which of the following statements best explains why a HCl solution of identical concentration should NOT be part of a buffer?

(A) The K_a value for benzoic acid is greater than the K_a value for HCl.

(B) HCl has stronger intermolecular forces than C_6H_5COOH.

(C) The Cl$^-$ ion is a less-effective conjugate base than $C_6H_5COO^-$.

(D) HCl has far few protons in it than C_6H_5COOH.

END OF SECTION I

INFORMATION IN THE TABLE BELOW AND ON THE FOLLOWING PAGES MAY BE USEFUL IN ANSWERING THE QUESTIONS IN THIS SECTION OF THE EXAMINATION

DO NOT DETACH FROM BOOK.

PERIODIC TABLE OF THE ELEMENTS

1	2	3	4	5	6	7	8	9	10	11	12	13	14	15	16	17	18
1 **H** 1.008																	2 **He** 4.00
3 **Li** 6.94	4 **Be** 9.01											5 **B** 10.81	6 **C** 12.01	7 **N** 14.01	8 **O** 16.00	9 **F** 19.00	10 **Ne** 20.18
11 **Na** 22.99	12 **Mg** 24.30											13 **Al** 26.98	14 **Si** 28.09	15 **P** 30.97	16 **S** 32.06	17 **Cl** 35.45	18 **Ar** 39.95
19 **K** 39.10	20 **Ca** 40.08	21 **Sc** 44.69	22 **Ti** 47.87	23 **V** 50.94	24 **Cr** 52.00	25 **Mn** 54.94	26 **Fe** 55.85	27 **Co** 58.93	28 **Ni** 58.69	29 **Cu** 63.55	30 **Zn** 65.38	31 **Ga** 69.72	32 **Ge** 72.63	33 **As** 74.92	34 **Se** 78.97	35 **Br** 79.90	36 **Kr** 83.80
37 **Rb** 85.47	38 **Sr** 87.62	39 **Y** 88.91	40 **Zr** 91.22	41 **Nb** 92.91	42 **Mo** 95.95	43 **Tc**	44 **Ru** 101.07	45 **Rh** 102.91	46 **Pd** 106.42	47 **Ag** 107.87	48 **Cd** 112.41	49 **In** 114.82	50 **Sn** 118.71	51 **Sb** 121.76	52 **Te** 127.60	53 **I** 126.90	54 **Xe** 131.29
55 **Cs** 132.91	56 **Ba** 137.33	57 57-71 *	72 **Hf** 178.49	73 **Ta** 180.95	74 **W** 183.94	75 **Re** 186.21	76 **Os** 190.23	77 **Ir** 192.22	78 **Pt** 195.08	79 **Au** 196.97	80 **Hg** 200.59	81 **Tl** 204.38	82 **Pb** 207.2	83 **Bi** 208.98	84 **Po**	85 **At**	86 **Rn**
87 **Fr**	88 **Ra**	89 89-103 †	104 **Rf**	105 **Db**	106 **Sg**	107 **Bh**	108 **Hs**	109 **Mt**	110 **Ds**	111 **Rg**	112 **Cn**	113 **Nh**	114 **Fl**	115 **Mc**	116 **Lv**	117 **Ts**	118 **Og**

*Lanthanoids	57 **La** 138.91	58 **Ce** 140.12	59 **Pr** 140.91	60 **Nd** 144.24	61 **Pm**	62 **Sm** 150.36	63 **Eu** 151.97	64 **Gd** 157.25	65 **Tb** 158.93	66 **Dy** 162.50	67 **Ho** 164.93	68 **Er** 167.26	69 **Tm** 168.93	70 **Yb** 173.05	71 **Lu** 174.97
†Actinoids	89 **Ac**	90 **Th** 232.04	91 **Pa** 231.04	92 **U** 238.03	93 **Np**	94 **Pu**	95 **Am**	96 **Cm**	97 **Bk**	98 **Cf**	99 **Es**	100 **Fm**	101 **Md**	102 **No**	103 **Lr**

GO ON TO THE NEXT PAGE.

AP® CHEMISTRY EQUATIONS & CONSTANTS

Throughout the exam the following symbols have the definitions specified unless otherwise noted.

L, mL = liter(s), milliliter(s)   mm Hg = millimeters of mercury
g = gram(s)   J, kJ = joule(s), kilojoule(s)
nm = nanometer(s)   V = volt(s)
atm = atmosphere(s)   mol = mole(s)

ATOMIC STRUCTURE

$$E = h\nu$$
$$c = \lambda\nu$$

E = energy
ν = frequency
λ = wavelength

Planck's constant, $h = 6.626 \times 10^{-34}$ J s

Speed of light, $c = 2.998 \times 10^8$ m s^{-1}

Avogadro's number = 6.022×10^{23} mol^{-1}

Electron charge, $e = -1.602 \times 10^{-19}$ coulomb

EQUILIBRIUM

$$K_c = \frac{[\text{C}]^c[\text{D}]^d}{[\text{A}]^a[\text{B}]^b}, \text{ where } a\,\text{A} + b\,\text{B} \rightleftarrows c\,\text{C} + d\,\text{D}$$

$$K_p = \frac{(P_\text{C})^c(P_D)^d}{(P_\text{A})^a(P_\text{B})^b}$$

$$K_a = \frac{[\text{H}^+][\text{A}^-]}{[\text{HA}]}$$

$$K_b = \frac{[\text{OH}^-][\text{HB}^+]}{[\text{B}]}$$

$$K_w = [\text{H}^+][\text{OH}^-] = 1.0 \times 10^{-14} \text{ at } 25°\text{C}$$
$$= K_a \times K_b$$

$$\text{pH} = -\log[\text{H}^+], \text{ pOH} = -\log[\text{OH}^-]$$

$$14 = \text{pH} + \text{pOH}$$

$$\text{pH} = \text{p}K_a + \log\frac{[\text{A}^-]}{[\text{HA}]}$$

$$\text{p}K_a = -\log K_a, \text{ p}K_b = -\log K_b$$

Equilibrium Constants

K_c (molar concentrations)
K_p (gas pressures)
K_a (weak acid)
K_b (weak base)
K_w (water)

KINETICS

$$[\text{A}]_t - [\text{A}]_0 = -kt$$

$$\ln[\text{A}]_t - \ln[\text{A}]_0 = -kt$$

$$\frac{1}{[\text{A}]_t} - \frac{1}{[\text{A}]_0} = kt$$

$$t_{1/2} = \frac{0.693}{k}$$

k = rate constant
t = time
$t_{1/2}$ = half-life

GO ON TO THE NEXT PAGE.

GASES, LIQUIDS, AND SOLUTIONS

$$PV = nRT$$

$$P_A = P_{total} \times X_A, \text{ where } X_A = \frac{\text{moles A}}{\text{total moles}}$$

$$P_{total} = P_A + P_B + P_C + \dots$$

$$n = \frac{m}{M}$$

$$K = {}^\circ C + 273$$

$$D = \frac{m}{V}$$

$$KE_{molecule} = \frac{1}{2}mv^2$$

Molarity, M = moles of solute per liter of solution

$$A = \varepsilon bc$$

P = pressure
V = volume
T = temperature
n = number of moles
m = mass
M = molar mass
D = density
KE = kinetic energy
v = velocity
A = absorbance
ε = molar absorptivity
b = path length
c = concentration

Gas constant, $R = 8.314 \text{ J mol}^{-1}\text{K}^{-1}$
$= 0.08206 \text{ L atm mol}^{-1}\text{K}^{-1}$
$= 62.36 \text{ L torr mol}^{-1}\text{K}^{-1}$
1 atm = 760 mm Hg = 760 torr
STP = 273.15 K and 1.0 atm
Ideal gas at STP = 22.4 L mol^{-1}

THERMOCHEMISTRY/ ELECTROCHEMISTRY

$$q = mc\Delta T$$

$$\Delta S^\circ = \sum S^\circ \text{ products} - \sum S^\circ \text{ reactants}$$

$$\Delta H^\circ = \sum \Delta H_f^\circ \text{ products} - \sum \Delta H_f^\circ \text{ reactants}$$

$$\Delta G^\circ = \sum \Delta G_f^\circ \text{ products} - \sum \Delta G_f^\circ \text{ reactants}$$

$$\Delta G^\circ = \Delta H^\circ - T\Delta S^\circ$$

$$= -RT \ln K$$

$$= -nFE^\circ$$

$$I = \frac{q}{t}$$

$$E_{cell} = E_{cell}^\circ - \frac{RT}{nF} \ln Q$$

q = heat
m = mass
c = specific heat capacity
T = temperature
S° = standard entropy
H° = standard enthalpy
G° = standard Gibbs free energy
n = number of moles
E° = standard reduction potential
I = current (amperes)
q = charge (coulombs)
t = time (seconds)
Q = reaction quotient

Faraday's constant, $F = 96{,}485$ coulombs per mole of electrons

$$1 \text{ volt} = \frac{1 \text{ joule}}{1 \text{ coulomb}}$$

GO ON TO THE NEXT PAGE.

CHEMISTRY
SECTION II
Time—1 hour and 45 minutes
7 Questions

Directions: Questions 1–3 are long free-response questions that require about 23 minutes each to answer and are worth 10 points each. Questions 4–7 are short free-response questions that require about 9 minutes each to answer and are worth 4 points each.

On test day, you will be asked to show your work for each part in the space provided after that part. For this practice test, you may use scrap paper. Examples and equations may be included in your responses where appropriate. For calculations, clearly show the method used and the steps involved in arriving at your answers. You must show your work to receive credit for your answer. Pay attention to significant figures.

Acetaldehyde, CH_3COH Chloroform, $CHCl_3$ Methanoic acid, $HCOOH$

BP = 293 K   BP = 334 K   BP = 374 K

1. The boiling points of three different compounds are listed above, along with their formulas.

   (a) In the blank box above, draw the Lewis diagram for methanoic acid. There are two carbon-oxygen bonds present, one of which is shorter than the other.

   (b) (i) What is the hybridization around atom C_2 in the acetaldehyde molecule?
       (ii) Chloroform is known to be polar. Draw the location of any positive and negative partial charges on a chloroform molecule.

   (c) Which of the three molecules would have the greatest degree of polarizability? Justify your answer.

GO ON TO THE NEXT PAGE.

Equal amounts of all three liquids are mixed together inside a beaker.

(d)   A sample of the mixture is placed on a piece of chromatography paper, and the bottom of the paper is immersed in water. After some time passes, the mixture separates. The locations of two components of the mixture are charted on the diagram below.

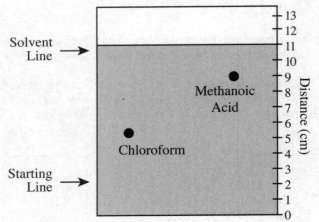

(i)   What is the R_f value for the methanoic acid?
(ii)   On the diagram above, draw and label a dot that would be a reasonable estimate of the distance the acetaldehyde would travel.

The beaker containing the mixture is then attached to a distillation apparatus and heated, as shown below.

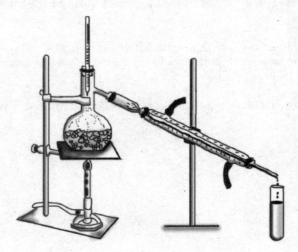

(e)   (i)   If the temperature of the beaker is gradually raised, which substance will separate out first? Justify your answer.
      (ii)   Which of the following temperatures would be the best option in order to create the purest distillate possible? Justify your answer.

210 K   300 K   330 K   390 K

GO ON TO THE NEXT PAGE.

2. A galvanic cell is set up according to the following diagram.

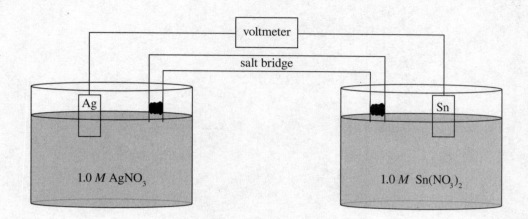

Half-Reaction	Standard Reduction Potential (V)
$Ag^+(aq) + e^- \rightarrow Ag(s)$	+0.80
$Sn^{2+}(aq) + 2\,e^- \rightarrow Sn(s)$	–0.14

(a) Write out the balanced net ionic equation that will occur.

(b) (i) Calculate E°_{cell}.
    (ii) Calculate ΔG° for the cell.

(c) If the solutions in each half-cell were replaced with 2.0 M solutions of $AgNO_3$ and $Sn(NO_3)_2$, how would that affect the cell potential? Justify your answer.

(d) (i) What is the purpose of the salt bridge?
    (ii) Suggest the formula for an aqueous solution that would be a good choice to fill the salt bridge. Justify your answer.

(e) The beaker of $Sn(NO_3)_2$ is disconnected from the cell, and the tin electrode is then connected to an outside source of current. Over the course of 10.0 minutes, 1.65 g of tin plates out onto the electrode. What is the amperage of the current source?

GO ON TO THE NEXT PAGE.

3.

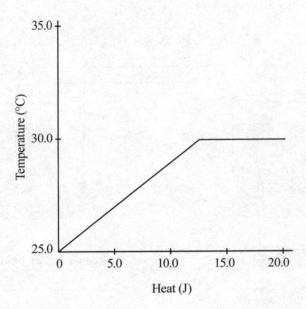

0.10 mol of solid gallium initially at room temperature is heated at a constant rate, and its temperature is tracked, leading to the above graph.

(a)   As heat is added, what is happening to the total entropy of the system? Justify your answer.

(b)   The horizontal portion of the graph indicates a phase change. Explain on a particulate level why the temperature is constant during a phase change.

(c)   (i)   Calculate the specific heat capacity of solid gallium in J $g^{-1}\,°C^{-1}$.
       (ii)   If the specific heat of the solid gallium were greater than what you calculated in (c)(i), how would the slope of the temperature versus heat line change during gallium's solid phase?

The gallium continues to be heated until it fully boils. Assume ideal behavior for the gallium gas.

$$Ga(l) \rightarrow Ga(g)$$

Substance	ΔH_f° (kJ mol^{-1})
Ga(l)	5.60
Ga(g)	277.1

(d)   (i)   Calculate the enthalpy of vaporization for gallium given the above data.
       (ii)   The enthalpy of vaporization for gallium is greater than its enthalpy of fusion. Explain why in terms of intermolecular forces.

(e)   Given your answer to (d)(i) and that $\Delta S° = 113.4$ J mol^{-1} K^{-1} for the boiling of gallium, what is the boiling point of the gallium?

(f)   After the gallium is fully converted to a gas, it continues to be heated. What would you expect to be true about the velocity distribution of the gaseous gallium atoms as the temperature increases?

GO ON TO THE NEXT PAGE.

4. Sulfurous acid, H_2SO_3, is a weak diprotic acid.

    (a) Write out the two acid-base dissociations that occur when H_2SO_3 is mixed with water.

    (b) A 0.50 M solution of H_2SO_3 has a pH of 1.10. Determine K_{a1} for H_2SO_3.

5. Hypobromous acid, HBrO, is a weak monoprotic acid with a K_a value of 2.0×10^{-9} at 25°C.

    (a) Write out the equilibrium reaction of hypobromous acid with water, identifying any conjugate acid/base pairs present.

    (b) (i) What would be the percent dissociation of a 0.50 M solution of hypobromous acid?
    (ii) If the 0.50 M solution were diluted, what would happen to the percent dissociation of the HBrO? Why?

6. Chlorofluorocarbons are byproducts of many different processes that are known to be dangerous to the environment as both a greenhouse gas, and as an agent for ozone (O_3) depletion. The accepted mechanism for the latter is:

$$\text{Step 1: } Cl(g) + O_3(g) \rightarrow ClO(g) + O_2(g)$$
$$\text{Step 2: } O(g) + ClO(g) \rightarrow Cl(g) + O_2(g)$$

    (a) Write out the full reaction with the above elementary steps, and identify all catalysts and intermediates.

    (b) Describe two ways by which a catalyst can reduce the activation energy of a reaction.

    (c) Both elementary steps in the reaction above are exothermic. On the axes provided, draw a potential reaction mechanism which supports this.

GO ON TO THE NEXT PAGE.

7. A stock solution of 2.0 M MgCl$_2$ is dissolved in water.

    (a)   (i)   In the beaker below, draw a particulate diagram that represents MgCl$_2$ dissolved in water. The approximate sizes of each atom/ion are provided for you. Your diagram should include at least four water molecules, which should be correctly oriented compared to the ions dissolved in solution.

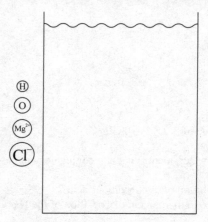

        (ii)   Why are the chloride ions from (a)(i) larger than the magnesium ions?

    (b)   (i)   A student wishes to make up 500 mL of 0.50 M MgCl$_2$ for an experiment. Explain the best method of doing so, starting with the 2.0 M MgCl$_2$ solution and utilizing a graduated cylinder and a volumetric flask. Assume MgCl$_2$ is fully soluble.

        (ii)   What are the concentrations of the Mg^{2+} and Cl$^-$ ions in the new solution?

STOP

END OF EXAM

Practice Test 2:
Answers and
Explanations

PRACTICE TEST 2: MULTIPLE-CHOICE ANSWER KEY

1.	A	21.	B	41.	D
2.	D	22.	B	42.	A
3.	B	23.	D	43.	D
4.	A	24.	B	44.	B
5.	B	25.	C	45.	A
6.	B	26.	C	46.	B
7.	C	27.	A	47.	C
8.	A	28.	C	48.	B
9.	C	29.	C	49.	A
10.	D	30.	C	50.	D
11.	D	31.	A	51.	B
12.	D	32.	B	52.	B
13.	B	33.	C	53.	C
14.	C	34.	A	54.	B
15.	A	35.	A	55.	A
16.	A	36.	C	56.	B
17.	B	37.	B	57.	A
18.	D	38.	D	58.	B
19.	A	39.	D	59.	D
20.	D	40.	A	60.	C

Section I—Multiple-Choice Answers and Explanations

1. **A**   Molar solubility describes the number of moles of a salt that will dissolve in one liter of water. Each unit of Na_3PO_4 that dissociates produces three Na^+ ions, which means that there are one-third as many units of Na_3PO_4 as there are of Na^+.

2. **D**   The overall dissolution of LiCl is exothermic; this is indicated by the temperature of the water rising. For that to be the case, the energy released when the ion-dipole attractions form has to exceed the amount of energy required to break the bonds in the LiCl lattice.

3. **B**   There is a 1:1 mole ratio between the HCl and the CH_3NH_2, meaning there will be the same number of moles of each present at the equivalence point. The equivalence point is located at 15.0 mL of HCl added. $(1.0\ M)(0.0150\ L) = 0.0150$ mol HCl $= 0.0150$ mol CH_3NH_2. Finally, divide that value by the volume of the CH_3NH_2, 0.0150 mol CH_3NH_2 / 0.020 L $= 0.75\ M\ CH_3NH_2$.

4. **A**   At the half-equivalence point (7.5 mL) of the titration, pK_b of methylamine = pOH of the solution. To determine the pOH, you simply have to see what the pH of the solution is at that half-equivalence point and use pH + pOH = 14. The pH at the half-equivalence is 10.5. So, 14 − 10.5 = 3.5.

5. **B**   A buffer solution is one that resists change in pH due to similar amounts of the base (CH_3NH_2) and its conjugate acid ($CH_3NH_3^+$). That occurs after the initial pH change, but prior to the equivalence region. The pH is also stable again starting at 16.5 mL, but that is due to the presence of the excess strong acid, and that does not create a buffer region.

6. **B**   Sodium hydroxide is a strong base, which dissociates completely in solution. Thus, it would initially have a higher pH (be more basic) than any weak base of an identical concentration. At equivalence of a strong acid/strong base titration, water is the only acid or base present, causing the solution to have a neutral pH of 7. Finally, in the post-equivalence region of the graph, the pH is driven by the excess H^+ from the HCl, and that would not change in the new titration.

7. **C**   The H–C bond is a single bond with a bond order of one. The C–O bonds display resonance, and the average bond order between them is (1 + 2)/2 = 1.5.

8. **A**   Diluting an aqueous system at equilibrium will cause a shift to the side with more aqueous species. In this case, that entails a shift to the left, which is what happens when the reverse reaction rate increases.

9. **C**   In order for the reaction to occur, one of the oxygen atoms from NO_2 must have collided directly with the carbon atom in CO. This will allow for the bonds in NO_2 to be broken, while a new C–O bond is also created.

10. **D** Given that both products are gases at the same temperature and pressure, their volumes are directly proportional to the number of moles of gas that are present. So, 1.5 times as much CO_2 was created as H_2O. However, you are interested in the C:H ratio, and there are two hydrogen atoms in each molecule of H_2O. The C:H ratio then becomes 1.5:2.0, which is the same as 3:4.

$$2\,H_2O_2(aq) \rightarrow 2\,H_2O(l) + O_2(g)$$

11. **D** In hydrogen peroxide, oxygen takes on an oxidation state of –1. Oxygen's oxidation state is –2 in water (a reduction) and 0 in oxygen gas (an oxidation). The oxidation state on hydrogen is a +1 in both compounds it appears in.

12. **D** Gases deviate from ideal behavior due to the strength of their IMFs. All four diatomic gases listed are non-polar, so the only difference in IMF strength arises from the polarizability of their electron clouds. Fluorine has the most electrons, and thus, experiences the strongest London dispersion forces and the greatest deviation from ideal behavior.

13. **B** With increasing temperature, the Maxwell-Boltzmann distribution curve shifts rightward and downward. A larger percentage of the curve will be past the activation energy barrier, as shown below, where the shaded areas under each curve are the particles that have enough energy to react.

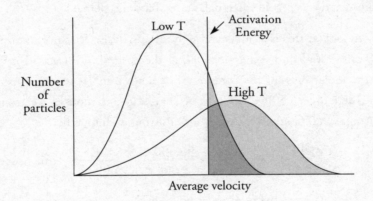

Choice (A) is a good distractor; however, just because the average kinetic energy present in the sample increases does not mean the velocity of ALL of the molecules increases (as shown in the distribution curves above).

14. **C** Phosphorus forms ions with a –3 charge (P^{3-}). Adding valence electrons increases the repulsion between the existing electrons, which increases the size of the ion. Zinc, on the other hand, loses electrons when forming ions, decreasing repulsion and reducing size.

15. **A** Phosphorus has electrons in both the $3s$ and $3p$ sublevels. These sublevels are located on the third energy level and are further away from the nucleus than the electrons in the first two energy levels, which are the only ones neon has electrons in.

16. **A** When transition metals form ions, they lose their outermost *s*-electrons first. Additionally, when filling the *d*-orbitals, you always need to subtract one from the period the element is found in to determine the principal energy level of its *d*-orbital.

17. **B** Both molecules have four electron groups around the central atom, so their base geometry is tetrahedral, which has bond angles of 109.5°. However, the PCl_3 has a lone pair as well, which exerts a slightly greater repulsive force than a bonded pair, bringing the bond angles in that molecule to slightly less than the ideal 109.5°.

18. **D** Stirring will only increase the rate of a reaction that is heterogeneous in nature. In this example, all reactants and products are in the liquid or aqueous phase, meaning the inherent speed of the molecules and ions is so great that any energy added via stirring would be negligible.

19. **A** Vapor pressure arises from the fact that molecules on the surface of a liquid may have enough kinetic energy to escape the intermolecular forces holding them together in liquid form. As temperature increases, the average overall kinetic energy of the molecules increases, meaning more of them will have enough energy to escape the surface of the liquid.

20. **D** Enthalpy of formation involves the formation of one mole of the substance from elements in their standard states. Both hydrogen and oxygen are diatomic elements in their standard state. Note that halves are considered acceptable when balancing diatomic elements.

21. **B** The best way to look at this is as an exercise in unit analysis. That means you need to start with the initial unit (grams), and find conversions to get the desired unit (seconds). First, go from grams to moles using the molar mass of the compound. It takes 2 moles of electrons to convert one mole of Fe^{2+} to $Fe(s)$. There are 96,500 Coulombs of charge in one mole of electrons (Faraday's constant). Finally, 5.0 Amps = 5.0 Coulombs per second. Putting it all together:

$$1.00 \text{ g Fe} \times \frac{1 \text{ mol Fe}}{55.85 \text{ g Fe}} \times \frac{2 \text{ mol } e^-}{1 \text{ mol Fe}} \times \frac{96,500 \text{ C}}{1 \text{ mol } e^-} \times \frac{1.0 \text{s}}{5.00 \text{ C}} = \text{Time in seconds}$$

22. **B** Pressure is directly related to the number of moles of a gas. CH_4 has the lowest molar mass out of the three gases, and therefore 10.0 grams of it represents the greatest number of moles, leading to the greatest pressure.

23. **D** Density is simply mass divided by volume. The mass and volume of each gas is identical; therefore, their densities are identical.

24. **B** The rate at which a gas effuses is inversely proportional with its molar mass. A gas with a lower molar mass (in this case, CH_4) will have molecules moving with a higher average velocity, meaning they are going to hit that hole more often and will be more likely to escape.

25. **C** To get the targeted equation, you have to flip the top one and multiple it by two before combining it with the bottom one. When manipulating equilibrium constants, flipping a reaction forces you to take the reciprocal of the original K, and doubling all coefficients leads to K being squared. So for $2\ CO(g) + 2\ H_2O(g) \rightleftharpoons 2\ CO_2(g) + 2\ H_2(g)$, $K_c = (1/1.4)^2$. After that, when you add the two equations together, to get the K value for their sum you have to multiple the two K values together.

26. **C** First, the limiting reactant is clearly the H_2, as twice as much of it is needed for the reaction to go to completion, yet there is not twice as much of it present. With that in mind, the ICE chart is as follows:

	2 H$_2$	O$_2$	2 H$_2$O
Initial	1.0	1.0	0
Change	−2x	−x	+2x
Ending	0		

A quick solve tells you x is 0.5, meaning that at the end of the reaction there are 0.5 mole of oxygen and 1.0 mole of H_2O. That is 1.5 total moles of gas left at the end, a 25% decrease in moles (and thus, pressure) from the 2.0 moles of gas that were present initially.

27. **A** An indicator changes color when the concentration of the deprotonated form exceeds the concentration of its protonated form. At the exact point of the color change, the concentrations of both the protonated and deprotonated forms will be equal.

28. **C** Ethanol displays hydrogen bonding, which is the strongest type of intermolecular force. However, London dispersion forces are based on the overall polarizability of a molecule's electron cloud, which in turn is based on the number of electrons that molecule has. Octane has significantly more electrons than ethanol, and the dispersion forces caused by that overwhelm even the hydrogen bonds in ethanol, causing octane to have a higher boiling point.

29. **C** When considering melting points for ionic compounds, a greater lattice energy means a greater melting point. Lattice energy is calculated using Coulomb's Law, which tells you that both greater charge and smaller size lead to higher energy. The oxygen anion has a charge of −2, while chlorine's anion charge is only −1. Note that oxygen is also smaller than chlorine, so that further reinforces the greater melting point of CaO.

30. **C** Noble gases don't usually form compounds because they have full outermost energy levels and are very stable. However, the larger noble gases (xenon and krypton) have valence electrons far enough away from the nucleus that they are able to share those with highly electronegative elements such as fluorine or oxygen.

31. **A** $pH = -\log[H^+]$. A 10 M solution of a strong acid such as nitric would have a pH of $-\log(10) = -1$. As 15.8 M is greater than 10 M, the pH of the 15.8 M solution must be more acidic (lower).

32. **B** Even though the spectator ions do not take part in the actual precipitation reaction, they can still "stick" to the surface of the precipitate itself. Rinsing the precipitate with water will carry those ions away as the dipoles in the water attract the free spectator ions.

33. **C** $Q = (P_{SO_2})^2(P_{O_2})/(P_{SO_3})^2$. Injecting neon into the tank will reduce the mole fraction, and thus the partial pressure, of all the gases. Reducing the partial pressures will decrease the numerator more than the denominator due to the ratios in Q. This, in turn, causes the value for Q to decrease, necessitating a shift to the right in order to re-establish equilibrium.

34. **A** The Na^+ cations will be attracted to the negative (oxygen) partial charges on the water molecules, and the NO_3^- anions will be attracted to the positive (hydrogen) partial charges.

35. **A** The temperature drops, which means the reaction absorbed energy from the water. This means the reaction itself is endothermic, giving it a positive value for $\Delta H°$.

36. **C** The formation of the ion-dipole attractions is an exothermic process during which energy is released.

37. **B** The equation here is $q = mc\Delta T$, where m is the combined mass of the water and the dissolved salt. The density of pure water is 1.0 g/mL, so 200.0 mL of water has a mass of 200.0 g, and the total mass of the solution is thus 210 g.

38. **D** If some of the temperature change of the water was due to heat flowing into the surroundings instead of into the dissolution itself, that would cause the overall magnitude of heat change, and thus the enthalpy of solution, to be artificially low.

39. **D** Via $c = \lambda v$, and given that c is a constant (3.0×10^8 m/s), the higher the wavelength, the lower the frequency. Via $E = hv$, with h being Planck's constant, a higher frequency means more energy.

40. **A** There is only one reactant in the rate law, and it is first order. This means the reaction is first order. Since the reaction is first order, the rate-determining step must be, too. Since there is only a single molecule in that step, it is unimolecular.

41. **D** The auto-ionization of water is represented by $H_2O(l) \rightleftharpoons H^+ + OH^-$. An increase in pH means a decrease in $[H^+]$, and as pure water is always neutral, a corresponding decrease in $[OH^-]$. This would occur only if the reaction experiences a permanent shift to the left, and that would occur only during a temperature decrease if the reaction were endothermic.

42. **A** Molecules that are square planar, like XeF_4, are symmetrical and thus, nonpolar.

43. **D** The most common isotope of strontium has an atomic mass of 88, and a strontium nucleus has 38 protons. As atomic mass = protons + neutrons, that means the most common isotope of strontium must have 50 neutrons.

44. **B** The manganese goes from an oxidation state of +7 in MnO_4^- to +2 in Mn^{2+}, necessitating a gain of 5 electrons. The carbon goes from an oxidation state of +3 in $C_2O_4^{2-}$ to +4 in CO_2, and in the balanced half-reaction that would happen twice (i.e., a 2 coefficient is needed in front of the CO_2 to balance it), necessitating a loss of 2 electrons. To balance the charge, the manganese half-reaction must be multiplied by 2 while the carbon half-reaction is multiplied by 5.

45. **A** $K_{sp} = [Zn^{2+}][IO_3^-]^2$

$4.0 \times 10^{-6} = (x)(2x)^2$

$4.0 \times 10^{-6} = 4x^3$

$1.0 \times 10^{-6} = x^3$

$x = 1.0 \times 10^{-2}\ M$

46. **B** Due to the common ion effect, zinc iodate would be least soluble in a solution with which it shared an ion. Of the options, only $NaIO_3$ shares an ion (IO_3^-) with $Zn(IO_3)_2$.

47. **C** In a saturated solution, the concentration of each component ion will always remain constant. For that to happen if water evaporates, some ions must "fall" out of solution, increasing the total amount of precipitate.

48. **B** Given that $\Delta G = \Delta H - T\Delta S$, for a reaction to only be favored at lower temperatures, ΔH must be negative, and the $-T\Delta S$ term must be positive (and smaller than ΔH). For the $-T\Delta S$ term to be positive, ΔS itself must be negative, which is the case with reaction (B), as the dispersion decreases when 1.5 moles of gaseous reactants turn into 1.0 mole of gaseous products.

49. **A** The reason that atoms bond is to minimize the potential energy in a system.

50. **D** The PES represents an electron configuration of $1s^2 2s^2 2p^6 3s^2$, that of magnesium. When forming an ion, the most stable pathway is to lose the two electrons in the third energy level, giving the ion a charge of +2.

51. **B** Using a battery with a larger voltage would lead to a larger current, which in turn would lead to a larger mass of copper being plated out of solution.

52. **B** $1.0 \text{ mol } NO_2 \times \dfrac{1 \text{ mol}_{rxn}}{3 \text{ mol } NO_2} \times \dfrac{-135 \text{ kJ}}{1 \text{ mol}_{rxn}} = -45 \text{ kJ}$ (the negative sign means energy is released).

53. **C**

$$\begin{array}{cccc} C\text{–}H & O\text{=}O & C\text{=}O & O\text{–}H \\ 4(413) & + \ 2(498) & - \ 2x & - \ 4(467) = -890 \text{ kJ/mol} \end{array}$$

$$x = +800 \text{ kJ/mol}$$

54. **B** Vapor pressure is inversely proportional with IMF strength, so the substance with the greatest vapor pressure (in this case, CH_3OH) would have the weakest IMF strength. That also means it has the lowest boiling point, and would boil first.

55. **A** The graph shows a constant half-life—that is, the concentration of the N_2O_5 decreases by half every 5 seconds. The rate law is thus rate = $k[N_2O_5]$; dimensional analysis gets us to the units on k: $M/s = k(M)$. $k = s^{-1}$.

56. **B** The reaction quotient is $Q = [Cu(NH_3)_4^{2+}]/[Cu^+][NH_3]^4$. Diluting the equilibrium will decrease the concentration of all ions by the same amount, but that will have a larger effect on the denominator in Q because there are more ions present. This, in turn, means Q will increase, causing a shift to the left, increasing the amount of reactants and decreasing the amount of products.

57. **A** Adding NaOH to the beaker would cause the following reaction to occur: $C_6H_5COOH + OH^- \rightleftharpoons C_6H_5COO^- + H_2O$. This creates sufficient conjugate base ($C_6H_5COO^-$) to create a buffer system with the benzoic acid.

58. **B** Diluting a buffer does not change its pH appreciatively, as the concentrations of both the acid and its conjugate change equally, which cancels out in Henderson-Hasselbalch. The pH will become a little bit closer to 7 (the pH of pure water), but not significantly so.

59. **D** For the pH to decrease by 1 from the pK_a value, via Henderson-Hasselbalch (pH = pK_a + log $[C_6H_5COO^-]/[C_6H_5COOH]$), the concentration of the acid must be 10 times greater than the concentration of the conjugate base.

60. **C** The conjugate bases of strong acids are ineffective, due to the fact that strong acids dissociate fully in solution. Without an effective conjugate base, a buffer cannot be formed.

Section II—Free-Response Answers and Explanations

1. (a) The Lewis diagram for methanoic acid:

$$
\begin{array}{c}
: \!\overset{\displaystyle :\text{O}:}{\underset{}{\|}} \\
\text{H} - \text{C} - \overset{..}{\underset{..}{\text{O}}} - \text{H}
\end{array}
$$

    (b) (i)   Three charge clouds means the hybridization is sp^2.

         (ii) The location of any positive and negative partial charges on a chloroform molecule:

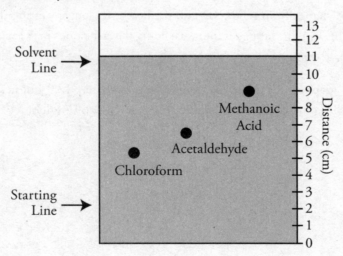

    (c) The polarizability is based on the number of electrons each molecule has. More electrons leads to stronger temporary dipoles, and of the three molecules, chloroform has the most electrons and is thus the most polarizable.

    (d) (i) $R_f = \dfrac{\text{distance methanoic acid}}{\text{distance solvent}} = \dfrac{9 \text{ cm}}{11 \text{ cm}} = 0.81$

         (ii) Below is the diagram including the labeled dot that would be a reasonable estimate of the distance the acetaldehyde would travel.

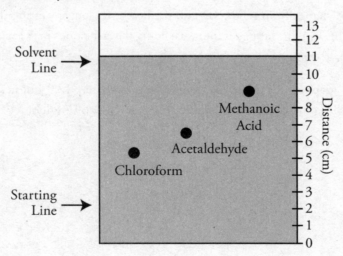

    (e) (i)   The acetaldehyde is the substance with the lowest boiling point and thus will boil first.

         (ii) 300 K is above the boiling point of acetaldehyde and will cause it to boil, but below the boiling points of the other substances, meaning they will not boil. 330 K also fits those guidelines; however, the vapor pressure of the other substances would increase with increasing temperature, and more of the methanoic acid and the chloroform would end up mixed in with the acetaldehyde distillate.

2. (a) Ag^+ has a higher Standard Reduction Potential than Sn^{2+}, so Ag^+ will be reduced ($Ag^+(aq) + e^- \rightarrow Ag(s)$) and $Sn(s)$ will be oxidized ($Sn(s) \rightarrow Sn^{2+}(aq)$). The reduction reaction must be multiplied by 2 to balance the charges when the half-reactions are combined, so the full net ionic equation becomes $2\,Ag^+(aq) + Sn(s) \rightarrow Sn^{2+}(aq) + 2\,Ag(s)$.

    (b) (i) $+0.80\text{ V} - (-0.14\text{ V}) = +0.94\text{ V}$

        (ii) $\Delta G° = -nFE°$. The number of moles of electrons transferred in this reaction is 2, so $\Delta G° = -(2\text{ mol } e^-)$ $(96{,}500\text{ C/mol } e^-)(+0.94\text{ V}) = -180{,}000$ J/mol or -180 kJ/mol.

    (c) Given that $Q = [Sn^{2+}]/[Ag^+]^2$, increasing both concentrations to 2.0 M would cause Q to decrease. As the cell is favored, the equilibrium constant is greater than 1, so a decrease in Q will bring the cell further from equilibrium (where the voltage is zero). Thus, the cell potential would increase.

    (d) (i)   The salt bridge allows ions to flow to each half-cell in order to maintain the electrical neutrality of the cell.

        (ii) Any aqueous solution that prevents the formation of a precipitate with the ions present in either half-cell would be ideal. As neither alkali metals nor nitrates can form precipitates, a solution of $NaNO_3$ or KNO_3 would work, as would any other alkali nitrate or ammonium nitrate.

    (e) $1.65\text{ g Sn} \times \dfrac{1\text{ mol Sn}}{118.71\text{ g}} \times \dfrac{2\text{ mol } e^-}{1\text{ mol Sn}} \times \dfrac{96{,}500\text{ C}}{1\text{ mol } e^-} = \dfrac{2{,}680\text{ C}}{600\text{ s}} = 4.47\text{ A}$

3. (a) The entropy of the system is increasing as it becomes more dispersed. Even in a solid phase, heating a substance causes the vibration of the particles to increase, meaning a greater level of dispersion in the system.

    (b) During a phase change from solid to liquid, the heat is no longer causing the atoms of the gallium to change speed. Instead, the heat goes into weakening the attractions between the gallium atoms in order to enact the phase change.

    (c) (i)   From the graph, you can see that it takes approximately 12.0 J of heat to raise the temperature of the gallium by 5.0°C. Additionally, $0.10\text{ mol Ga} \times 69.7\text{ g/mol} = 6.97$ g of gallium.

$$q = mc\Delta T$$

$$12.0\text{ J} = (6.97\text{ g})(c)(5.0°C)$$

$$c = 0.344\text{ J g}^{-1}\,°C^{-1}$$

(ii) If the specific heat were actually greater, it would take more energy to raise the temperature of the gallium, meaning the slope of the line would be smaller, as shown below. Note that the melting point of the gallium would be unchanged; it would just take more heat to raise the temperature of the solid gallium to that point.

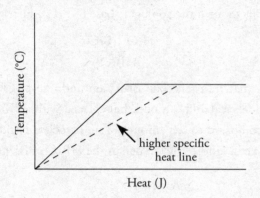

(d) (i) This is just products–reactants, so $277.1 - 5.60 = 271.5$ kJ mol^{-1}.

(ii) Enthalpy of vaporization describes the amount of heat necessary to completely break any attractions between gallium atoms, as ideal gas particles have no attractions between each other. That amount of energy will be greater than the amount necessary to simply weaken them, which is what occurs in the solid-to-liquid phase change (quantified by the enthalpy of fusion).

(e) During any phase change, $\Delta G° = 0$. With that information, and converting the units to match:

$$\Delta G° = \Delta H° - T\Delta S°$$

$$0 = 271.5 \text{ kJ mol}^{-1} - T(0.1134 \text{ kJ mol}^{-1} \text{ K}^{-1})$$

$$T = 2{,}394 \text{ K}$$

(f) As the gas is heated, the average energy in the gas will increase, causing a greater distribution of potential velocities for the gaseous atoms.

4. (a) $H_2SO_3(aq) + H_2O(l) \rightleftharpoons HSO_3^-(aq) + H_3O^+(aq)$

$HSO_3^-(aq) + H_2O(l) \rightleftharpoons SO_3^{2-}(aq) + H_3O^+(aq)$

(b) For a diprotic acid, only the first dissociation contributes significantly to the pH of the resulting solution. In this case, if pH = 1.10, then $-\log[H_3O^+] = 1.10$ and $[H_3O^+] = 0.079$ M. Keeping in mind that $[H_3O^+] = [HSO_3^-]$, you get:

$$K_{a1} = \frac{[H_3O^+][HSO_3^-]}{[H_2SO_3]} \qquad K_{a1} = \frac{(0.079)^2}{(0.50 - 0.079)} \qquad K_{a1} = 1.5 \times 10^{-2}$$

Note that if the -0.079 is ignored in the denominator, it leads to the incorrect answer. The fact that some of the H_2SO_3 dissociates must be considered!

5. (a) $HBrO(aq) + H_2O(l) \rightleftharpoons H_3O^+(aq) + BrO^-(aq)$

    Conjugate pairs: HBrO and BrO⁻, H₂O and H₃O⁺

    (b) (i) Percent dissociation describes what percentage of the acid ions will dissociate—that is, give up their hydrogen ions to form the conjugate base of BrO⁻. First, you need the K_a expression:

    $$K_a = \frac{[H_3O^+][BrO^-]}{[HBrO]}$$

    For every proton that transfers, one BrO⁻ ion and one H₃O⁺ ion are created, meaning those numbers will be identical and can both be replaced with x. The amount of dissociated ions will be insignificant compared to the amount of undissociated acid molecules in solution with such a small K_a, so you can ignore any change in the HBrO concentration.

    $$2.0 \times 10^{-9} = \frac{x^2}{0.50\ M}$$

    $$x = 3.2 \times 10^{-5}\ M$$

    Finally, to calculate percent dissociation, take the concentration of the conjugates created and divide it by the original concentration of the acid.

    $$\% \text{ Dissociation} = \frac{3.2 \times 10^{-5}\ M}{0.50\ M} \times 100\% = 6.4 \times 10^{-3}\%$$

    (ii) The forward reaction, HBrO donating a proton to water, can occur easily regardless of the acid's concentration. This is because the vast majority of the HBrO remains undissociated, and there is always a very large amount of water molecules available to accept a proton from the acid.

    However, as the acid becomes less concentrated via dilution, less and less of the conjugates (BrO⁻ and H₃O⁺) are created, making the reverse reaction increasingly less likely to happen, especially as the conjugates are surrounded by HBrO and water molecules. This means more of the conjugates will stay dissociated (they have a hard time "finding" each other to make the reaction occur), which increases the percent dissociation of the acid.

6. (a) Full reaction: $O_3(g) + O(g) \rightarrow 2\ O_2(g)$

    Catalyst: Cl(g) Intermediate: ClO(g)

    (b) A catalyst can either stabilize an existing transition state, or it can cause the creation of a new intermediate. Both of these can lower the overall activation energy for the reaction, speeding it up.

(c)   The most important thing in the diagram below is that the energy level decreases after each step.

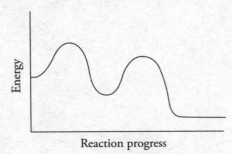

Reaction progress

7.   (a)   (i)   Note that the negative (oxygen) partial charges on the water are attracted to the positive magnesium cations, while the positive (hydrogen) partial charges on the water attract to the negative chloride anions.

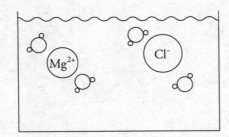

(ii)   The chlorine atom gains a single electron to become the chloride ion, filling its third energy level. The magnesium atom loses two electrons to form the magnesium ion, leaving it with only two full energy levels.

(b)   (i)   For dilutions, use $M_1V_1 = M_2V_2$ to figure out the necessary volume of stock solution.

$$(2.0\ M)(V_1) = (0.50\ M)(500.\ \text{mL})$$

$$V_1 = 125\ \text{mL}$$

Using a graduated cylinder, add 125 mL of 2.0 M MgCl$_2$ to the volumetric flask. Next, dilute to the mark in the neck of the flask with distilled water before capping and inverting the flask several times to ensure a fully mixed solution.

(ii)   The equation for the dissociation is $MgCl_2(s) \rightarrow Mg^{2+}(aq) + 2\ Cl^-(aq)$.

Thus, the magnesium ion concentration will be the same as the overall solution, and the chloride ion concentration will be twice that: $[Mg^{2+}] = 0.50\ M$, $[Cl^-] = 1.0\ M$.

Practice Test 3

The Exam

AP® Chemistry Exam

SECTION I: Multiple-Choice Questions

DO NOT OPEN THIS BOOKLET UNTIL YOU ARE TOLD TO DO SO.

At a Glance

Total Time
1 hour and 30 minutes
Number of Questions
60
Percent of Total Grade
50%
Writing Instrument
Pencil required

Instructions

Section I of this examination contains 60 multiple-choice questions. Fill in only the ovals for numbers 1 through 60 on your answer sheet.

Indicate all of your answers to the multiple-choice questions on the answer sheet. No credit will be given for anything written in this exam booklet, but you may use the booklet for notes or scratch work. After you have decided which of the suggested answers is best, completely fill in the corresponding oval on the answer sheet. Give only one answer to each question. If you change an answer, be sure that the previous mark is erased completely. Here is a sample question and answer.

Sample Question   Sample Answer

Chicago is a
(A) state
(B) city
(C) country
(D) continent

Use your time effectively, working as quickly as you can without losing accuracy. Do not spend too much time on any one question. Go on to other questions and come back to the ones you have not answered if you have time. It is not expected that everyone will know the answers to all the multiple-choice questions.

About Guessing

Many candidates wonder whether or not to guess the answers to questions about which they are not certain. Multiple-choice scores are based on the number of questions answered correctly. Points are not deducted for incorrect answers, and no points are awarded for unanswered questions. Because points are not deducted for incorrect answers, you are encouraged to answer all multiple-choice questions. On any questions you do not know the answer to, you should eliminate as many choices as you can, and then select the best answer among the remaining choices.

GO ON TO THE NEXT PAGE.

CHEMISTRY
SECTION I
Time—1 hour and 30 minutes

INFORMATION IN THE TABLE BELOW AND ON THE FOLLOWING PAGES MAY BE USEFUL IN
ANSWERING THE QUESTIONS IN THIS SECTION OF THE EXAMINATION

DO NOT DETACH FROM BOOK.

PERIODIC TABLE OF THE ELEMENTS

1																	18
1 H 1.008	2											13	14	15	16	17	2 He 4.00
3 Li 6.94	4 Be 9.01											5 B 10.81	6 C 12.01	7 N 14.01	8 O 16.00	9 F 19.00	10 Ne 20.18
11 Na 22.99	12 Mg 24.30	3	4	5	6	7	8	9	10	11	12	13 Al 26.98	14 Si 28.09	15 P 30.97	16 S 32.06	17 Cl 35.45	18 Ar 39.95
19 K 39.10	20 Ca 40.08	21 Sc 44.69	22 Ti 47.87	23 V 50.94	24 Cr 52.00	25 Mn 54.94	26 Fe 55.85	27 Co 58.93	28 Ni 58.69	29 Cu 63.55	30 Zn 65.38	31 Ga 69.72	32 Ge 72.63	33 As 74.92	34 Se 78.97	35 Br 79.90	36 Kr 83.80
37 Rb 85.47	38 Sr 87.62	39 Y 88.91	40 Zr 91.22	41 Nb 92.91	42 Mo 95.95	43 Tc	44 Ru 101.07	45 Rh 102.91	46 Pd 106.42	47 Ag 107.87	48 Cd 112.41	49 In 114.82	50 Sn 118.71	51 Sb 121.76	52 Te 127.60	53 I 126.90	54 Xe 131.29
55 Cs 132.91	56 Ba 137.33	57 57-71 *	72 Hf 178.49	73 Ta 180.95	74 W 183.94	75 Re 186.21	76 Os 190.23	77 Ir 192.22	78 Pt 195.08	79 Au 196.97	80 Hg 200.59	81 Tl 204.38	82 Pb 207.2	83 Bi 208.98	84 Po	85 At	86 Rn
87 Fr	88 Ra	89-103 †	104 Rf	105 Db	106 Sg	107 Bh	108 Hs	109 Mt	110 Ds	111 Rg	112 Cn	113 Nh	114 Fl	115 Mc	116 Lv	117 Ts	118 Og

*Lanthanoids	57 La 138.91	58 Ce 140.12	59 Pr 140.91	60 Nd 144.24	61 Pm	62 Sm 150.36	63 Eu 151.97	64 Gd 157.25	65 Tb 158.93	66 Dy 162.50	67 Ho 164.93	68 Er 167.26	69 Tm 168.93	70 Yb 173.05	71 Lu 174.97
†Actinoids	89 Ac	90 Th 232.04	91 Pa 231.04	92 U 238.03	93 Np	94 Pu	95 Am	96 Cm	97 Bk	98 Cf	99 Es	100 Fm	101 Md	102 No	103 Lr

GO ON TO THE NEXT PAGE.

AP® CHEMISTRY EQUATIONS & CONSTANTS

Throughout the exam the following symbols have the definitions specified unless otherwise noted.

L, mL	=	liter(s), milliliter(s)	mm Hg	= millimeters of mercury
g	=	gram(s)	J, kJ	= joule(s), kilojoule(s)
nm	=	nanometer(s)	V	= volt(s)
atm	=	atmosphere(s)	mol	= mole(s)

ATOMIC STRUCTURE

$E = h\nu$

$c = \lambda\nu$

E = energy

ν = frequency

λ = wavelength

Planck's constant, $h = 6.626 \times 10^{-34}$ J s

Speed of light, $c = 2.998 \times 10^8$ m s^{-1}

Avogadro's number $= 6.022 \times 10^{23}$ mol^{-1}

Electron charge, $e = -1.602 \times 10^{-19}$ coulomb

EQUILIBRIUM

$K_c = \dfrac{[C]^c[D]^d}{[A]^a[B]^b}$, where $a\,A + b\,B \rightleftarrows c\,C + d\,D$

$K_p = \dfrac{(P_C)^c(P_D)^d}{(P_A)^a(P_B)^b}$

$K_a = \dfrac{[H^+][A^-]}{[HA]}$

$K_b = \dfrac{[OH^-][HB^+]}{[B]}$

$K_w = [H^+][OH^-] = 1.0 \times 10^{-14}$ at 25°C

$\quad = K_a \times K_b$

$pH = -\log[H^+],\ pOH = -\log[OH^-]$

$14 = pH + pOH$

$pH = pK_a + \log\dfrac{[A^-]}{[HA]}$

$pK_a = -\log K_a,\ pK_b = -\log K_b$

Equilibrium Constants

K_c (molar concentrations)

K_p (gas pressures)

K_a (weak acid)

K_b (weak base)

K_w (water)

KINETICS

$[A]_t - [A]_0 = -kt$

$\ln[A]_t - \ln[A]_0 = -kt$

$\dfrac{1}{[A]_t} - \dfrac{1}{[A]_0} = kt$

$t_{1/2} = \dfrac{0.693}{k}$

k = rate constant

t = time

$t_{1/2}$ = half-life

GO ON TO THE NEXT PAGE.

GASES, LIQUIDS, AND SOLUTIONS

$$PV = nRT$$

$$P_A = P_{total} \times X_A, \text{ where } X_A = \frac{\text{moles A}}{\text{total moles}}$$

$$P_{total} = P_A + P_B + P_C + \ldots$$

$$n = \frac{m}{M}$$

$$K = {}^\circ C + 273$$

$$D = \frac{m}{V}$$

$$KE_{molecule} = \frac{1}{2}mv^2$$

Molarity, M = moles of solute per liter of solution

$$A = \varepsilon bc$$

P = pressure
V = volume
T = temperature
n = number of moles
m = mass
M = molar mass
D = density
KE = kinetic energy
v = velocity
A = absorbance
ε = molar absorptivity
b = path length
c = concentration

Gas constant, R = 8.314 J mol^{-1}K^{-1}

$\qquad$ = 0.08206 L atm mol^{-1}K^{-1}

$\qquad$ = 62.36 L torr mol^{-1}K^{-1}

1 atm = 760 mm Hg = 760 torr

STP = 273.15 K and 1.0 atm

Ideal gas at STP = 22.4 L mol^{-1}

THERMOCHEMISTRY/ ELECTROCHEMISTRY

$$q = mc\Delta T$$

$$\Delta S^\circ = \sum S^\circ \text{ products} - \sum S^\circ \text{ reactants}$$

$$\Delta H^\circ = \sum \Delta H_f^\circ \text{ products} - \sum \Delta H_f^\circ \text{ reactants}$$

$$\Delta G^\circ = \sum \Delta G_f^\circ \text{ products} - \sum \Delta G_f^\circ \text{ reactants}$$

$$\Delta G^\circ = \Delta H^\circ - T\Delta S^\circ$$

$$= -RT \ln K$$

$$= -nFE^\circ$$

$$I = \frac{q}{t}$$

$$E_{cell} = E_{cell}^\circ - \frac{RT}{nF} \ln Q$$

q = heat
m = mass
c = specific heat capacity
T = temperature
S° = standard entropy
H° = standard enthalpy
G° = standard Gibbs free energy
n = number of moles
E° = standard reduction potential
I = current (amperes)
q = charge (coulombs)
t = time (seconds)
Q = reaction quotient

Faraday's constant, F = 96,485 coulombs per mole of electrons

$$1 \text{ volt} = \frac{1 \text{ joule}}{1 \text{ coulomb}}$$

GO ON TO THE NEXT PAGE.

$$NH_3(l) \rightarrow NH_3(g) \qquad \Delta H_{vap} = 23.9 \text{ kJ/mol}$$

1. NH_3 has a boiling point of 239 K. Which of the following values would be closest to the entropy of vaporization for NH_3?

   (A) 0.100 J/mol × K
   (B) 100 J/mol × K
   (C) 200 J/mol × K
   (D) 260 J/mol × K

Substance	Boiling Point (°C)
C_6H_6	80.2
C_2H_5OH	78.4

2. Given the data in the above table, which substance would have a lower vapor pressure at 298 K, and why?

   (A) C_6H_6, due to its more polarizable electron cloud
   (B) C_6H_6, due to its lack of permanent dipoles
   (C) C_2H_5OH, due to its hydrogen bonding
   (D) C_2H_5OH, due to the presence of lone pairs on the oxygen

3. Which of the following 1.0 M aqueous solutions would experience the highest % ionization?

   (A) HClO
   (B) $HClO_2$
   (C) HBrO
   (D) $HBrO_2$

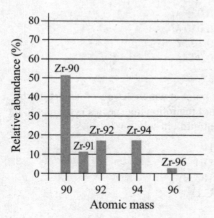

Atomic mass

4. Based on the mass spectrum shown above, which of the following can be concluded about zirconium?

   (A) The most common charge on a zirconium ion is +2.
   (B) Zirconium nuclei can have different numbers of protons.
   (C) The average atomic mass of a zirconium atom is 90 amu.
   (D) The most common isotope of zirconium has 50 neutrons.

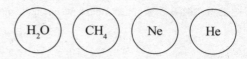

5. As shown above, four identical containers hold the same number of moles of four different gases at 298 K. If ideal behavior is NOT assumed, in which container would the pressure be the lowest?

   (A) H_2O
   (B) CH_4
   (C) Ne
   (D) He

Use the following information to answer questions 6–8.

The <u>unbalanced</u> reaction below occurs when a solution of potassium dichromate, $K_2Cr_2O_7$, is titrated into a solution containing aqueous Fe^{2+} ions.

$$\_\_\_ Fe^{2+}(aq) + Cr_2O_7^{2-}(aq) + 14\ H^+ \rightarrow$$
$$7\ H_2O(l) + 2\ Cr^{3+}(aq) + \_\_\_ Fe^{3+}(aq)$$

6. Which species is being oxidized, and which is being reduced?

	Oxidized	Reduced
(A)	$Cr_2O_7^{2-}$	H^+
(B)	Fe^{2+}	H^+
(C)	$Cr_2O_7^{2-}$	Fe^{2+}
(D)	Fe^{2+}	$Cr_2O_7^{2-}$

7. What must the coefficient in front of the iron on both sides of the reaction be in order to balance the reaction?

   (A) 1
   (B) 3
   (C) 4
   (D) 6

8. Which of the following corresponds to the electron configuration for Fe^{3+}?

   (A) $[Ar]4s^23d^3$
   (B) $[Ar]3d^5$
   (C) $[Ar]4s^14d^3$
   (D) $[Ar]4s^13d^4$

GO ON TO THE NEXT PAGE.

9. Which of the below molecules would have no dipole moment?

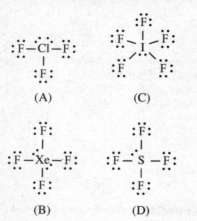

(A)

(C)

(B)

(D)

Sub-stance	Conductivity as solid	Conductivity as liquid	Conductivity in water
A	High	High	Chemical Reaction Occurs
B	Low	High	High
C	Low	Low	Does not dissolve
D	Low	Low	Low

10. Data considering the conductivity of four different substances in their various phases is given in the table above. Of the four options, which substance is most likely to be NaCl?

(A) Substance A
(B) Substance B
(C) Substance C
(D) Substance D

Co^{2+} Concentration (M)	Absorbance
0	0
0.025	0.13
0.050	0.25
0.075	0.38
0.100	0.50

11. The absorbance of Co^{2+} at several different concentrations was tested, yielding the above data. What is the molar absorptivity value for Co^{2+} under the given conditions if a cuvette with a 1.0 cm path length was used?

(A) $0.05\ M^{-1}cm^{-1}$
(B) $0.20\ M^{-1}cm^{-1}$
(C) $5.0\ M^{-1}cm^{-1}$
(D) $20.0\ M^{-1}cm^{-1}$

12. Which of the following correctly pairs the reaction showing the electron affinity for an aluminum atom?

(A) $Al(g) + e^- \rightarrow Al^-(g)$
(B) $Al(g) \rightarrow Al^{3+}(g) + 3\ e^-$
(C) $Al(g) \rightarrow Al^+(g) + e^-$
(D) $Al(g) + 5\ e^- \rightarrow Ar(g)$

Use the following information to answer questions 13–16.

A 0.10 M solution of NaOH is titrated into 20 mL of $H_2C_2O_4$, a diprotic acid, of an unknown concentration. The pH of the $H_2C_2O_4$ solution is monitored as the NaOH is added to it, resulting in the below graph.

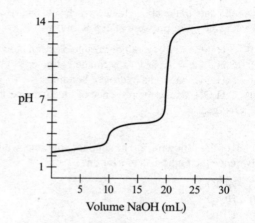

13. What is the concentration of the $H_2C_2O_4$ solution?

(A) 0.025 M
(B) 0.050 M
(C) 0.10 M
(D) 0.20 M

14. At the point at which 20 mL of NaOH has been added, which of the following species is present in the greatest concentration in solution?

(A) H^+
(B) OH^-
(C) $HC_2O_4^-$
(D) $C_2O_4^{2-}$

GO ON TO THE NEXT PAGE.

15. Phenolphthalein is an acid-base indicator with a pK_a of 9.1. Its protonated form is often abbreviated as HIn, while its conjugate base is abbreviated as In⁻. At the following volumes of NaOH added, select the option that accurately describes which form of the indicator will be present in a greater concentration.

	5 mL	15 mL	25 mL
(A)	HIn	HIn	In⁻
(B)	HIn	In⁻	In⁻
(C)	In⁻	In⁻	HIn
(D)	In⁻	HIn	HIn

16. If the $H_2C_2O_4$ were to be replaced with an identical volume of H_2SO_4, what volume of NaOH would be required to fully neutralize the acid?

    (A) 10 mL
    (B) 20 mL
    (C) 40 mL
    (D) 60 mL

$$2\,Ag^+(aq) + Fe(s) \rightarrow Fe^{2+}(aq) + 2\,Ag(s) \quad E^\circ_{cell} = +1.24\ V$$

17. The above reaction takes places in a galvanic cell at 25°C. Which of the following would decrease the voltage for the cell?

    (A) Doubling the mass of the Fe(s) electrode
    (B) Adding a catalyst
    (C) Increasing the concentration of Ag⁺
    (D) Adding water

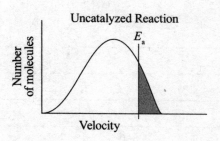

Uncatalyzed Reaction

$$2\,H_2O_2(aq) \rightarrow 2\,H_2O(l) + O_2(g)$$

18. A sample of H_2O_2 is present in a flask. As time passes, the molecules may collide to form the indicated products. The shaded area under the graph represents the number of effective collisions which create products under standard conditions. Given the energy distributions curve for the uncatalyzed reaction, which curve would best represent the catalyzed reaction?

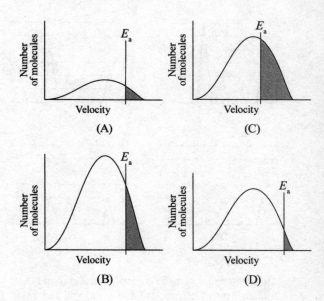

$$2\,CO(g) + 2\,NO(g) \rightarrow 2\,CO_2(g) + N_2(g)$$

19. CO(g) at a partial pressure of 2.0 atm and NO(g) at a partial pressure of 1.0 atm are mixed in an evacuated and sealed container where they react via the above equation. What is the total pressure of all gases present in the flask after the reaction goes to completion?

    (A) 1.5 atm
    (B) 2.5 atm
    (C) 3.0 atm
    (D) 2.0 atm

GO ON TO THE NEXT PAGE.

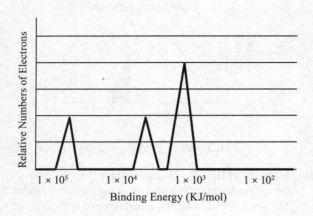

20. The photoelectron spectrum for an oxygen atom is shown above. Which of the diagrams below would be the correct spectrum for the oxide ion (O^{2-})?

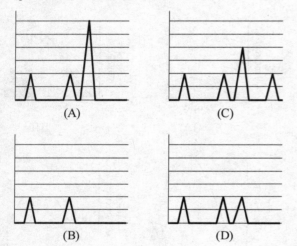

21. An unknown substance was analyzed and found to contain 9.0 g of carbon, 2.0 g of hydrogen, and 4.0 g of oxygen. What is the empirical formula of the substance?

    (A) CH_5O
    (B) C_3H_8O
    (C) CH_6O_2
    (D) C_2H_3O

22. Which of the following molecules is most likely to behave as an acid?

    (A)   (C)

    (B)   (D)

23. Which of the following ways can a catalyst function in order to increase reaction rate?

    (A) They can create intermediates with a higher activation energy
    (B) By reversing the role of the products and reactants
    (C) Through increasing the thermodynamic favorability of a reaction
    (D) By causing reactant molecules to be oriented more favorably for a successful collision

    $$4\ Fe(s) + 3\ O_2(g) \rightarrow 2\ Fe_2O_3(s)$$

24. How many grams of Fe must react with excess oxygen to create 8.0 g of Fe_2O_3 (MM = 160 g/mol)?

    (A) 2.79 g
    (B) 5.58 g
    (C) 8.37 g
    (D) 11.16 g

GO ON TO THE NEXT PAGE.

Use the following information to answer questions 25–29.

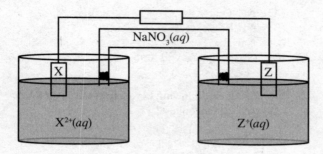

A galvanic cell is constructed as shown above. Metals X and Z serve as the electrodes, and the following reaction occurs as the cell operates:

$$X^{2+}(aq) + 2\ Z(s) \rightarrow 2\ Z^+(aq) + X(s)$$

25. Which of the following options correctly identifies which substance is being reduced, as well as the half-reaction occurring at the cathode?

	Reduced	Cathode Half-Reaction
(A)	X^{2+}	$X(s) \rightarrow X^{2+}(aq) + 2\ e^-$
(B)	X^{2+}	$X^{2+}(aq) + 2\ e^- \rightarrow X(s)$
(C)	Z^+	$Z(s) \rightarrow Z^+(aq) + e^-$
(D)	Z^+	$Z^+(aq) + e^- \rightarrow Z(s)$

26. What is true about the value for K and ΔG for this reaction?

	K	ΔG
(A)	>1	<0
(B)	>1	>0
(C)	<1	>0
(D)	<1	<0

27. Which of the following describes what is happening in the salt bridge?

(A) Electrons are flowing from the X half-cell to the Z half-cell

(B) Electrons are flowing from the Z half-cell to the X half-cell

(C) Na^+ ions are moving towards the Z half-cell, and NO_3^- ions are flowing towards the X half-cell

(D) NO_3^- ions are flowing towards the Z half-cell, and Na^+ ions are flowing towards the X half-cell

28. A student claims that the mass gained by the X electrode will be equal to the mass lost by the Z electrode. Is this correct? Why or why not?

(A) No, because the mass of electrode X is gained from the reduced X^{2+} ions, not the Z electrode.

(B) No, because there are no interactions between the two half-cells.

(C) Yes, because the electrons carry the mass from the X half-cell to the Z half-cell.

(D) Yes, because the salt bridge ensures the mass of both electrodes remains equal.

29. If the concentration of both X^{2+} and Z^+ is doubled, how would that affect the voltage of the cell, if at all?

(A) The voltage would increase because Q decreases.

(B) The voltage would decrease because Q increases.

(C) The voltage would remain unchanged as the doubled concentrations cancel out.

(D) The voltage would remain unchanged as the reduction potential of both half-reactions would not change.

---

30. $Ca(OH)_2$ would be the most soluble in a solution at which pH value?

(A) 4.0

(B) 7.0

(C) 10.0

(D) 13.0

GO ON TO THE NEXT PAGE.

31. Spinach leaves contain three visible dyes: carotene, xanthophyll, and chlorophyll. A spinach sample is subjected to paper chromatography using a nonpolar solvent, and the final chromatogram looks as follows:

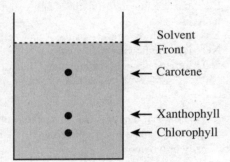

Which of the following options orders the dyes from the most to the least polar?

(A) Carotene > xanthophyll > chlorophyll
(B) Carotene > chlorophyll > xanthophyll
(C) Chlorophyll > xanthophyll > carotene
(D) Chlorophyll > carotene > xanthophyll

32. Some noble gases, such as helium and neon, are never found as part of compounds. However, other noble gases, such as krypton and xenon, can be found bonded to other elements in various compounds. Which of the following best explains the above observations?

(A) Noble gases with smaller radii have lower electronegativity values
(B) Noble gases with smaller radii have smaller ionization energies and can more easily transfer electrons
(C) Noble gases with more filled energy levels have empty d-subshells that can participate in bonding
(D) Noble gases with a lower molar mass will experience weaker London dispersion forces

$$2\ SO_3(g) \rightleftharpoons 2\ SO_2(g) + O_2(g)$$

33. The above system is at equilibrium in a sealed container. Some He (g) is added to the container and the container expands, maintaining a constant total pressure. In order to re-establish equilibrium, which of the following must occur?

(A) The partial pressure of all three gases must decrease
(B) P_{SO2} and P_{O2} will increase, while P_{SO3} will decrease
(C) P_{SO3} will increase, while P_{SO2} and P_{O2} will decrease
(D) The system remains at equilibrium, so no changes will occur

34. Solutions of NaBr and NH_4NO_3 are mixed. Which precipitate will form, if any?

(A) $NaNO_3$
(B) NH_4Br
(C) $NaNH_4$
(D) No precipitate will form

35. In which of the following Lewis diagrams is the sulfur-oxygen bond the longest?

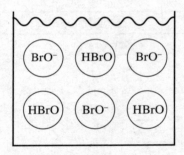

36. HBrO has a pK_a of 8.7. A particulate view of a BrO^-/HBrO buffer is shown below:

What would the pH of the buffer solution be?

(A) 5.7
(B) 6.7
(C) 8.7
(D) 9.7

GO ON TO THE NEXT PAGE.

$$A + D \rightarrow E$$

37. Reactants A and D combine to form product E via the above reaction. Given the data below that charts initial concentration of each reactant vs. reaction rate, what must be the initial concentration of reactant D in trial 4?

Trial	[A] (M)	[D] (M)	Rate (M/s)
1	0.10	0.10	5.0×10^{-4}
2	0.10	0.20	2.0×10^{-3}
3	0.20	0.10	1.0×10^{-3}
4	0.050	???	1.0×10^{-3}

(A) 0.050 M
(B) 0.10 M
(C) 0.15 M
(D) 0.20 M

Use the following information to answer questions 38–41.

10.0 g of an unknown metal is heated to 98.0°C, then dropped into a Styrofoam cup containing 100. mL of water ($c = 4.2$ J/g°C) originally at a temperature of 13.0°C. The final temperature of the water stabilizes at 14.0°C.

38. What the specific heat capacity of the metal?

(A) 0.10 J/g°C
(B) 0.25 J/g°C
(C) 0.50 J/g°C
(D) 2.0 J/g°C

39. Which of the following is true at thermal equilibrium?

(A) The amount of thermal energy contained in both the metal and water is identical
(B) The temperature of the metal must be the same as that of the water
(C) All energy transfer between the metal and the water has stopped
(D) No more energy will be lost to the surroundings

40. Why is the specific heat capacity for the water greater than that of the metal?

(A) The sea of electrons in the metal causes internal repulsion
(B) The hydrogen bonding present in the water requires a lot of energy to overcome
(C) The covalent O-H bonds in the metal are stronger than the bonds between the metallic nuclei
(D) The metal experiences stronger London dispersion forces than the water

41. If some of the energy lost by the metal is absorbed by the cup, how would that affect the calculated specific heat of the metal?

(A) It would be artificially low because the water would not absorb all the energy lost by the metal
(B) It would be artificially low because the rate of energy transfer between the metal and water would decrease
(C) It would be artificially high because the amount of energy lost by the metal would decrease
(D) It would be artificially high because the cup has a lower specific heat capacity than water

Use the following information to answer questions 42–45.

Element	Atomic Radius (pm)	First Ionization Energy (kJ/mol)	Electronegativity
Phosphorus	98	1,012	2.19
Sulfur	88	999	2.59

42. Which of the following is the best explanation for why a sulfur atom has a lower ionization energy than a phosphorus atom?

(A) Sulfur's electrons are, on average, further from the nucleus.
(B) Sulfur's valence electrons experience greater shielding.
(C) Sulfur commonly forms an ion with a lower charge.
(D) Sulfur has paired p-electrons, providing additional repulsion.

43. The main isotope of sulfur, ^{32}S, has a natural abundance of approximately 97%. Which of the following isotopes is most likely to make up the remaining 3% of naturally occurring sulfur atoms?

(A) ^{30}S
(B) ^{31}S
(C) ^{33}S
(D) ^{34}S

GO ON TO THE NEXT PAGE.

44. If phosphorus and sulfur form a bond, which element would have a partial negative charge? Why?

    (A) Phosphorus, because it has a larger atomic radius
    (B) Sulfur, because it has a larger electronegativity value
    (C) Neither, because both elements have the same valence subshell
    (D) Neither, because both elements are in the same period

45. Which option below represents the reaction describing the first electron affinity of sulfur?

    (A) $S(g) + e^- \rightarrow S^-(g)$
    (B) $S(g) + 2 e^- \rightarrow S^{2-}(g)$
    (C) $S(g) \rightarrow S^+(g) + e^-$
    (D) $S(g) + S(g) \rightarrow S_2(g)$

46. Choose the correct Lewis diagram for the nitrite ion, NO_2^-.

    (A) $\left[\ddot{:O} - \ddot{N} - \ddot{O:} \right]^-$

    (B) $\left[\ddot{O} = N = \ddot{O} \right]^-$

    (C) $\left[\ddot{O} = \ddot{N} - \ddot{O:} \right]^- \leftrightarrow \left[\ddot{:O} - \ddot{N} = \ddot{O} \right]^-$

    (D) $\left[\ddot{O} = \ddot{N} = \ddot{O:} \right]^-$

47. Which of the following substances would be the best conductor of electricity?

    (A) $1.0\ M\ C_6H_{12}O_6(aq)$
    (B) $LiNO_3(s)$
    (C) $1.0\ M\ CaCl_2(aq)$
    (D) $CH_3OH(l)$

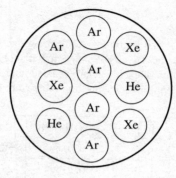

48. In the above diagram, $P_{Xe} = 0.40$ atm. What is the total pressure in the container?

    (A) 1.33 atm
    (B) 1.20 atm
    (C) 1.10 atm
    (D) 0.95 atm

49. In which of the following compounds is the oxidation state on chlorine the most positive?

    (A) Cl_2
    (B) $MgCl_2$
    (C) $NaClO_3$
    (D) $KClO_2$

Use the following information to answer questions 50–52.

$$2\ ClO_2(aq) + 2\ OH^-(aq) \rightarrow ClO_2^-(aq) + ClO_3^-(aq) + H_2O(l)$$

Chlorine dioxide is a disinfectant used to treat water at municipal water plants. It dissolves in a basic solution, causing the redox reaction above.

50. Which species is oxidized, and which is reduced?

	Oxidized	Reduced
(A)	Cl	O
(B)	O	H
(C)	H	O
(D)	Cl	Cl

GO ON TO THE NEXT PAGE.

51. Which of the following would cause the activation energy of the reaction to decrease?

    (A) Stirring the reaction
    (B) Increasing the temperature
    (C) Diluting the reactants
    (D) Adding a catalyst

Trial	$[ClO_2]$ (M)	$[OH^-]$ (M)	Initial Rate (M/s)
1	0.100	0.100	0.0400
2	0.050	0.100	0.0100
3	0.200	0.050	0.0800

52. Three trials were done by examining the rate of reaction when various concentrations of each reactant are mixed, and the data above was gathered. What is the rate law for the reaction?

    (A) Rate = $k[ClO_2][OH^-]$
    (B) Rate = $k[ClO_2]^2[OH^-]$
    (C) Rate = $k[ClO_2][OH^-]^2$
    (D) Rate = $k[ClO_2]^2[OH^-]^2$

$$2\,HI(g) + S(s) \rightleftharpoons H_2S(g) + I_2(g) \qquad K_c = 2.3 \times 10^{-5}$$

53. A container with some solid sulfur in it has $HI(g)$ injected into it. If the initial concentration of the HI is 0.15 M, what is the concentration of $H_2S(g)$ at equilibrium?

    (A) $5.2 \times 10^{-7}\ M$
    (B) $2.3 \times 10^{-5}\ M$
    (C) $7.2 \times 10^{-4}\ M$
    (D) $6.8 \times 10^{-3}\ M$

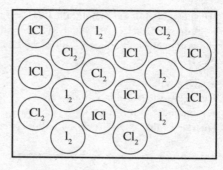

$$Cl_2\,(g) + I_2\,(g) \rightleftharpoons 2\,ICl\,(g)$$

54. The container represented above contains three different species involved in the given reaction at equilibrium. The total pressure in the flask is 1.7 atm. What is K_p for the reaction under these conditions?

    (A) 0.35
    (B) 0.51
    (C) 0.71
    (D) 2.0

Use the following information to answer questions 55–57.

$$2\,CrO_4^{2-}(aq) + 2\,H^+(aq) \rightleftharpoons Cr_2O_7^{2-}(aq) + H_2O(l)$$
$$\text{(yellow)} \qquad\qquad\qquad \text{(orange)}$$

The chromate (CrO_4^{2-}) and dichromate ($Cr_2O_7^{2-}$) exist in a dynamic equilibrium as demonstrated by the reaction above. 100. mL of 0.10 M K_2CrO_4 is present in a beaker, and the system is allowed to come to equilibrium.

55. Which of the following changes would cause the solution to turn orange?

    (A) Diluting the solution
    (B) Adding some $HCl(aq)$
    (C) Adding some $NaNO_3(aq)$
    (D) Stirring the solution vigorously

56. If the system is heated, the solution turns orange. Which of the following must be true about the reaction?

    (A) The reaction is exothermic and the value for K increases.
    (B) The reaction is exothermic and the value for K does not change.
    (C) The reaction is endothermic and the value for K decreases.
    (D) The reaction is endothermic and the value for K increases.

GO ON TO THE NEXT PAGE.

57. If NaOH(s) is added to the system, the OH⁻ will react with the H⁺ in solution to form water. Which of the following will occur immediately following the addition of the NaOH?

   (A) $[Cr_2O_7^{2-}]$ will increase.
   (B) The pH of the solution will decrease.
   (C) A precipitate will form.
   (D) The rate of the reverse reaction will increase.

---

58. The indicator thymol blue is yellow in a solution with a pH of 8 but turns blue when the pH of the solution is increased to 9. What is the approximate pK_b range for the conjugate base of thymol blue?

   (A) 2–3
   (B) 5–6
   (C) 8–9
   (D) 11–12

59. How long would it take to plate 0.100 g of Cu(s) out of a solution containing 1.00 M Cu^{2+} ions if a current of 0.500 A is run through the solution?

   (A) 608 s
   (B) 304 s
   (C) 152 s
   (D) 76.0 s

Substance	ΔH_f° (kJ/mol)	S° (J/mol·K)
$Br_2(l)$	0	152.2
$Br_2(g)$	30.9	245.5

60. Given the above data, what is the boiling point of $Br_2(l)$?

   (A) 288 K
   (B) 301 K
   (C) 310 K
   (D) 331 K

END OF SECTION I

INFORMATION IN THE TABLE BELOW AND ON THE FOLLOWING PAGES MAY BE USEFUL IN
ANSWERING THE QUESTIONS IN THIS SECTION OF THE EXAMINATION

DO NOT DETACH FROM BOOK.

PERIODIC TABLE OF THE ELEMENTS

1	2	3	4	5	6	7	8	9	10	11	12	13	14	15	16	17	18
1 **H** 1.008																	2 **He** 4.00
3 **Li** 6.94	4 **Be** 9.01											5 **B** 10.81	6 **C** 12.01	7 **N** 14.01	8 **O** 16.00	9 **F** 19.00	10 **Ne** 20.18
11 **Na** 22.99	12 **Mg** 24.30											13 **Al** 26.98	14 **Si** 28.09	15 **P** 30.97	16 **S** 32.06	17 **Cl** 35.45	18 **Ar** 39.95
19 **K** 39.10	20 **Ca** 40.08	21 **Sc** 44.69	22 **Ti** 47.87	23 **V** 50.94	24 **Cr** 52.00	25 **Mn** 54.94	26 **Fe** 55.85	27 **Co** 58.93	28 **Ni** 58.69	29 **Cu** 63.55	30 **Zn** 65.38	31 **Ga** 69.72	32 **Ge** 72.63	33 **As** 74.92	34 **Se** 78.97	35 **Br** 79.90	36 **Kr** 83.80
37 **Rb** 85.47	38 **Sr** 87.62	39 **Y** 88.91	40 **Zr** 91.22	41 **Nb** 92.91	42 **Mo** 95.95	43 **Tc**	44 **Ru** 101.07	45 **Rh** 102.91	46 **Pd** 106.42	47 **Ag** 107.87	48 **Cd** 112.41	49 **In** 114.82	50 **Sn** 118.71	51 **Sb** 121.76	52 **Te** 127.60	53 **I** 126.90	54 **Xe** 131.29
55 **Cs** 132.91	56 **Ba** 137.33	57-71 *	72 **Hf** 178.49	73 **Ta** 180.95	74 **W** 183.94	75 **Re** 186.21	76 **Os** 190.23	77 **Ir** 192.22	78 **Pt** 195.08	79 **Au** 196.97	80 **Hg** 200.59	81 **Tl** 204.38	82 **Pb** 207.2	83 **Bi** 208.98	84 **Po**	85 **At**	86 **Rn**
87 **Fr**	88 **Ra**	89-103 †	104 **Rf**	105 **Db**	106 **Sg**	107 **Bh**	108 **Hs**	109 **Mt**	110 **Ds**	111 **Rg**	112 **Cn**	113 **Nh**	114 **Fl**	115 **Mc**	116 **Lv**	117 **Ts**	118 **Og**

*Lanthanoids	57 **La** 138.91	58 **Ce** 140.12	59 **Pr** 140.91	60 **Nd** 144.24	61 **Pm**	62 **Sm** 150.36	63 **Eu** 151.97	64 **Gd** 157.25	65 **Tb** 158.93	66 **Dy** 162.50	67 **Ho** 164.93	68 **Er** 167.26	69 **Tm** 168.93	70 **Yb** 173.05	71 **Lu** 174.97
†Actinoids	89 **Ac**	90 **Th** 232.04	91 **Pa** 231.04	92 **U** 238.03	93 **Np**	94 **Pu**	95 **Am**	96 **Cm**	97 **Bk**	98 **Cf**	99 **Es**	100 **Fm**	101 **Md**	102 **No**	103 **Lr**

GO ON TO THE NEXT PAGE.

AP® CHEMISTRY EQUATIONS & CONSTANTS

Throughout the exam the following symbols have the definitions specified unless otherwise noted.

L, mL	=	liter(s), milliliter(s)	mm Hg	= millimeters of mercury
g	=	gram(s)	J, kJ	= joule(s), kilojoule(s)
nm	=	nanometer(s)	V	= volt(s)
atm	=	atmosphere(s)	mol	= mole(s)

ATOMIC STRUCTURE

$$E = h\nu$$
$$c = \lambda\nu$$

E = energy
ν = frequency
λ = wavelength

Planck's constant, $h = 6.626 \times 10^{-34}$ J s

Speed of light, $c = 2.998 \times 10^8$ m s^{-1}

Avogadro's number $= 6.022 \times 10^{23}$ mol^{-1}

Electron charge, $e = -1.602 \times 10^{-19}$ coulomb

EQUILIBRIUM

$$K_c = \frac{[C]^c[D]^d}{[A]^a[B]^b}, \text{ where } a\,A + b\,B \rightleftarrows c\,C + d\,D$$

$$K_p = \frac{(P_C)^c (P_D)^d}{(P_A)^a (P_B)^b}$$

$$K_a = \frac{[H^+][A^-]}{[HA]}$$

$$K_b = \frac{[OH^-][HB^+]}{[B]}$$

$$K_w = [H^+][OH^-] = 1.0 \times 10^{-14} \text{ at } 25°C$$
$$= K_a \times K_b$$

$$pH = -\log[H^+], \ pOH = -\log[OH^-]$$

$$14 = pH + pOH$$

$$pH = pK_a + \log\frac{[A^-]}{[HA]}$$

$$pK_a = -\log K_a, \ pK_b = -\log K_b$$

Equilibrium Constants

K_c (molar concentrations)

K_p (gas pressures)

K_a (weak acid)

K_b (weak base)

K_w (water)

KINETICS

$$[A]_t - [A]_0 = -kt$$

$$\ln[A]_t - \ln[A]_0 = -kt$$

$$\frac{1}{[A]_t} - \frac{1}{[A]_0} = kt$$

$$t_{1/2} = \frac{0.693}{k}$$

k = rate constant
t = time
$t_{1/2}$ = half-life

GO ON TO THE NEXT PAGE.

GASES, LIQUIDS, AND SOLUTIONS

$$PV = nRT$$

$$P_A = P_{total} \times X_A, \text{ where } X_A = \frac{\text{moles A}}{\text{total moles}}$$

$$P_{total} = P_A + P_B + P_C + \ldots$$

$$n = \frac{m}{M}$$

$$K = {}^\circ C + 273$$

$$D = \frac{m}{V}$$

$$KE_{molecule} = \frac{1}{2}mv^2$$

Molarity, M = moles of solute per liter of solution

$$A = \varepsilon bc$$

P = pressure
V = volume
T = temperature
n = number of moles
m = mass
M = molar mass
D = density
KE = kinetic energy
v = velocity
A = absorbance
ε = molar absorptivity
b = path length
c = concentration

Gas constant, R = 8.314 J mol^{-1} K^{-1}
$\qquad$ = 0.08206 L atm mol^{-1} K^{-1}
$\qquad$ = 62.36 L torr mol^{-1} K^{-1}
1 atm = 760 mm Hg = 760 torr
STP = 273.15 K and 1.0 atm
Ideal gas at STP = 22.4 L mol^{-1}

THERMOCHEMISTRY/ ELECTROCHEMISTRY

$$q = mc\Delta T$$

$$\Delta S^\circ = \sum S^\circ \text{ products} - \sum S^\circ \text{ reactants}$$

$$\Delta H^\circ = \sum \Delta H_f^\circ \text{ products} - \sum \Delta H_f^\circ \text{ reactants}$$

$$\Delta G^\circ = \sum \Delta G_f^\circ \text{ products} - \sum \Delta G_f^\circ \text{ reactants}$$

$$\Delta G^\circ = \Delta H^\circ - T\Delta S^\circ$$

$$= -RT \ln K$$

$$= -nFE^\circ$$

$$I = \frac{q}{t}$$

$$E_{cell} = E_{cell}^\circ - \frac{RT}{nF} \ln Q$$

q = heat
m = mass
c = specific heat capacity
T = temperature
S° = standard entropy
H° = standard enthalpy
G° = standard Gibbs free energy
n = number of moles
E° = standard reduction potential
I = current (amperes)
q = charge (coulombs)
t = time (seconds)
Q = reaction quotient

Faraday's constant, F = 96,485 coulombs per mole
$\qquad$ of electrons

$$1 \text{ volt} = \frac{1 \text{ joule}}{1 \text{ coulomb}}$$

GO ON TO THE NEXT PAGE.

CHEMISTRY
SECTION II
Time—1 hour and 45 minutes
7 Questions

Directions: Questions 1–3 are long free-response questions that require about 23 minutes each to answer and are worth 10 points each. Questions 4–7 are short free-response questions that require about 9 minutes each to answer and are worth 4 points each.

On test day, you will be asked to show your work for each part in the space provided after that part. For this practice test, you may use scrap paper. Examples and equations may be included in your responses where appropriate. For calculations, clearly show the method used and the steps involved in arriving at your answers. You must show your work to receive credit for your answer. Pay attention to significant figures.

1. 5.00 g of $PbCl_2$ is added to 300 mL of water in a 400 mL beaker, which is then heated for 10 minutes. At the end of the heating period, some solid $PbCl_2$ is still present at the bottom of the beaker, and the solution is cooled to room temperature before being left out overnight.

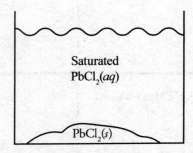

Saturated
$PbCl_2(aq)$

$PbCl_2(s)$

(a) If 50 mL of water evaporates overnight at constant temperature, what would happen to the following values? Justify your answer.
   (i) The concentration of the Pb^{2+} and Cl^- ions in solution
   (ii) The mass of $PbCl_2(s)$ on the bottom of the beaker

The next day, 100 mL of the saturated solution is decanted into a separate 250 mL beaker, taking care not to transfer any remaining solid. 100 mL of 0.75 *M* KI solution is added, causing the following precipitation reaction to go to completion.

$$Pb^{2+} + 2\ I^- \rightarrow PbI_2(s)$$

(b) Given the following equipment, describe how to make up 100 mL of 0.75 *M* KI solution. You need not use all of the equipment listed.

250 mL Erlenmeyer flask	Stir station
50 mL buret	Hot plate
100 mL volumetric flask	Analytical balance
Solid KI	Weigh boats
100 mL graduated cylinder	Filter paper

GO ON TO THE NEXT PAGE.

The PbI$_2$ is filtered out, dried, and massed. The mass of the precipitate is found to be 0.747 g.

(c)   (i)   How many moles of Pb^{2+} are in the PbI$_2$ precipitate?

       (ii)   What is the concentration of Pb^{2+} in the saturated solution that was decanted from the beaker?

     (iii)   Calculate the solubility product constant, K_{sp}, for PbCl$_2$.

(d)   If the PbI$_2$ precipitate was not completely dried, how would that affect your calculated value for the K_{sp} of PbCl$_2$ in (c)(iii)? Justify your answer.

(e)   Which salt would have a greater melting point: PbCl$_2$ or PbI$_2$? Justify your answer.

GO ON TO THE NEXT PAGE.

H   H
| |
H–C–H   H–C–H
:O: H | H | H
‖ | | | | |
H– C — Ö–H   H — C — C — C — C — C — H
·· | | | | |
                                 H   H | H   H
                                      H–C–H
                                         |
                                         H

Formic Acid   Trimethylpentane
Boiling Point = 100°C   Boiling Point = 114°C

2. Use the above Lewis diagrams of formic acid and trimethylpentane to answer the following questions.

(a) For trimethylpentane, identify the bond angle and the hybridization around any of the carbon atoms.

(b) (i) Identify all types of IMFs present in both molecules.
    (ii) For any molecule that is polar, draw the location of any partial charges on the Lewis diagrams.

(c) The boiling point of the trimethylpentane is greater than that of formic acid. Explain why.

(d) Which liquid would have a higher vapor pressure at room temperature? Justify your answer.

(e) Some of each liquid is injected into a column, where they saturate the stationary phase. Then, the column is eluted with a polar eluent.

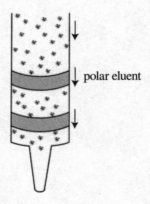

polar eluent

Which liquid will leave the column first? Justify your answer.

GO ON TO THE NEXT PAGE.

3. 40. mL of a 0.500 M solution of sodium bicarbonate, $NaHCO_3$, is present in a 1 L flask.

(a) How many grams of solid $NaHCO_3$ were necessary to create the above solution?

(b) Write the reaction between $NaHCO_3$ and water that causes the creation of a basic environment.

(c) The K_b of the HCO_3^- ion is 2.00×10^{-8}.
   (i) What is the pH of the 0.500 M solution?
   (ii) What is the percent ionization of the HCO_3^- ion in solution?
   (iii) The percent ionization of a 0.100 M solution of $NaHCO_3$ is higher than your answer to part (c)(ii). Explain why utilizing equilibrium principles.

The $NaHCO_3$ solution is then titrated with some 0.750 M HCl, causing the following reaction to occur:

$$HCO_3^-(aq) + H^+(aq) \rightarrow H_2CO_3\ (aq)$$

(d) What volume of HCl will be needed to reach the equivalence point?

(e) What would the pH of the solution be at the half-equivalence point?

(f) Using your answers to (c)(i) and (e), sketch the titration curve for this reaction on the below axes.

You need not calculate the exact pH at any other point.

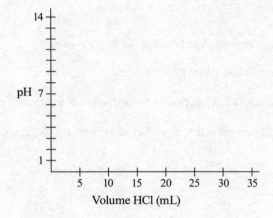

GO ON TO THE NEXT PAGE.

4. A single magnesium atom will be ionized when exposed to energy with a frequency of 1.86×10^{15} s^{-1}.

   (a) What wavelength of light, in nm, would be required to ionize a magnesium atom?

   (b) What is the first ionization energy, in kJ/mol, for magnesium?

   (c) How would the required frequency to ionize an atom of magnesium compare to the required frequency to ionize an atom of sodium? Justify your answer in terms of Coulombic attractions.

5. $$C_6H_{12}O_6(s) \rightarrow 2\ C_2H_5OH(l) + 2\ CO_2(g) \qquad \Delta H = -68\ kJ/mol_{rxn}$$

   At 298 K, glucose will convert into ethanol and carbon dioxide, as shown above.

   (a) Predict the sign of ΔS for the reaction at 298 K.

   (b) Based on your answer to (a), would this reaction be favored at low temperatures, high temperatures, or all temperatures?

   (c) Even though this reaction will proceed at 298 K, it does so very slowly. Suggest one factor that may contribute to the slow speed of the reaction.

GO ON TO THE NEXT PAGE.

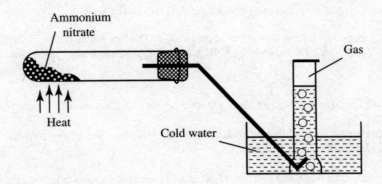

6. A student wants to experimentally determine the value for the ideal gas constant, R. In order to do so, 0.27 g of ammonium nitrate is placed in a test tube and heated, causing the below reaction to occur.

$$NH_4NO_3(g) \rightarrow N_2O(g) + 2\, H_2O(l)$$

The gaseous N_2O is collected over water in the graduated cylinder. The following data is gathered:

Air/Water Temperature	23°C
P_{H_2O} at 23°C	21 torr
Volume of gas	79.8 mL
Atmospheric Pressure	0.97 atm

(a) What is the partial pressure of the N_2O gas collected in the graduated cylinder?

(b) Calculate the experimental value for R. Don't forget to include units!

(c) If some of the N_2O dissolves in the water as it is bubbling up through it, what effect would this have on the student's calculated value of R?

GO ON TO THE NEXT PAGE.

$$NH_4CNO(aq) \rightarrow H_2NCONH_2(aq)$$

7. Ammonium cyanate isomerizes via the reaction above to create urea. The rate law is rate = $k[NH_4CNO]^2$, and the magnitude of the rate constant is 9.81×10^{-4} at 25°C. A sample of NH_4CNO is dissolved in water, placed into a flask, and the reaction is allowed to commence.

(a) Assuming time is measured in seconds, what are the units on the rate constant, k?

(b) Suggest one possible way to increase the rate constant for this reaction. Give a brief physical description of how this would increase the rate constant.

(c) If the initial concentration of NH_4CNO is 0.350 M, what will the concentration be at t = 250. s?

STOP

END OF EXAM

Practice Test 3:
Answers and
Explanations

PRACTICE TEST 3: MULTIPLE-CHOICE ANSWER KEY

1.	B	21.	B	41.	A
2.	A	22.	B	42.	D
3.	B	23.	D	43.	D
4.	D	24.	B	44.	B
5.	A	25.	B	45.	A
6.	D	26.	A	46.	C
7.	D	27.	D	47.	C
8.	B	28.	A	48.	A
9.	B	29.	B	49.	C
10.	B	30.	A	50.	D
11.	C	31.	C	51.	D
12.	A	32.	C	52.	B
13.	B	33.	B	53.	C
14.	D	34.	D	54.	D
15.	A	35.	D	55.	B
16.	B	36.	C	56.	D
17.	D	37.	D	57.	D
18.	C	38.	C	58.	C
19.	B	39.	B	59.	A
20.	A	40.	B	60.	D

Section I—Multiple-Choice Answers and Explanations

1. **B** During a phase change, $\Delta G = 0$. Plugging that into the $\Delta G = \Delta H - T\Delta S$ equation, you get $0 = \dfrac{23.9 \text{ kJ}}{\text{mol}} - (239 \text{ K})(\Delta S)$. Solving that yields $\Delta S = \dfrac{0.10 \text{ kJ}}{\text{mol} \times \text{K}}$, which converted to Joules is $\dfrac{100 \text{ J}}{\text{mol} \times \text{K}}$.

2. **A** As C_6H_6 has the higher boiling point, it must have stronger IMFs. This also means it has a lower vapor pressure. As C_6H_6 is completely nonpolar, the only way it could have stronger IMFs than C_2H_5OH is if the LDFs in C_6H_6 are significantly stronger than any IMFs present in C_2H_5OH. Stronger LDFs = a more polarizable electron cloud.

3. **B** For oxoacids, the more electronegative the halogen is, the stronger the acid is due to the fact that the electrons are pulled away from the O-H bond, causing the H^+ to ionize more frequently. So, any oxoacid containing Cl is a stronger acid than an analogous one containing Br. Between the two acids containing Cl, the more oxygens that are present, the smaller the magnitude of the partial negative charge is on each oxygen in the conjugate base. That means the oxygens in ClO_2^- are less likely than the oxygen in ClO^- to attract protons, which in turn means the protons are more likely to stay ionized in $HClO_2$, creating a stronger acid.

4. **D** From the graph, you can see that the most common isotope of zirconium has a mass number of 90. Mass number is equal to the number of protons plus the number of neutrons, and all zirconium atoms have 40 protons. $90 - 40 = 50$ neutrons.

5. **A** Gases deviate from ideal behavior when their IMFs are significant enough to affect their behavior. When IMFs cause gas molecules to attract each other, that means they will hit the sides of the container less often and decrease the pressure. In this case, H_2O has the strongest IMFs due to hydrogen bonding, and thus H_2O molecules are most likely to deviate from ideal behavior.

6. **D** Chromium goes from +6 to +3, meaning it gains electrons and is reduced. Fe goes from +2 to +3, meaning it loses electrons and it is oxidized.

7. **D** The chromium goes from an oxidation state of +6 to +3, and there are two chromium atoms. That means chromium gains a total of six electrons. The iron only loses one electron every time it goes from an oxidation state of +2 to +3, and thus it must happen six times for the charge to balance.

8. **B** When losing electrons, transition metals will lose their outermost s electrons first, followed by their outermost d electrons. The electron configuration of an iron atom is $[Ar]4s^23d^6$, so after losing three electrons it will be $[Ar]3d^5$.

9. **B** XeF_4 has a molecular geometry that is square planar, and because all of the terminal atoms are identical it will have no dipole moment. All of the other options have asymmetrical geometries.

10. **B** Ionic substances such as NaCl do not conduct electricity in their solid form, but do when melted and/or dissolved in water.

11. **C** Beer's law states that $A = abc$, where the small a is molar absorptivity. When plotting absorbance on the y-axis and concentration on the x-axis, the slope of the line would be equal to ab. As $b = 1.0$ cm, the slope of the line is equal to the molar absorptivity constant. Use data from two points: $\dfrac{(0.25 - 0.13)}{(0.050 - 0.025)} = \dfrac{1.2 \times 10^{-1}}{2.5 \times 10^{-2}} = 4.8 \ M^{-1}\cdot cm^{-1}$. Choice (C) is the only answer in the ballpark.

12. **A** Electron affinity is defined as a neutral atom gaining a single electron.

13. **B** $(0.10 \ M \ NaOH)(0.020 \ L) = 0.0020 \ \text{mol OH}^- \times \dfrac{1 \ \text{mol} \ H_2C_2O_4}{2 \ \text{mol NaOH}} = 0.0010 \ \text{mol} \ H_2C_2O_4$

    $\dfrac{0.0010 \ \text{mol} \ H_2C_2O_4}{0.020 \ L} = \dfrac{1.0 \times 10^{-3} \ \text{mol}}{2.0 \times 10^{-2} \ L} = 0.50 \times 10^{-1} \ M = 0.050 \ M.$

14. **D** At a volume of 20 mL, the OH^- has fully reacted with the $H_2C_2O_4$, removing all of its protons. Other than water, the only species present in significant concentrations at that point is the $C_2O_4^{2-}$.

15. **A** Below a pH of 9.1, the predominant form of the indicator will be its protonated state, HIn. Above 9.1, there will be more of the conjugate base In^-.

16. **B** The strength of the acid is irrelevant. H_2SO_4 has the same number of protons as $H_2C_2O_4$, and thus it will require the same amount of base to fully neutralize.

17. **D** The voltage of the cell is dependent on how close it is to equilibrium. The closer a cell is to equilibrium, the lower the voltage will become. Because all cells with a positive E_{cell} are favored, K is very large. In this case, $Q = \dfrac{[Fe^{2+}]}{[Ag^+]^2}$. If water is added, it will decrease the concentration of both Ag^+ and Fe^{2+}. However, doing so affects the value of the denominator more because it is squared. Consequently, Q will end up getting bigger, bringing the reaction closer to equilibrium and decreasing the overall voltage.

18. **C** Adding a catalyst reduces the activation energy for the reaction. This means a larger percentage of collisions will have enough energy to successfully react.

19. **B** The best way to solve this problem is by using an ICE chart. Start by entering the initial pressure values, as well as the change in each (determined by the coefficients in the balanced equation).

$$2 \ CO(g) + 2 \ NO(g) \rightarrow 2 \ CO_2(g) + N_2(g)$$

I	2.0	1.0	0	0
C	$-2x$	$-2x$	$+2x$	$+x$
E				

The NO limits, given that it reacts at the same rate as the CO, but started with half as much of it. So, $1.0 - 2x = 0$, and thus $x = 0.50$. Using that, finish the chart.

$$2 CO(g) + 2 NO(g) \rightarrow 2 CO_2(g) + N_2(g)$$

I	2.0	1.0	0	0
C	$-2x$	$-2x$	$+2x$	$+x$
E	1.0	0	1.0	0.50

$1.0 + 1.0 + 0.5 = 2.5$ atm

20. **A** The oxygen atom has a configuration of $1s^2 2s^2 2p^4$. The oxide ion would be $1s^2 2s^2 2p^6$. So, there will be two additional electrons in the $2p$ subshells, which corresponds with the rightmost peak. That peak thus grows taller.

21. **B** First, divide the mass of each element by its molar mass to convert to moles. $9.0/12.0 = 0.75$ mol of C, $2.0/1.0 = 2.0$ mol of H, and $4.0/16.0 = 0.25$ mol O. If you divide each number of moles by the lowest number (that of oxygen), you get carbon: $0.75/0.25 = 3$, hydrogen: $2.0/0.25 = 8$, and oxygen $0.25/0.25 = 1$. Those ratios become the subscripts for each element in the empirical formula.

22. **B** For acids, the hydrogen atoms that are directly bonded to an oxygen atom are dissociable. The only structure with that is B, where it actually happens twice. That also means B is likely a diprotic acid.

23. **D** Reactant molecules must collide with the correct orientation to support the breaking of their bonds, as well as the formation of new product bonds. One way that a catalyst can work is by creating more favorable orientations for the reaction to occur.

24. **B** $8.0 \text{ g Fe}_2O_3 \times \dfrac{1 \text{ mol Fe}_2O_3}{160 \text{ g}} \times \dfrac{4 \text{ mol Fe}}{2 \text{ mol Fe}_2O_3} \times \dfrac{55.8 \text{ g Fe}}{1 \text{ mol Fe}} = 5.58 \text{ g Fe}$

25. **B** Reduction means gaining electrons, and the only substance to do that is $X^{2+} \rightarrow X$, with the oxidation number changing from +2 to 0 (becoming more negative). Reduction reactions take place at the cathode.

26. **A** As this reaction occurs/is favored, K must be greater than 1 and ΔG must be negative.

27. **D** The role of the salt bridge is to make sure the cell remains electrically neutral. The X half-cell is becoming less positive (by losing X^{2+}), and so Na^+ would flow to that. The Z half-cell is becoming more positive (by gaining Z^+), so NO_3^- must flow to that.

28. **A** Conservation of mass is certainly a thing, but the only interaction between the half-cells is the flow of electrons. When X^{2+} gains two electrons, it turns into the solid X electrode. As the mass of the gained electrons is insignificant, the mass of the X electrode will increase as much as the mass of the X^{2+} in solution decreases.

29. **B**   Given that $Q = [Z^+]^2/[X^{2+}]$, doubling the concentration of both species will have a larger effect on the numerator, causing Q to increase. An increase in Q brings the cell closer to equilibrium, and thus closer to zero volts.

30. **A**   Any hydroxide salt will be more soluble in a lower-pH solution. This is because a lower pH means more H^+ ions to react with the OH^- in the salt, causing more of it to dissolve.

31. **C**   If the solvent is nonpolar, the most polar substance will travel the shortest distance, and the most nonpolar substance will travel the furthest ("like dissolves like").

32. **C**   For the central atom in a molecule to have more than eight electrons, it must have empty d-orbitals available for bonding. No elements in energy level 1 or 2 (such as He or Ne) have d-orbitals, and thus they cannot expand their octet.

33. **B**   The total pressure remains constant so the partial pressure of the other gases must decrease to accommodate the added partial pressure of helium. This will cause a shift to the side with more gas molecules—the right, in this reaction.

34. **D**   Via solubility rules, no compounds with NH_4^+, NO_3^-, or any alkali ion (such as Na^+) will be insoluble/present in a precipitate.

35. **D**   In both structures with only double bonds (A and C), the bond order is 2. In structure B, the bond order is 1.5 (3 bonds/2 locations), and in structure D it is 1.33 (4 bonds/3 locations). A lower bond order is equivalent to longer bonds.

36. **C**   Via Henderson-Hasselbalch, $pH = pK_a + \log [A^-]/[HA]$. In the diagrams, $[A^-] = [HA]$, so the pH of the solution will be equal to the pK_a of HBrO.

37. **D**   Using trials 1 and 2, we can see when A is constant and D doubles, the rate goes up by a factor of four. So, the reaction is second order with respect to D. Using trials 1 and 3, when D is constant and A doubles, the rate doubles. So, the reaction is first order with respect to A. Thus, rate = $k[A][D]^2$. Since A decreased but the rate stays the same from trial 3 to 4, D must have increased to exactly compensate for the loss of A. Since A decreased by a factor of four and is first order, D has to increase by a factor of 2 to compensate because it is second order. $2(0.10\ M) = 0.20\ M$.

38. **C**   The heat gained by the water is $q = (100.)(4.2)(1.0) = 420$ J. The metal must have lost the same amount of heat, so $-420 = (10.0)(c)(-84)$. $c = 0.50$ J/g·°C.

39. **B**   At thermal equilibrium, the average kinetic energy of the particles in both the metal and water will be identical. Energy transfer will continue to occur as water and metal particles collide, but the rate of transfer will be constant and the temperature will remain unchanged.

40. **B**   Specific heat measures how much energy is needed to change the speed of the particles in a substance. To increase the speed of the water molecules, the H-bonds between the molecules must be overcome, which requires a lot of energy. In the metal, there are no strong attractive forces between the various particles, so it is easier to change their speeds.

41. **A**   If some energy is absorbed by the cup, the water will experience an artificially low temperature change, which would lead to an artificially low calculated value for q and an artificially low value for the specific heat of the metal.

42. **D**   Phosphorus' $3p$ subshell electrons are each in a separate orbital due to Hund's Rule. Two of sulfur's $3p$ electrons are paired in the same orbital, and as they both have a negative charge that adds some extra repulsive forces between them.

43. **D**   The average atomic mass is a weighted average of all isotopes of an element. We can obtain the average atomic mass of sulfur from the Periodic Table, leading to the following calculation:

$$32.06 = 32(0.97) + x(0.03) \qquad x = 34$$

44. **B**   Electronegativity describes an atom's ability to attract electrons in a bond. As sulfur has the higher electronegativity value, it would attract the shared electrons more effectively, creating a partial negative charge.

45. **A**   Electron affinity is defined as the energy change when a gaseous atom of an element gains an electron.

46. **C**   Nitrite has $5 + 6(2) + 1 = 18$ valence electrons. The pair of resonance structures are the only option where the correct number of valence electrons are included.

47. **C**   Aqueous ionic solutions conduct electricity. None of the other options represent conductors as (A) and (D) are both covalent substances, and (B) is an ionic solid.

48. **A**   Partial pressure = Total pressure × mole fraction. In the diagram, Xe represent 3/10 particles, so its mole fraction is 0.3. Thus, 0.40 atm = P_{total} × 0.30 and P_{total} = 1.33 atm.

49. **C**   In $NaClO_3$, the oxidation state calculation is $+1 + Cl + (-2)3 = 0$, so $Cl = +5$. That is the highest out of all the options. Other Cl oxidation states: (A) = 0, (B) = −1, (D) = +3.

50. **D**   Cl starts as +4 in ClO_2, and reduces to +3 in ClO_2^- while also oxidizing to +5 in ClO_3^-. None of the other atoms change oxidation states.

51. **D**   Adding a catalyst can create an alternative pathway for a chemical reaction with a lower activation energy. Thus, the reaction proceeds more quickly.

52. **B** Between trial 1 and 2, the [ClO$_2$] halves while [OH$^-$] does not change. The rate decreases by a factor of four, which is consistent for a reaction that is second order with respect to ClO$_2$. Between trial 1 and 3 [ClO$_2$] doubles, which by itself would make the rate go up by a factor of four. However, [OH$^-$] gets cut in half and as a result the overall reaction rate only goes up by a factor of two. Logically this means that reducing [OH$^-$] by half would by itself cause the reaction rate to decrease by half, making the reaction first order with respect to OH$^-$.

53. **C** We start by writing out the equilibrium expression for the reaction and plugging in some variables.

$$K_c = \frac{[H_2S][I_2]}{[HI]^2} \qquad 2.3\times10^{-5} = \frac{(x)(x)}{(0.15-2x)^2}$$

As $K \ll 1$, we can assume any change in [HI] in the denominator would be insignificant.

$$2.3\times10^{-5} = \frac{x^2}{(0.15)^2} \qquad x = 7.2\times10^{-4}\ M = [H_2S]$$

54. **D** As there are seven circles of ICl and five circles of both Cl$_2$ and I$_2$, we can use ICl = $7x$ and Cl$_2$ = I$_2$ = $5x$ to describe their mole ratios. That would also be their partial pressure ratio, as the two are directly proportional. Thus, 1.70 atm = $7x + 5x + 5x$, and $x = 0.10$ atm. That means $P_{ICl} = 0.70$ atm and $P_{Cl_2} = P_{I_2} = 0.50$ atm. Plugging those into the K_p expression:

$$K_p = \frac{(P_{ICl})^2}{(P_{I_2})(P_{Cl_2})} = \frac{(0.70)^2}{(0.50)(0.50)} = 2.0$$

55. **B** The solution turning orange is caused by a shift to the right. Adding HCl will increase [H$^+$], and that will cause the reaction to shift away from H$^+$—so, to the right.

56. **D** The solution turning orange means a shift to the right. When heated, an endothermic equilibrium system will shift to the right. Additionally, temperature shifts change the value for K. By shifting to the right, there will be more products and less reactants when equilibrium re-establishes, causing an increase in K.

57. **D** Reducing [H$^+$] causes a shift to the left, which is synonymous with an increase in the reverse reaction rate.

58. **C** An indicator changes color when the pK_a of the indicator = pH of the solution. Thus, the pK_a of thymol blue is between 8 and 9. Given that pK_a + pK_b = 14 for any conjugate pair, that makes the pK_b of thymol blue's conjugate base between 5 and 6.

59. **A** First we need to find the amount of charge necessary to plate out the Cu.

$$0.100\ g\ Cu \times \frac{1\ mol\ Cu}{63.55\ g} \times \frac{2\ mol\ e^-}{1\ mol\ Cu} \times \frac{96,500\ C}{1\ mol\ e^-} = 304\ C$$

Then, we can use the current equation to figure out the necessary time.

$$0.500 \, \text{A} = \frac{304 \, \text{C}}{t} \qquad t = 608 \, \text{s}$$

60. **D**   We can find ΔH and ΔS for this reaction by doing products – reactants:

$$\Delta H = 30.9 \, \text{kJ/mol} - 0 \, \text{kJ/mol} = 30.9 \, \text{kJ/mol} \qquad \Delta S = 245.5 \, \text{J/mol·K} - 152.2 \, \text{J/mol·K} = 93.3 \, \text{J/mol·K} =$$
$$0.0933 \, \text{kJ/mol·K}$$

Then, at the point of phase change for any substance, $\Delta G = 0 \, \text{kJ/mol}$. Using $\Delta G = \Delta H - T\Delta S$:

$$0 \, \text{kJ/mol} = 30.9 \, \text{kJ/mol} - T(0.0933 \, \text{kJ/mol·K}) \qquad T = 331 \, \text{K}$$

Section II—Free-Response Answers and Explanations

1. (a)   (i and ii) The concentration of both ions would remain constant, but the mass of $PbCl_2$ on the bottom of the beaker would increase. As long as the temperature remains unchanged, K_{sp} will also stay the same. While some water evaporates and reduces the volume, some Pb^{2+} and Cl^- ions "fall" out of solution and combine to precipitate out as solid $PbCl_2$. Adding more $PbCl_2$ does not affect the equilibrium calculations, as it is a solid and does not appear in the K_{sp} equation.

    (b)   First, figure out the mass of KI required.

$$\text{M} = \frac{n}{v} \qquad 0.75 \, M = \frac{n}{0.100 \, \text{L}} \qquad n = 0.075 \, \text{mol KI} \times \frac{166 \, \text{g}}{1 \, \text{mol KI}} = 12.5 \, \text{g KI}$$

Mass out 12.5 g of KI in a weigh boat using the analytical balance. Transfer the KI to the 100 mL volumetric flask and fill it halfway with distilled water. Swirl the water until the KI dissolves, and then fill the flask to the mark. Cap the flask and invert it a few times to ensure thorough mixing.

    (c)   (i) $0.747 \, \text{g PbI}_2 \times \dfrac{1 \, \text{mol PbI}_2}{461 \, \text{g PbI}_2} \times \dfrac{1 \, \text{mol Pb}}{1 \, \text{mol PbI}_2} = 0.00162 \, \text{moles}$

          (ii) $\dfrac{1.62 \times 10^{-3} \, \text{mol Pb}}{0.100 \, \text{L}} = 0.0162 \, M$

          (iii) When the $PbCl_2$ solution is saturated, the reaction is $PbCl_2 \, (s) \rightleftharpoons Pb^{2+} + 2 \, Cl^-$.
                 So, for every one Pb^{2+} ion, there will be two Cl^- ions.
                 If $[Pb^{2+}] = 0.0162 \, M$, then $[Cl^-] = 0.0324 \, M$.
                 $K_{sp} = [Pb^{2+}][Cl^-]^2 = (0.0162 \, M)(0.0324 \, M)^2 = 1.70 \times 10^{-5}$

    (d)   If the PbI_2 precipitate were not fully dried, that would artificially increase the mass of precipitate. That in turn would give an artificially high mass of lead in the PbI_2, leading to a higher apparent concentration of lead in the saturated $PbCl_2$ solution. All of this would lead to an artificially high value for the K_{sp} of $PbCl_2$.

(e) Both salts have the same magnitude of charge, but Cl^- is smaller than I^-. Via Coulomb's Law, a smaller radius will lead to greater attraction and lattice energy, meaning the $PbCl_2$ will have the higher melting point.

2. (a) All of the carbon atoms have four charge clouds, which means a bond angle of 109.5° and a hybridization of sp^3.

(b) (i) Formic acid has LDFs, hydrogen-bonding, and permanent dipoles. Trimethylpentane has only LDFs.

(ii)

Oxygen is more electronegative than either carbon or hydrogen, so the negative partial charges go on the oxygen atoms.

(c) Even though trimethylpentane is fully nonpolar, it has a significantly higher number of electrons than formic acid. This means it will have a much more polarizable electron cloud and stronger LDFs, which must be stronger than any permanent-dipole attractions present in formic acid.

(d) As formic acid has a lower boiling point, that means it has weaker IMFs, and thus a higher vapor pressure.

(e) The formic acid is more polar, and thus will be more attracted to the polar eluent ("like dissolves like"). This will cause the formic acid to leave the column first. Remember that IMF strength is NOT the same thing as polarity, and only polarity matters when looking at chromatography!

3. (a) $0.500\ M \times 0.040\ L = 0.020$ mol $NaHCO_3 \times 84.01$ g/mol $= 1.7$ g

(b) To create a basic solution, the bicarbonate anion must accept a proton from water.
$HCO_3^-(aq) + H_2O(l) \rightleftharpoons H_2CO_3(aq) + OH^-(aq)$

(c) (i) $K_b = \dfrac{[H_2CO_3][OH^-]}{[HCO_3^-]}$

$$2.00 \times 10^8 = \frac{x^2}{(0.50-x)}$$

Note that due to the K_b value being so small, we can ignore the "-x" term in the denominator. Thus:

$x = [OH^-] = 1.0 \times 10^{-4}\ M$

$pOH = -\log(1.0 \times 10^{-4}) = 4.0$

$pH + 4.0 = 14.0$

$pH = 10.0$

(ii) $\%\text{ ionization} = \dfrac{\left[OH^-\right]}{\left[HCO_3^-\right]} \times 100\%$ $\dfrac{1.0 \times 10^{-4}\,M}{0.50\,M} \times 100\% = 0.020\%$

(iii) If $[HCO_3^-]$ decreases, that is due to a dilution of the solution. Anytime you dilute an aqueous equilibrium such as this one, it will shift to the side with more ions—so, to the right. This increases the overall $[OH^-]$ relative to $[HCO_3^-]$ once equilibrium reestablishes, which means the percent ionization will also increase.

(d) We start with 0.020 mol of HCO_3^- (see part a). Thus:

$$0.020 \text{ mol } HCO_3^- \times \frac{1 \text{ mol HCl}}{1 \text{ mol } HCO_3^-} = 0.020 \text{ mol HCl}$$

$$\frac{0.020 \text{ mol HCl}}{V} = 0.750\,M \qquad V = 0.0267 \text{ L} = 27 \text{ mL}$$

(e) At half-equivalence, the pOH of the solution will be equal to the pK_b of the HCO_3^- ion.
$-\log(2.00 \times 10^{-8}) = \text{pOH}$
$7.70 = \text{pOH}$
$\text{pH} = 14 - 7.70 = 6.30$

(f)

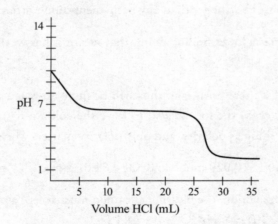

We know that prior to any HCl being added, the pH was 10. We also know that at the half-equivalence point ($V = 13.5$ mL), the pH should be 6.30. Beyond those two points, there is also a large drop in the equivalence region ($V = 24$–28 mL) because all of the original HCO_3^- has been neutralized and the pH shifts to being very acidic due to excess HCl.

4. (a) $c = \lambda v$ 3.0×10^8 m/s = $\lambda(1.86 \times 10^{15}$ s$^{-1})$ $\lambda = 1.61 \times 10^{-7}$ m $\times \dfrac{1.0 \times 10^9 \text{ nm}}{1.0 \text{ m}} = 161$ nm

(b) $E = hv$ $(6.626 \times 10^{-34}$ J·s$^{-1})(1.86 \times 10^{15}$ s$^{-1}) = 1.23 \times 10^{-18}$ J (for a single atom)

$$\frac{1.23 \times 10^{-18} \text{ J}}{1 \text{ atom}} \times \frac{1 \text{ kJ}}{1{,}000 \text{ J}} \times \frac{6.02 \times 10^{23} \text{ atoms}}{1 \text{ mol}} = 740 \text{ kJ/mol}$$

(c) A sodium atom is slightly larger than a magnesium atom due to the fact that there are fewer protons in the nucleus of sodium and both atoms have the same number of occupied principal energy levels. As the valence electrons in sodium are further from the nucleus, less energy is required to remove them. Since energy and frequency are directly proportional ($E = hv$), the frequency of light needed to ionize a magnesium atom would be greater than the frequency of light needed to ionize a sodium atom.

5. (a) The solid reactant becomes more dispersed when it converts into a liquid and gas. This increase in dispersion means ΔS will be positive.

(b) Using $\Delta G = \Delta H - T\Delta S$, if ΔS is positive the entire $-T\Delta S$ term is negative. Coupled with a negative ΔH, the value for ΔG will be negative at all temperatures, meaning the reaction is always favored.

(c) Just because a reaction is favored does not mean it will be fast! As this reaction consists of only a single reactant molecule, the most likely cause of a slow reaction is a high activation energy.

6. (a) There are two gases in the tube: the N_2O, and the water vapor that arises from the evaporation of the liquid water. The partial pressure of those two gases will equal the total atmospheric pressure.

$$P_{H_2O} = \frac{21 \text{ torr}}{760 \text{ torr} \cdot \text{atm}^{-1}} = 0.028 \text{ atm}$$

$P_T = P_{H_2O} + P_{N_2O}$

0.97 atm $= 0.028$ atm $+ P_{N_2O}$

$P_{N_2O} = 0.94$ atm

(b) We first need to calculate the moles of gas evolved.

$$0.27 \text{ g NH}_4\text{NO}_3 \times \frac{1 \text{ mol NH}_4\text{NO}_3}{80.05 \text{ g NH}_4\text{NO}_3} \times \frac{1 \text{ mol N}_2\text{O}}{1 \text{ mol NH}_4\text{NO}_3} = 0.0034 \text{ N}_2\text{O}$$

Then, using the ideal gas law:

$PV = nRT$

$(0.94$ atm$)(0.0798$ L$) = (0.0034$ mol$)R(296$ K$)$

$R = 0.0745$ L·atm/mol·K

(c) Using 0.0034 moles of N_2O in the calculation of R assumes that all the gas that was produced was collected over the water. If some of it dissolved instead, then this value would be artificially high. Additionally, the measured 0.0798 L would artificially low if some of the gas had dissolved; it would occupy more volume if none was lost in dissolution. Since $R = PV/nT$, an artificially high n and artificially low V will both contribute to an artificially low calculated value of R.

7. (a) Given the rate law, we can perform a dimensional analysis to determine the units on k.

Rate = $k[NH_4CNO]^2$

$M/s = k(M)^2$

$k = M^{-1} \cdot s^{-1}$

(b) Adding a catalyst or increasing the temperature at which the reaction takes place will increase the value of the rate constant. Adding a catalyst lowers the energy needed for the reaction to occur (activation energy). Increasing the temperature increases the average kinetic energy of the particles. In either case, there will be a greater number of effective collisions, ones with sufficient energy for the reaction to occur. Thus, the rate constant will increase.

(c) As this is a second-order reaction, the integrated rate law is:

$$\frac{1}{[NH_4CNO]_t} = kt + \frac{1}{[NH_4CNO]_0}$$

$$\frac{1}{[NH_4CNO]_{250}} = (9.81 \times 10^{-4}\ M^{-1} \cdot s^{-1})(250.\ s) + \frac{1}{0.350\ M}$$

$$\frac{1}{[NH_4CNO]_{250}} = 0.245\ M^{-1} + 2.86\ M^{-1}$$

$$[NH_4NCO]_{250} = 0.322\ M$$

Part VII
Additional Practice Tests

Practice Test 4

The Exam

AP® Chemistry Exam

SECTION I: Multiple-Choice Questions

DO NOT OPEN THIS BOOKLET UNTIL YOU ARE TOLD TO DO SO.

At a Glance

Total Time
1 hour and 30 minutes
Number of Questions
60
Percent of Total Grade
50%
Writing Instrument
Pencil required

Instructions

Section I of this examination contains 60 multiple-choice questions. Fill in only the ovals for numbers 1 through 60 on your answer sheet.

Indicate all of your answers to the multiple-choice questions on the answer sheet. No credit will be given for anything written in this exam booklet, but you may use the booklet for notes or scratch work. After you have decided which of the suggested answers is best, completely fill in the corresponding oval on the answer sheet. Give only one answer to each question. If you change an answer, be sure that the previous mark is erased completely. Here is a sample question and answer.

Sample Question   Sample Answer

Chicago is a
(A) state
(B) city
(C) country
(D) continent

Use your time effectively, working as quickly as you can without losing accuracy. Do not spend too much time on any one question. Go on to other questions and come back to the ones you have not answered if you have time. It is not expected that everyone will know the answers to all the multiple-choice questions.

About Guessing

Many candidates wonder whether or not to guess the answers to questions about which they are not certain. Multiple-choice scores are based on the number of questions answered correctly. Points are not deducted for incorrect answers, and no points are awarded for unanswered questions. Because points are not deducted for incorrect answers, you are encouraged to answer all multiple-choice questions. On any questions you do not know the answer to, you should eliminate as many choices as you can, and then select the best answer among the remaining choices.

GO ON TO THE NEXT PAGE.

CHEMISTRY
SECTION I
Time—1 hour and 30 minutes

INFORMATION IN THE TABLE BELOW AND ON THE FOLLOWING PAGES MAY BE USEFUL IN ANSWERING THE QUESTIONS IN THIS SECTION OF THE EXAMINATION

DO NOT DETACH FROM BOOK.

PERIODIC TABLE OF THE ELEMENTS

1	2	3	4	5	6	7	8	9	10	11	12	13	14	15	16	17	18
1 **H** 1.008																	2 **He** 4.00
3 **Li** 6.94	4 **Be** 9.01											5 **B** 10.81	6 **C** 12.01	7 **N** 14.01	8 **O** 16.00	9 **F** 19.00	10 **Ne** 20.18
11 **Na** 22.99	12 **Mg** 24.30											13 **Al** 26.98	14 **Si** 28.09	15 **P** 30.97	16 **S** 32.06	17 **Cl** 35.45	18 **Ar** 39.95
19 **K** 39.10	20 **Ca** 40.08	21 **Sc** 44.69	22 **Ti** 47.87	23 **V** 50.94	24 **Cr** 52.00	25 **Mn** 54.94	26 **Fe** 55.85	27 **Co** 58.93	28 **Ni** 58.69	29 **Cu** 63.55	30 **Zn** 65.38	31 **Ga** 69.72	32 **Ge** 72.63	33 **As** 74.92	34 **Se** 78.97	35 **Br** 79.90	36 **Kr** 83.80
37 **Rb** 85.47	38 **Sr** 87.62	39 **Y** 88.91	40 **Zr** 91.22	41 **Nb** 92.91	42 **Mo** 95.95	43 **Tc**	44 **Ru** 101.07	45 **Rh** 102.91	46 **Pd** 106.42	47 **Ag** 107.87	48 **Cd** 112.41	49 **In** 114.82	50 **Sn** 118.71	51 **Sb** 121.76	52 **Te** 127.60	53 **I** 126.90	54 **Xe** 131.29
55 **Cs** 132.91	56 **Ba** 137.33	57-71 *	72 **Hf** 178.49	73 **Ta** 180.95	74 **W** 183.94	75 **Re** 186.21	76 **Os** 190.23	77 **Ir** 192.22	78 **Pt** 195.08	79 **Au** 196.97	80 **Hg** 200.59	81 **Tl** 204.38	82 **Pb** 207.2	83 **Bi** 208.98	84 **Po**	85 **At**	86 **Rn**
87 **Fr**	88 **Ra**	89-103 †	104 **Rf**	105 **Db**	106 **Sg**	107 **Bh**	108 **Hs**	109 **Mt**	110 **Ds**	111 **Rg**	112 **Cn**	113 **Nh**	114 **Fl**	115 **Mc**	116 **Lv**	117 **Ts**	118 **Og**

*Lanthanoids

57 **La** 138.91	58 **Ce** 140.12	59 **Pr** 140.91	60 **Nd** 144.24	61 **Pm**	62 **Sm** 150.36	63 **Eu** 151.97	64 **Gd** 157.25	65 **Tb** 158.93	66 **Dy** 162.50	67 **Ho** 164.93	68 **Er** 167.26	69 **Tm** 168.93	70 **Yb** 173.05	71 **Lu** 174.97

†Actinoids

89 **Ac**	90 **Th** 232.04	91 **Pa** 231.04	92 **U** 238.03	93 **Np**	94 **Pu**	95 **Am**	96 **Cm**	97 **Bk**	98 **Cf**	99 **Es**	100 **Fm**	101 **Md**	102 **No**	103 **Lr**

GO ON TO THE NEXT PAGE.

AP® CHEMISTRY EQUATIONS & CONSTANTS

Throughout the exam the following symbols have the definitions specified unless otherwise noted.

L, mL	= liter(s), milliliter(s)	mm Hg	= millimeters of mercury
g	= gram(s)	J, kJ	= joule(s), kilojoule(s)
nm	= nanometer(s)	V	= volt(s)
atm	= atmosphere(s)	mol	= mole(s)

ATOMIC STRUCTURE

$E = h\nu$

$c = \lambda\nu$

E = energy

ν = frequency

λ = wavelength

Planck's constant, $h = 6.626 \times 10^{-34}$ J s

Speed of light, $c = 2.998 \times 10^8$ m s^{-1}

Avogadro's number $= 6.022 \times 10^{23}$ mol^{-1}

Electron charge, $e = -1.602 \times 10^{-19}$ coulomb

EQUILIBRIUM

$K_c = \dfrac{[C]^c[D]^d}{[A]^a[B]^b}$, where $a\,A + b\,B \rightleftarrows c\,C + d\,D$

$K_p = \dfrac{(P_C)^c(P_D)^d}{(P_A)^a(P_B)^b}$

$K_a = \dfrac{[H^+][A^-]}{[HA]}$

$K_b = \dfrac{[OH^-][HB^+]}{[B]}$

$K_w = [H^+][OH^-] = 1.0 \times 10^{-14}$ at 25°C

$\quad = K_a \times K_b$

$pH = -\log[H^+]$, $pOH = -\log[OH^-]$

$14 = pH + pOH$

$pH = pK_a + \log\dfrac{[A^-]}{[HA]}$

$pK_a = -\log K_a$, $pK_b = -\log K_b$

Equilibrium Constants

K_c (molar concentrations)

K_p (gas pressures)

K_a (weak acid)

K_b (weak base)

K_w (water)

KINETICS

$[A]_t - [A]_0 = -kt$

$\ln[A]_t - \ln[A]_0 = -kt$

$\dfrac{1}{[A]_t} - \dfrac{1}{[A]_0} = kt$

$t_{1/2} = \dfrac{0.693}{k}$

k = rate constant

t = time

$t_{1/2}$ = half-life

GO ON TO THE NEXT PAGE.

GASES, LIQUIDS, AND SOLUTIONS

$$PV = nRT$$

$$P_A = P_{total} \times X_A, \text{ where } X_A = \frac{\text{moles A}}{\text{total moles}}$$

$$P_{total} = P_A + P_B + P_C + \ldots$$

$$n = \frac{m}{M}$$

$$K = {}^{\circ}C + 273$$

$$D = \frac{m}{V}$$

$$KE_{molecule} = \frac{1}{2}mv^2$$

Molarity, M = moles of solute per liter of solution

$$A = \varepsilon bc$$

P = pressure
V = volume
T = temperature
n = number of moles
m = mass
M = molar mass
D = density
KE = kinetic energy
v = velocity
A = absorbance
ε = molar absorptivity
b = path length
c = concentration

Gas constant, R = 8.314 J mol^{-1}K^{-1}
$\quad$ = 0.08206 L atm mol^{-1}K^{-1}
$\quad$ = 62.36 L torr mol^{-1}K^{-1}
1 atm = 760 mm Hg = 760 torr
STP = 273.15 K and 1.0 atm
Ideal gas at STP = 22.4 L mol^{-1}

THERMOCHEMISTRY/ ELECTROCHEMISTRY

$$q = mc\Delta T$$

$$\Delta S^{\circ} = \sum S^{\circ} \text{ products} - \sum S^{\circ} \text{ reactants}$$

$$\Delta H^{\circ} = \sum \Delta H_f^{\circ} \text{ products} - \sum \Delta H_f^{\circ} \text{ reactants}$$

$$\Delta G^{\circ} = \sum \Delta G_f^{\circ} \text{ products} - \sum \Delta G_f^{\circ} \text{ reactants}$$

$$\Delta G^{\circ} = \Delta H^{\circ} - T\Delta S^{\circ}$$

$$= -RT \ln K$$

$$= -n F E^{\circ}$$

$$I = \frac{q}{t}$$

$$E_{cell} = E_{cell}^{\circ} - \frac{RT}{nF} \ln Q$$

q = heat
m = mass
c = specific heat capacity
T = temperature
S° = standard entropy
H° = standard enthalpy
G° = standard Gibbs free energy
n = number of moles
E° = standard reduction potential
I = current (amperes)
q = charge (coulombs)
t = time (seconds)
Q = reaction quotient

Faraday's constant, F = 96,485 coulombs per mole
of electrons

$$1 \text{ volt} = \frac{1 \text{ joule}}{1 \text{ coulomb}}$$

GO ON TO THE NEXT PAGE.

$$2 \text{ Ag}^+(aq) + \text{CO}_3^{2-}(aq) \rightarrow \text{Ag}_2\text{CO}_3(s)$$

1. A student mixes equimolar amounts of silver nitrate and sodium carbonate in a beaker, causing the above reaction to occur. Which of the following steps could the student take to make more precipitate form?

   (A) Adding more silver nitrate
   (B) Adding more sodium carbonate
   (C) Heating the solution
   (D) No more precipitate can be formed as the reaction is already complete

2. Approximately how many grams of oxygen are present in a 10.0 g sample of NaOH?

   (A) 2.00 g
   (B) 4.00 g
   (C) 6.00 g
   (D) 8.00 g

$$2 \text{ Li}(s) + 2 \text{ H}_2\text{O}(l) \rightarrow 2 \text{ Li}^+(aq) + \text{H}_2(g) + 2 \text{ OH}^-(aq)$$

3. A 3.50 g sample of lithium metal is dropped into a beaker of water, causing the above reaction to occur. How many liters of hydrogen gas are produced if this reaction takes place at STP?

   (A)   5.60 L
   (B)   11.2 L
   (C)   22.4 L
   (D)   28.0 L

Metal Initial Temperature	98.0°C
Water Initial Temperature	26.0°C
Water + Metal Final Temperature	28.0°C
Mass Metal	10.0 g
Mass Water	100. g

4. A piece of an unknown metal is heated to a high temperature, and then dropped into a cup of water. Given the data above and assuming no heat is lost to the environment, what is the approximate specific heat of the metal? (Water has a specific heat of 4.2 J/g°C.)

   (A)   0.60 J/g°C
   (B)   0.80 J/g°C
   (C)   1.00 J/g°C
   (D)   1.20 J/g°C

$$\text{Fe}_2\text{O}_3(s) + 3 \text{ H}_2(g) \rightleftharpoons 3 \text{ H}_2\text{O}(l) + 2 \text{ Fe}(s)$$

5. What would the equilibrium constant expression, K_c, be for the above reaction?

   (A) $K_c = \dfrac{[\text{Fe}]^2[\text{H}_2\text{O}]^3}{[\text{Fe}_2\text{O}_3][\text{H}_2]^3}$

   (B) $K_c = \dfrac{[\text{H}_2\text{O}]^3}{[\text{H}_2]^3}$

   (C) $K_c = \dfrac{[\text{H}_2]^3}{[\text{H}_2\text{O}]^3}$

   (D) $K_c = \dfrac{1}{[\text{H}_2]^3}$

GO ON TO THE NEXT PAGE.

Questions 6–10 refer to the following standard reduction potentials.

Half Reaction	Reduction Potential
$O_2(g) + 4\,H^+(aq) + 4\,e^- \rightarrow 2\,H_2O(l)$	1.23 V
$Ag^+(aq) + e^- \rightarrow Ag(s)$	0.80 V
$Cu^{2+}(aq) + 2\,e^- \rightarrow Cu(s)$	0.34 V
$Fe^{2+}(aq) + 2\,e^- \rightarrow Fe(s)$	−0.44 V
$2\,H_2O(l) + 2\,e^- \rightarrow H_2(g) + 2\,OH^-$	−0.83 V
$Al^{3+}(aq) + 3\,e^- \rightarrow Al(s)$	−1.66 V

6. What would the cell potential be for a galvanic cell constructed of Ag/Ag^+ and Fe/Fe^{2+} half-cells?

   (A) 0.36 V
   (B) 1.16 V
   (C) 1.24 V
   (D) 2.04 V

7. What would the net ionic reaction be for the galvanic cell described in question #6?

   (A) $2\,Ag^+(aq) + Fe(s) \rightarrow 2\,Ag(s) + Fe^{2+}(aq)$
   (B) $Fe^{2+}(aq) + 2\,Ag(s) \rightarrow Fe(s) + 2\,Ag^+(aq)$
   (C) $Ag^+(aq) + Fe^{2+}(aq) \rightarrow Ag(s) + Fe(s)$
   (D) $2\,Ag(s) + Fe(s) \rightarrow 2\,Ag^+(aq) + Fe^{2+}(aq)$

8. A 2.0 A current is applied to beakers containing identical volumes of each of the following 1.0 M solutions. In which beaker will the greatest mass of metal plate out?

   (A) $AgNO_3(aq)$
   (B) $CuSO_4(aq)$
   (C) $FeCl_2(aq)$
   (D) $AlBr_3(aq)$

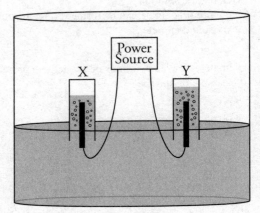

9. A current is run through a solution of pure water, and gases are collected as shown in the above diagram. Which of the following correctly identifies the cathode and anode during the hydrolysis, along with identifying the gas present at each?

	Electrode X Identity	Gas collected at Electrode X
(A)	Cathode	Hydrogen
(B)	Anode	Hydrogen
(C)	Cathode	Oxygen
(D)	Anode	Oxygen

10. Which of the following would NOT cause a reaction to occur?

   (A) Adding some aluminum pellets to a solution containing Fe^{2+} ions
   (B) Placing a copper strip into a solution containing Ag^+ ions
   (C) Adding powdered iron to a solution containing Cu^{2+} ions
   (D) Dropping a silver coin into a solution containing Al^{3+} ions

GO ON TO THE NEXT PAGE.

Use the following information to answer questions 11–13.

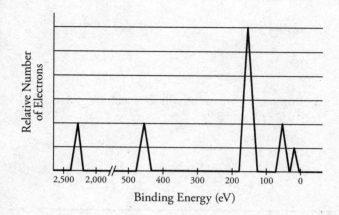

Peak 1	Peak 2	Peak 3	Peak 4	Peak 5
2300 eV	450 eV	150 eV	30 eV	5.0 eV

The photoelectron spectrum for a neutral aluminum atom is located above.

11. The amount of energy necessary to remove an electron from the $2p$ subshell is closest to which value?

    (A) 450 eV
    (B) 150 eV
    (C) 30 eV
    (D) 5.0 eV

12. On a spectrum of an aluminum ion

    (A) all peaks would be identical
    (B) the peak furthest to the right would be twice as tall
    (C) the two peaks furthest to the right would be missing
    (D) all peaks would be half as tall

13. A different aluminum atom is exposed to incoming radiation with an energy of 200 eV. Ejected electrons that were originally in which orbital would have the lowest kinetic energy?

    (A) $1s$
    (B) $2s$
    (C) $2p$
    (D) $3p$

14. A certain process is favored at 250 K, but becomes thermodynamically unfavored at 500 K. Which of the following accurately predicts the signs for ΔH and ΔS for the process?

	ΔH	ΔS
(A)	+	+
(B)	+	−
(C)	−	+
(D)	−	−

15. Which of the following orbital diagrams shows the correct electron configuration for an oxygen atom?

    (A)
    $2p$ [↑↓] [↑] [↑]
    $2s$ [↑↓]
    $1s$ [↑↓]

    (C)
    $2p$ [↑] [↑] [↑] [↑]
    $2s$ [↑↓]
    $1s$ [↑↓]

    (B)
    $2p$ [↑↓] [↑↓] [ ]
    $2s$ [↑↓]
    $1s$ [↑↓]

    (D)
    $2p$ [↑↓] [↑↓] [↑↓]
    $2s$ [ ]
    $1s$ [↑↓]

$$C_xH_y(g) + O_2(g) \rightarrow CO_2(g) + H_2O(g)$$

16. A hydrocarbon of an unknown formula consisting of only hydrogen and carbon is combusted in excess oxygen, producing 44.0 g of carbon dioxide and 18.0 g of water via the above **unbalanced** reaction. Which of the following is a potential formula for the hydrocarbon?

    (A) CH_4
    (B) C_2H_3
    (C) C_2H_4
    (D) C_3H_5

GO ON TO THE NEXT PAGE.

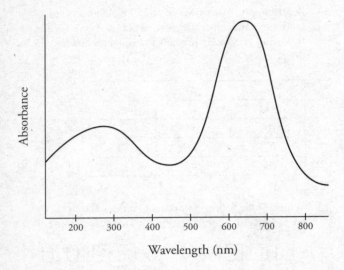

Wavelength (nm)

17. Cu^{2+} ions appear blue in solution, and the NO_3^- ion is colorless. The absorbance of a $Cu(NO_3)_2$ solution is measured at various wavelengths using a spectrophotometer. To determine the concentration of a $Cu(NO_3)_2$ solution of unknown concentration, what wavelength of light would produce the best results?

(A) 350 nm
(B) 500 nm
(C) 640 nm
(D) 800 nm

Use the following information to answer questions 18–20.

$$Cr_2O_7^{2-}(aq) + 3\ Sn(s) + 14\ H^+(aq) \rightarrow$$
$$2\ Cr^{3+}(aq) + 7\ H_2O(l) + 3\ Sn^{2+}(aq)$$

A solution of potassium dichromate, $K_2Cr_2O_7$, is poured over a piece of solid tin, causing the oxidation-reduction reaction above to occur.

18. Which substance is oxidized, and which is reduced?

	Oxidized	Reduced
(A)	Cr_2O_7	Sn
(B)	H^+	Cr_2O_7
(C)	Sn	Cr_2O_7
(D)	Cr_2O_7	H^+

19. How many moles of electrons are transferred in the balanced equation?

(A) 2
(B) 3
(C) 4
(D) 6

20. How many grams of tin will react with 100. mL of 0.033 $M\ K_2Cr_2O_7$?

(A) 0.40 g
(B) 1.2 g
(C) 2.4 g
(D) 3.6 g

Salt	$\Delta H_{soln}°$ (kJ/mol)
$LiClO_4$	−26.55
$NaClO_4$	13.88
$KClO_4$	41.38

21. The enthalpy of solution for three different perchlorate compounds is listed in the table above. Based on the data, which of the following statements is correct?

(A) As cation size increases, the rate at which the lattice energy is changing exceeds the rate at which the hydration energy is changing.
(B) The smaller the cation, the smaller the magnitude of the lattice energy will be.
(C) As cation size increases, the magnitude of the temperature change that occurs in the surrounding solution consistently increases.
(D) Of the three salts, only lithium perchlorate is fully soluble in water, which is demonstrated by its negative enthalpy of solution.

$$Au^+(aq) + 2\ CN^-(aq) \rightleftharpoons Au(CN)_2^-(aq) \qquad \Delta H° = 242\ kJ/mol$$

22. The above system is at equilibrium. Which of the following would cause an immediate increase in the rate of the reverse reaction?

(A) Adding some water
(B) Adding some $AuNO_3(aq)$
(C) Increasing the pressure
(D) Increasing the temperature

GO ON TO THE NEXT PAGE.

$$2\,Ag^+(aq) + Cu(s) \rightarrow 2\,Ag(s) + Cu^{2+}(aq) \qquad E_{cell} = +0.56\ V$$

23. A silver/copper galvanic cell is set up, and the above reaction occurs. If all other variables were to remain the same, what is one way to increase the cell potential?

    (A) Increase the mass of the copper electrode
    (B) Increase the mass of the silver electrode
    (C) Increase $[Ag^+]$
    (D) Increase $[Cu^{2+}]$

Use the following information to answer questions 24–27.

Titration of HN$_3$ with NaOH

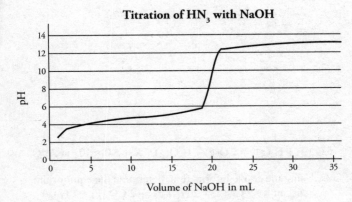

Volume of NaOH in mL

$$HN_3(aq) + OH^-(aq) \rightarrow N^{3-}(aq) + H_2O(l)$$

25.0 mL of hydrazoic acid, HN$_3$ ($K_a = 1.9 \times 10^{-5}$), is titrated with 0.50 M NaOH. The pH is measured during the titration, producing the above graph. The equivalence point is reached at a volume of 20.0 mL of NaOH added.

24. What is the concentration of the HN$_3$?

    (A) 0.40 M
    (B) 0.50 M
    (C) 0.60 M
    (D) 0.80 M

25. At what volume of NaOH added would $[N^{3-}] = [HN_3]$?

    (A) 5.0 mL
    (B) 10.0 mL
    (C) 20.0 mL
    (D) 25.0 mL

26. Which indicator would be the best to use to determine the endpoint of the titration?

	Indicator	pK_a
(A)	Methyl Violet	0.80
(B)	Congo Red	4.0
(C)	Thymol Blue	8.9
(D)	Indigo Carmine	12.2

27. If the hydrazoic acid were replaced with hydrocyanic acid ($K_a = 6.2 \times 10^{-10}$) of identical concentration, which of the following changes would occur?

    (A) Less NaOH would be required to reach the titration endpoint.
    (B) After NaOH is initially added, the change in pH will be greater than that of the original titration.
    (C) The pH of the solution at the 30.0 mL mark would be higher.
    (D) The pH at the equivalence point would be decreased.

Ion	Ionization energy (kJ/mol)
Al^{3+}	11,500
Mg^{2+}	7,732
Na$^+$	4,562

28. The next ionization energy values for different ions of three period 3 metals are listed above. The reason that the aluminum ion has the greatest next ionization energy is

    (A) aluminum metal is less chemically active than magnesium or sodium
    (B) the radius of the aluminum ion is the smallest
    (C) the aluminum ion has the greatest number of unpaired electrons
    (D) the aluminum ion experiences the smallest amount of shielding

GO ON TO THE NEXT PAGE.

$$NaCl(s) \rightarrow Na^+(aq) + Cl^-(aq)$$

29. A sample of sodium chloride, NaCl, is dissolved fully in water. Which diagram below is an accurate representation of how the solution would look on the particulate level?

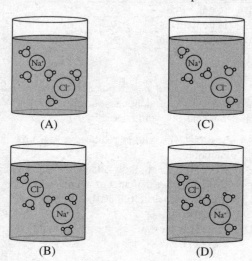

(A)

(C)

(B)

(D)

30. Which of the following expressions would accurately give the magnitude for the density of a sample of helium gas at STP?

(A) $\dfrac{(4.0)}{(0.0821)(273)}$

(B) $\dfrac{(273)(0.0821)}{4.0}$

(C) $\dfrac{(273)(4.0)}{(0.0821)}$

(D) $(4.0)(273)(0.0821)$

31. In which of the four molecules below is the bond angle between the terminal fluorine atoms and the central atom the smallest?

$$:\!\overset{\displaystyle ..}{\underset{\displaystyle ..}{F}}\!:$$

(A) $:\!\ddot{F}\!-\!C\!-\!\ddot{F}\!:$ with $:\ddot{F}:$ above and below central C

(C) B bonded to three F atoms

(B) $:\!\ddot{F}\!-\!\ddot{N}\!-\!\ddot{F}\!:$ with $:\ddot{F}:$ below N

(D) O bonded to two F atoms

$$CO_2(g) + H_2(g) \rightleftharpoons CO(g) + H_2O(g) \quad K = 5.0 \times 10^{-3} \text{ at } 25^\circ C$$

32. Using the above information, determine the equilibrium constant for $2\,CO(g) + 2\,H_2O(g) \rightleftharpoons 2\,CO_2(g) + 2\,H_2(g)$.

(A) -1.0×10^{-4}
(B) 2.0×10^3
(C) 1.0×10^4
(D) 4.0×10^4

GO ON TO THE NEXT PAGE.

Use the following information to answer questions 33–37.

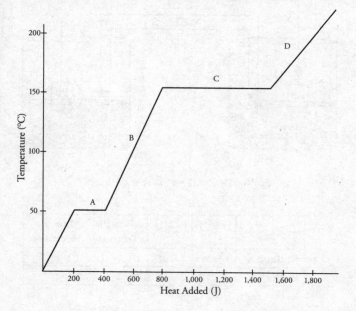

A 2.0 g sample of a solid covalent substance is placed inside an enclosed chamber, and heat is added at a constant rate. The temperature of the substance is tracked as heat is added, producing the above graph.

33. At which point on the graph will the substance exist in both its solid and liquid phase simultaneously?

(A) Point A
(B) Point B
(C) Point C
(D) Point D

34. At which point(s) on the graph will intermolecular forces be the weakest?

(A) Point A only
(B) Points B and D
(C) Point D only
(D) Intermolecular forces are present at all times.

35. Which of the following statements is true, according to the graph?

(A) The intramolecular forces are strongest during phase A.
(B) The heat of vaporization exceeds the heat of fusion.
(C) The strength of the intermolecular forces is unaffected by the addition of energy.
(D) The average molecular speed increases throughout the heating process.

36. How much energy would be required to melt a separate 1.0 g sample of the substance?

(A) 100 J
(B) 200 J
(C) 300 J
(D) 400 J

37. What is the approximate specific heat of the substance in its liquid phase?

(A) 2.0 J/g°C
(B) 4.0 J/g°C
(C) 6.0 J/g°C
(D) 8.0 J/g°C

GO ON TO THE NEXT PAGE.

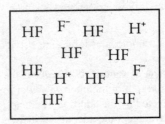

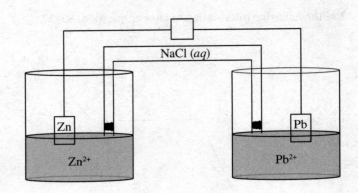

$$HF(aq) \rightleftharpoons H^+(aq) + F^-(aq)$$

38. HF is a weak acid with a K_a value of 7.2×10^{-4}. The box above represents a particulate-level view of an HF solution of known concentration. If some water is added, which of the following options would represent a particulate representation of the diluted solution?

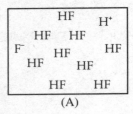

(A)   (C)

(B)   (D)

$$Pb^{2+}(aq) + Zn(s) \rightarrow Pb(s) + Zn^{2+}(aq)$$

	$[Pb^{2+}]$	$[Zn^{2+}]$	Mass Pb electrode	Mass Zn electrode
Battery X	1.0 M	1.0 M	10. g	10. g
Battery Y	2.0 M	2.0 M	20. g	20. g

39. Two batteries are constructed via the above diagram. The concentration of the solutions and the mass of the electrodes are identified in the data table below the diagram. How would the voltage and the battery life for the two batteries compare?

	Voltage	Battery Life
(A)	X < Y	X < Y
(B)	X < Y	X = Y
(C)	X = Y	X = Y
(D)	X = Y	X < Y

40. Which of the four substances listed below would have the highest melting point?

(A) KCl
(B) $CaBr_2$
(C) MgO
(D) $C_6H_{12}O_6$

GO ON TO THE NEXT PAGE.

Use the following information to answer questions 41–43.

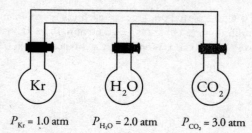

$P_{Kr} = 1.0$ atm $\quad P_{H_2O} = 2.0$ atm $\quad P_{CO_2} = 3.0$ atm

Three identical 1.0 L containers filled with different gases are connected using glass tubes. The stopcocks at the top of each flask are initially closed and all three containers are held at 125°C.

41. Which gas particles are moving the fastest on average?

    (A) Kr
    (B) H_2O
    (C) CO_2
    (D) All gas particles are moving at the same speed.

42. Assuming the volume of the connecting tubes is negligible and that temperature remains constant, what would the total pressure of the system be after all three stopcocks are opened and the gases fully mix?

    (A) 2.0 atm
    (B) 4.0 atm
    (C) 6.0 atm
    (D) 8.0 atm

43. After the stopcocks have been opened, the entire system is heated. Choose the graph below that correctly shows how the pressure would change as the temperature is increased.

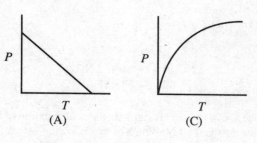

(A)

(C)

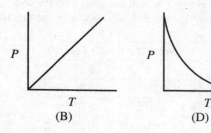

(B)

(D)

GO ON TO THE NEXT PAGE.

44. A 1.0 M solution of which of the following would have the highest conductivity?

(A) $C_2H_5OH(aq)$
(B) $NH_4Cl(aq)$
(C) $PF_3(aq)$
(D) $Na_2CO_3(aq)$

$$2 NO(g) + O_2(g) \rightarrow 2 NO_2(g)$$
$$rate = k[NO]^2[O_2]$$

45. The above reaction is run with the concentration of both reactants at 0.10 M. Which of the following values for the initial concentration of both reactants would lead to an initial reaction rate which is double that of the first trial?

	[NO] (M)	[O_2] (M)
(A)	0.15	0.15
(B)	0.20	0.20
(C)	0.20	0.050
(D)	0.050	0.20

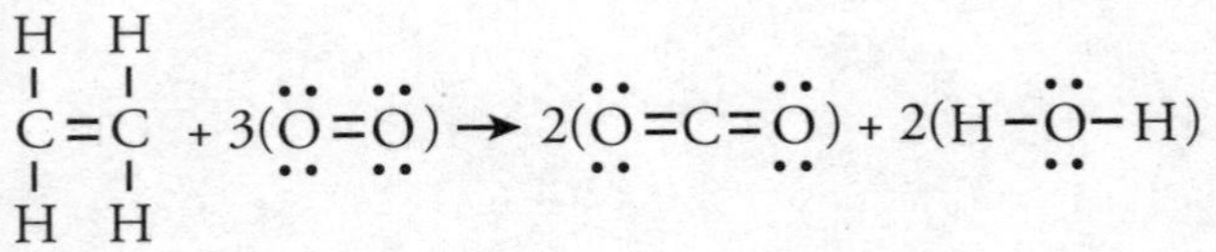

$$\underset{\substack{|\ \ |\\ H\ H}}{\overset{\substack{H\ H\\|\ \ |}}{C=C}} + 3(\ddot{O}=\ddot{O}) \rightarrow 2(\ddot{O}=C=\ddot{O}) + 2(H-\ddot{O}-H)$$

Bond	Enthalpy (kJ/mol)	Bond	Enthalpy (kJ/mol)
C–H	410	C=O	800
C=C	720	H–O	470
O=O	500		

46. Given the bond enthalpy values in the data table above, determine the enthalpy change for the reaction above it.

(A) –1,220 kJ/mol
(B) –400 kJ/mol
(C) 400 kJ/mol
(D) 1,220 kJ/mol

$$2 NO_2(g) \rightleftharpoons 2 NO(g) + O_2(g)$$

47. At 0°C, the partial pressures of each gas above in an equilibrium system are $P_{NO_2} = 0.60$ atm, $P_{NO} = 0.30$ atm, and $P_{O_2} = 0.20$ atm. What is K_p for this reaction at 0°C?

(A) 0.05
(B) 0.10
(C) 10.
(D) 20.

Use the following information to answer questions 48–50.

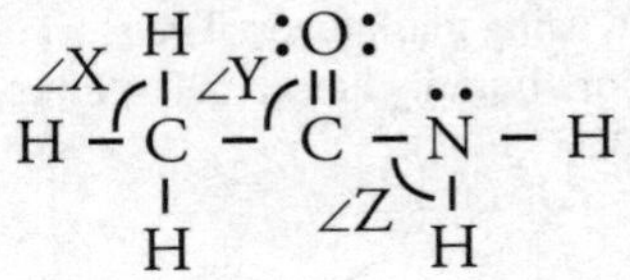

A Lewis diagram of the acetamide molecule, CH_3CONH_2, is drawn above. Note the bond angles are not drawn to scale.

48. Identify the bond angles on the diagram from greatest to smallest.

(A) X > Y > Z
(B) X > Z > Y
(C) Z > X > Y
(D) Y > X > Z

49. How many π bonds are present in an acetamide molecule?

(A) 0
(B) 1
(C) 2
(D) 3

50. What is the strongest type of intermolecular force present between acetamide molecules in an aqueous solution?

(A) London dispersion forces
(B) Hydrogen bonding
(C) Ionic bonding
(D) Covalent bonding

GO ON TO THE NEXT PAGE.

$$CO(g) + 2\,H_2(g) \rightleftharpoons CH_3OH(g) \qquad K_c = 14.5 \text{ @ } 298 \text{ K}$$

51. Three gases are injected into a sealed container. Initially, all three gases are at a concentration of 0.10 M. Which of the following must occur in order for the reaction to reach equilibrium?

    (A) The concentration of all three species must increase.
    (B) The concentration of all three species must decrease.
    (C) The concentration of the reactants must increase, while the concentration of the product will decrease.
    (D) The concentration of the reactants must decrease, while the concentration of the product will increase.

$$SO_2(g) + O_3(g) \rightarrow SO_3(g) + O_2(g)$$

Trial	$[SO_2](M)$	$[O_3](M)$	Initial Rate (M/s)
1	0.10	0.10	0.075
2	0.10	0.20	0.075
3	0.20	0.10	0.300

52. The above reaction is run three separate times, varying the initial concentration of both reactants as shown. The initial rate is shown for each reaction. What is the rate law for the reaction?

    (A) Rate = $k[SO_2][O_3]$
    (B) Rate = $k[SO_2]$
    (C) Rate = $k[SO_2]^2$
    (D) Rate = $k[O_3]^2$

Use the following information to answer questions 53–56.

$$NH_4HS(s) \rightleftharpoons H_2S(g) + NH_3(g) \quad \text{At 298 K, } K_c = 9.0 \times 10^{-2}$$
$$\Delta H^\circ = +91 \text{ kJ/mol}_{rxn}$$

Some solid NH_4HS is placed in a sealed and evacuated 250 mL flask at 298 K, and the above decomposition reaction is allowed to reach equilibrium.

53. Calculate the concentration of the $NH_3(g)$ at equilibrium.

    (A) $9.0 \times 10^{-2}\,M$
    (B) $3.0 \times 10^{-1}\,M$
    (C) $1.0 \times 10^{2}\,M$
    (D) $9.0 \times 10^{2}\,M$

54. If some additional $NH_3(g)$ is injected into the flask, which of the following options correctly predicts the reaction shift and what would happen to the concentration of NH_4HS?

	Shift	$[NH_4HS]$
(A)	Left	Increase
(B)	Left	No Change
(C)	Right	Decrease
(D)	None	No Change

55. If the temperature of the flask were to be increased to 313 K, which of the following identifies what would happen to the amount of NH_4HS present at equilibrium and correctly explains why?

    (A) Increase because both reaction rates have been permanently increased
    (B) Decrease because the value for K has increased
    (C) Decrease because the reaction is becoming more dispersed
    (D) Remain the same because it is a solid and unaffected by reaction shifts

56. If the reaction were to be repeated at 298 K using the same initial amount of NH_4HS, but in a 500 mL flask, how would that affect the concentration of NH_3 at equilibrium compared to the trial in the 250 mL flask?

    (A) It would be unchanged.
    (B) It would decrease by a factor of four.
    (C) It would decrease by a factor of two.
    (D) It would increase by a factor of two.

GO ON TO THE NEXT PAGE.

57. A sample of lithium hydroxide, LiOH, would be the most soluble in which of the following solutions?

   (A) Pure water
   (B) 0.10 M LiOH(aq)
   (C) 1.0 M Sr(OH)$_2$(aq)
   (D) 1.0 M HCl(aq)

58. What is the oxidation state on the sulfur atom in the compound S$_2$Cl$_2$?

   (A) −2
   (B) −1
   (C) 0
   (D) +1

59. Which of the following diagrams correctly demonstrates the permanent dipole attractions between two molecules of oxygen difluoride, OF$_2$? Note that the dashed lines are representative of attractive intermolecular forces.

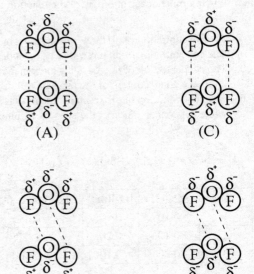

60. Which of the following reactions would have the most negative value for $\Delta S°$?

   (A) 2 NH$_3$(g) → N$_2$(g) + 3 H$_2$(g)
   (B) H$_2$O(l) → H$_2$O(g)
   (C) CaO(s) + CO$_2$(g) → CaCO$_3$(s)
   (D) MgCl$_2$(s) + H$_2$O(l) → MgO(s) + 2 HCl(g)

END OF SECTION I

INFORMATION IN THE TABLE BELOW AND ON THE FOLLOWING PAGES MAY BE USEFUL IN ANSWERING THE QUESTIONS IN THIS SECTION OF THE EXAMINATION

DO NOT DETACH FROM BOOK.

PERIODIC TABLE OF THE ELEMENTS

1	2	3	4	5	6	7	8	9	10	11	12	13	14	15	16	17	18
1 **H** 1.008																	2 **He** 4.00
3 **Li** 6.94	4 **Be** 9.01											5 **B** 10.81	6 **C** 12.01	7 **N** 14.01	8 **O** 16.00	9 **F** 19.00	10 **Ne** 20.18
11 **Na** 22.99	12 **Mg** 24.30											13 **Al** 26.98	14 **Si** 28.09	15 **P** 30.97	16 **S** 32.06	17 **Cl** 35.45	18 **Ar** 39.95
19 **K** 39.10	20 **Ca** 40.08	21 **Sc** 44.69	22 **Ti** 47.87	23 **V** 50.94	24 **Cr** 52.00	25 **Mn** 54.94	26 **Fe** 55.85	27 **Co** 58.93	28 **Ni** 58.69	29 **Cu** 63.55	30 **Zn** 65.38	31 **Ga** 69.72	32 **Ge** 72.63	33 **As** 74.92	34 **Se** 78.97	35 **Br** 79.90	36 **Kr** 83.80
37 **Rb** 85.47	38 **Sr** 87.62	39 **Y** 88.91	40 **Zr** 91.22	41 **Nb** 92.91	42 **Mo** 95.95	43 **Tc**	44 **Ru** 101.07	45 **Rh** 102.91	46 **Pd** 106.42	47 **Ag** 107.87	48 **Cd** 112.41	49 **In** 114.82	50 **Sn** 118.71	51 **Sb** 121.76	52 **Te** 127.60	53 **I** 126.90	54 **Xe** 131.29
55 **Cs** 132.91	56 **Ba** 137.33	57-71 *	72 **Hf** 178.49	73 **Ta** 180.95	74 **W** 183.94	75 **Re** 186.21	76 **Os** 190.23	77 **Ir** 192.22	78 **Pt** 195.08	79 **Au** 196.97	80 **Hg** 200.59	81 **Tl** 204.38	82 **Pb** 207.2	83 **Bi** 208.98	84 **Po**	85 **At**	86 **Rn**
87 **Fr**	88 **Ra**	89-103 †	104 **Rf**	105 **Db**	106 **Sg**	107 **Bh**	108 **Hs**	109 **Mt**	110 **Ds**	111 **Rg**	112 **Cn**	113 **Nh**	114 **Fl**	115 **Mc**	116 **Lv**	117 **Ts**	118 **Og**

*Lanthanoids	57 **La** 138.91	58 **Ce** 140.12	59 **Pr** 140.91	60 **Nd** 144.24	61 **Pm**	62 **Sm** 150.36	63 **Eu** 151.97	64 **Gd** 157.25	65 **Tb** 158.93	66 **Dy** 162.50	67 **Ho** 164.93	68 **Er** 167.26	69 **Tm** 168.93	70 **Yb** 173.05	71 **Lu** 174.97
†Actinoids	89 **Ac**	90 **Th** 232.04	91 **Pa** 231.04	92 **U** 238.03	93 **Np**	94 **Pu**	95 **Am**	96 **Cm**	97 **Bk**	98 **Cf**	99 **Es**	100 **Fm**	101 **Md**	102 **No**	103 **Lr**

GO ON TO THE NEXT PAGE.

AP® CHEMISTRY EQUATIONS & CONSTANTS

Throughout the exam the following symbols have the definitions specified unless otherwise noted.

L, mL	=	liter(s), milliliter(s)	mm Hg	= millimeters of mercury
g	=	gram(s)	J, kJ	= joule(s), kilojoule(s)
nm	=	nanometer(s)	V	= volt(s)
atm	=	atmosphere(s)	mol	= mole(s)

ATOMIC STRUCTURE

$E = h\nu$

$c = \lambda\nu$

E = energy

ν = frequency

λ = wavelength

Planck's constant, $h = 6.626 \times 10^{-34}$ J s

Speed of light, $c = 2.998 \times 10^8$ m s^{-1}

Avogadro's number $= 6.022 \times 10^{23}$ mol^{-1}

Electron charge, $e = -1.602 \times 10^{-19}$ coulomb

EQUILIBRIUM

$K_c = \dfrac{[C]^c[D]^d}{[A]^a[B]^b}$, where $a\,A + b\,B \rightleftarrows c\,C + d\,D$

$K_p = \dfrac{(P_C)^c(P_D)^d}{(P_A)^a(P_B)^b}$

$K_a = \dfrac{[H^+][A^-]}{[HA]}$

$K_b = \dfrac{[OH^-][HB^+]}{[B]}$

$K_w = [H^+][OH^-] = 1.0 \times 10^{-14}$ at 25°C

$\quad = K_a \times K_b$

$pH = -\log[H^+]$, $pOH = -\log[OH^-]$

$14 = pH + pOH$

$pH = pK_a + \log\dfrac{[A^-]}{[HA]}$

$pK_a = -\log K_a$, $pK_b = -\log K_b$

Equilibrium Constants

K_c (molar concentrations)

K_p (gas pressures)

K_a (weak acid)

K_b (weak base)

K_w (water)

KINETICS

$[A]_t - [A]_0 = -kt$

$\ln[A]_t - \ln[A]_0 = -kt$

$\dfrac{1}{[A]_t} - \dfrac{1}{[A]_0} = kt$

$t_{1/2} = \dfrac{0.693}{k}$

k = rate constant

t = time

$t_{1/2}$ = half-life

GO ON TO THE NEXT PAGE.

GASES, LIQUIDS, AND SOLUTIONS

$$PV = nRT$$

$$P_A = P_{\text{total}} \times X_A, \text{ where } X_A = \frac{\text{moles A}}{\text{total moles}}$$

$$P_{total} = P_A + P_B + P_C + \ldots$$

$$n = \frac{m}{M}$$

$$K = {}^{\circ}C + 273$$

$$D = \frac{m}{V}$$

$$KE_{\text{molecule}} = \frac{1}{2}mv^2$$

Molarity, M = moles of solute per liter of solution

$$A = \varepsilon bc$$

P = pressure
V = volume
T = temperature
n = number of moles
m = mass
M = molar mass
D = density
KE = kinetic energy
v = velocity
A = absorbance
ε = molar absorptivity
b = path length
c = concentration

Gas constant, $R = 8.314 \text{ J mol}^{-1}\text{K}^{-1}$
$= 0.08206 \text{ L atm mol}^{-1}\text{K}^{-1}$
$= 62.36 \text{ L torr mol}^{-1}\text{K}^{-1}$
1 atm = 760 mm Hg = 760 torr
STP = 273.15 K and 1.0 atm
Ideal gas at STP = 22.4 L mol^{-1}

THERMOCHEMISTRY/ ELECTROCHEMISTRY

$$q = mc\Delta T$$

$$\Delta S^{\circ} = \sum S^{\circ} \text{ products} - \sum S^{\circ} \text{ reactants}$$

$$\Delta H^{\circ} = \sum \Delta H_f^{\circ} \text{ products} - \sum \Delta H_f^{\circ} \text{ reactants}$$

$$\Delta G^{\circ} = \sum \Delta G_f^{\circ} \text{ products} - \sum \Delta G_f^{\circ} \text{ reactants}$$

$$\Delta G^{\circ} = \Delta H^{\circ} - T\Delta S^{\circ}$$
$$= -RT \ln K$$
$$= -n F E^{\circ}$$

$$I = \frac{q}{t}$$

$$E_{\text{cell}} = E_{\text{cell}}^{\circ} - \frac{RT}{nF} \ln Q$$

q = heat
m = mass
c = specific heat capacity
T = temperature
S° = standard entropy
H° = standard enthalpy
G° = standard Gibbs free energy
n = number of moles
E° = standard reduction potential
I = current (amperes)
q = charge (coulombs)
t = time (seconds)
Q = reaction quotient

Faraday's constant, $F = 96,485$ coulombs per mole of electrons

1 volt = $\dfrac{1 \text{ joule}}{1 \text{ coulomb}}$

GO ON TO THE NEXT PAGE.

CHEMISTRY

SECTION II

Time—1 hour and 45 minutes

7 Questions

Directions: Questions 1–3 are long free-response questions that require about 23 minutes each to answer and are worth 10 points each. Questions 4–7 are short free-response questions that require about 9 minutes each to answer and are worth 4 points each.

On test day, you will be asked to show your work for each part in the space provided after that part. For this practice test, you may use scrap paper. Examples and equations may be included in your responses where appropriate. For calculations, clearly show the method used and the steps involved in arriving at your answers. You must show your work to receive credit for your answer. Pay attention to significant figures.

$$CaCl_2(aq) + (NH_4)_2C_2O_4(aq) \rightarrow CaC_2O_4(s) + 2\ NH_4Cl(aq)$$

1. Most vitamin tablets consist of not only the active ingredient, but also other ingredients that are there to bind the vitamin together or coat it for easier swallowing. A calcium tablet with a mass of 1.49 g is crushed, and then dissolved in some concentrated HCl, creating aqueous calcium chloride. The tablet solution is then mixed with excess ammonium oxalate in a basic environment, causing calcium oxalate to precipitate.

   (a) Describe a method to fully separate the precipitate from the solution and accurately measure its mass.

   (b) The final mass of the CaC_2O_4 is determined to be 1.64 g.
      (i) How many moles of calcium are present in the precipitate?
      (ii) What is the mass percent of calcium in the original vitamin tablet?

   (c) CaC_2O_4 has a K_{sp} value of 1.9×10^{-9} and dissociates in solution as follows.
      $$CaC_2O_4(s) \rightleftharpoons Ca^{2+}(aq) + C_2O_4^{2-}(aq)$$
      (i) What is the concentration of Ca^{2+} ions and $C_2O_4^{2-}$ ions in a saturated solution of CaC_2O_4?
      (ii) If some of the water evaporated from a saturated solution of CaC_2O_4, how would that affect the concentration of the ions you determined in part (c)(i)? Justify your answer.

   The results do not match the accepted value of calcium in the tablet provided by the manufacturer. Upon reviewing the reaction, it was determined that the final product was not, in fact, pure calcium oxalate, but was instead a hydrate of calcium oxalate.

   (d) Would not accounting for the water of hydration cause the calcium value in the tablet to be artificially high or artificially low? Why?

   (e) A 5.00 g sample of the calcium oxalate hydrate is heated several times, resulting in a final mass for the anhydrous salt of 4.38 g.
      (i) How many moles of water were present in the sample?
      (ii) What is the hydration coefficient?

GO ON TO THE NEXT PAGE.

$$NaF(s) \rightarrow Na^+(aq) + F^-(aq)$$

2. Sodium fluoride is the active ingredient in toothpaste that helps prevent tooth decay. One of the important aspects of sodium fluoride is that it is completely soluble in water, and will dissociate fully.

(a) What are the forces that exist between sodium fluoride and water molecules that allow for this dissociation to occur?

(b) Which ion has the larger size—Na^+ or F^-? Justify your answer.

(c) The below beaker is filled with water molecules. In the beaker, correctly draw the location of the Na^+ and F^- ions in a dissolved solution of sodium fluoride. Make sure each ion is labeled clearly.

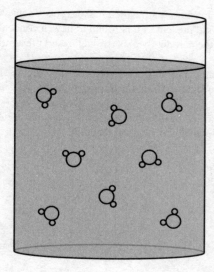

(d) 5.00 g of NaF is dissolved into 50.0 mL of water, and the temperature of the water decreases by 0.52°C. The specific heat and density of solution are identical to that of pure water: 4.18 J/g°C and 1.0 g/mL, respectively.
   (i)   How much energy is lost by the solution during the dissolution process?
   (ii) What is the enthalpy of solution, in kJ/mol, for the dissolution of sodium fluoride?
   (iii) For the dissolution of sodium fluoride, which value has a greater magnitude—the lattice energy of sodium fluoride, or the hydration energy between sodium fluoride and water? How do you know?

(e) How would you expect the magnitude of the hydration energy for sodium fluoride to compare to that of potassium chloride, KCl? Justify your answer.

GO ON TO THE NEXT PAGE.

$$HC_3H_5O_3(aq) \rightleftharpoons H^+(aq) + C_3H_5O_3^-(aq) \qquad K_a = 1.38 \times 10^{-4}$$

3. Lactic acid is an acid found in many dairy-related products.

   (a) Write the equilibrium constant expression, K_a, for lactic acid.

   (b) (i) The pH of an 0.50 M solution of lactic acid is found to be 2.08. Calculate the percent dissociation of the 0.50 M lactic acid solution.

       (ii) If the 0.50 M solution from part (i) were to be diluted, how would that affect the percent dissociation, if at all? Justify your answer.

   (c) Lactic acid is the active ingredient in sour cream which makes it taste sour. A sour cream sample with a volume of 500. mL is titrated, and the lactic acid present in it is found to have a concentration of 0.113 M. If the sour cream sample has a density of 1.0 g/mL, what is the percent by mass of lactic acid present in the sour cream?

   (d) To control the pH level of lactic acid containing substances, sodium lactate, $NaC_3H_5O_3$, is often used.
       (i) How many moles of sodium lactate would need to be added to 100. mL of 0.10 M lactic acid to create a buffer with a pH of 4.00?
       (ii) Adding a small amount of which of the following chemicals to a solution of lactic acid would also create a buffer solution? Justify your answer using a chemical reaction.

   HCl   NaOH   KCl

GO ON TO THE NEXT PAGE.

4. Many foods are dyed various colors to make them more aesthetically pleasing. The dyes are often complex organic compounds, including one that is commercially called Brilliant Blue FCF, which can be abbreviated as BB. When BB reacts with the active ingredient in bleach, the hypochlorite ion (ClO^-), the color will fade over time. Thus, the rate of reaction can be determined by using a colorimeter to measure the rate at which the BB fades.

To determine the reaction order for both BB and ClO^- in this reaction, three trials are run at 25°C and the rate of reaction is measured in each.

Trial	[BB] (M)	[ClO⁻] (M)	Rate (M/s)
1	1.5×10^{-3}	1.5×10^{-3}	3.3×10^{-3}
2	2.0×10^{-3}	1.5×10^{-3}	4.4×10^{-3}
3	3.0×10^{-3}	2.5×10^{-3}	1.1×10^{-2}

(a) What is the reaction order with respect to:
(i) BB
(ii) ClO^-

(b) Calculate the rate constant for the reaction between BB and ClO^- at 25°C. Include units.

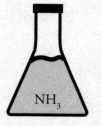

P_{NH_3} = 7,600 mmHg $P_{C_4H_{10}}$ = 2,200 mmHg $P_{CH_3OCH_3}$ = 200 mmHg

5. Three different liquids are placed in three sealed flasks of identical volume as shown above. All of the flasks are held at a constant temperature of 25°C, and the vapor pressure of each gas is listed below the flask.

(a) Which liquid has the strongest intermolecular forces? Justify your answer on a particular level.

(b) If the temperature of each of the flasks were to be increased, what effect would that have on the vapor pressure, if any? Why?

(c) (i) Are the gases present in each flask least likely to behave ideally at very low or very high temperatures? Explain your reasoning.
(ii) Which substance would show the greatest deviation from ideal behaviors under the conditions you chose in (c)(i)? Justify your answer.

GO ON TO THE NEXT PAGE.

6. Use the below diagram of the chlorate ion, ClO_3^-, to answer the following questions as needed.

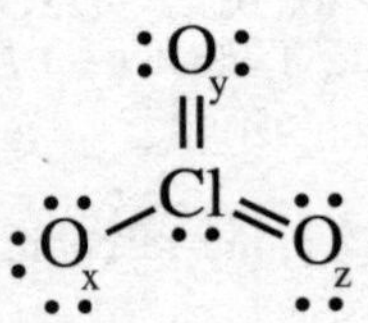

(a) What is the oxidation state on the chlorine atom in the chlorate ion? Show any necessary calculations.

(b) Calculate the formal charge on each atom in the chlorate ion. Note that each oxygen atom should be clearly labeled to match the subscripts in the diagram.

(c) Is the $Cl-O_x$ bond shorter than, longer than, or the same length as the $Cl-O_y$ bond? Justify your answer.

(d) Chlorine atoms can expand their octet, but oxygen atoms cannot. Why?

7. The Br-Br bond in a bromine molecule has a bond energy of 193 kJ/mol.

(a) (i) How much energy, in Joules, is required to break the bond in a single Br molecule?
    (ii) What wavelength of light is necessary to break a single Br-Br bond?

(b) Is the wavelength of light necessary to break an F-F bond shorter than, longer than, or the same as the wavelength of light necessary to break the Br-Br bond? Justify your answer.

STOP

END OF EXAM

Practice Test 4:
Answers and
Explanations

PRACTICE TEST 4: MULTIPLE-CHOICE ANSWER KEY

1.	A	21.	A	41.	B
2.	B	22.	A	42.	A
3.	A	23.	C	43.	B
4.	D	24.	A	44.	D
5.	D	25.	B	45.	C
6.	C	26.	C	46.	A
7.	A	27.	B	47.	A
8.	A	28.	B	48.	D
9.	A	29.	A	49.	B
10.	D	30.	A	50.	B
11.	B	31.	D	51.	C
12.	C	32.	D	52.	C
13.	C	33.	A	53.	B
14.	D	34.	C	54.	B
15.	A	35.	B	55.	B
16.	C	36.	A	56.	A
17.	C	37.	A	57.	D
18.	C	38.	C	58.	D
19.	D	39.	D	59.	D
20.	B	40.	C	60.	C

Section I—Multiple-Choice Answers and Explanations

1. **A** As an equal number of moles of both reactants were mixed together, the reactant with the greater stoichiometry ratio (in this case, the Ag^+) will run out first and limit the formation of additional precipitate, at the same time leaving excess CO_3^{2-} ions in solution. Adding more $AgNO_3$ will add more Ag^+ ions to react with those CO_3^{2-} ions in solution, allowing more precipitate to form.

2. **B** NaOH has a molar mass of 40 g/mol. The oxygen present in the NaOH has a molar mass of 16 g/mol, and so makes up 16/40 = 40% of the molar mass of NaOH. Oxygen would also represent 40% of the mass of the 10.0 g NaOH sample, which is 4.00 g.

3. **A** Remember that Avogadro's Law states that one mole of ANY gas at STP takes up 22.4 L of space. With that in mind:

$$3.5 \text{ g Li} \times \frac{1 \text{ mol Li}}{7.0 \text{ g Li}} \times \frac{1 \text{ mol H}_2}{2 \text{ mol Li}} \times \frac{22.4 \text{ L}}{1 \text{ mol H}_2} = 5.60 \text{ L}$$

4. **D** The amount of heat gained by the water is equal to the amount of heat lost by the metal. This is mathematically represented by $q_{water} = -q_{metal}$. Using the calorimetry equation,

$$m_w c_w \Delta T_w = -m_m c_m \Delta T_m$$

$$(100)(4.2)(2) = -(10) \, c_m \, (-70)$$

$$840 = 700 \, c_m$$

$$c_m = 1.2 \text{ J/g}°\text{C}$$

5. **D** In an equilibrium constant expression, it is products over reactants and neither solids nor liquids are included due to their unchanging concentration.

6. **C** Silver has the higher SRP and would be the cathode. To calculate cell potential, subtract the reduction potential of the anode from the reduction potential of the cathode.

$$0.80 \text{ V} - (-0.44 \text{ V}) = 1.24 \text{ V}$$

7. **A** Electrons must be balanced in a full reaction, and to do that you must multiply the Ag half reaction by 2 before combining it with the Fe half-reaction. The Fe half-reaction will flip and become an oxidation because it has the lower SRP.

8. **A** For metallic M ions with a charge of +Y:

$$\text{Current} \times \text{time} = \text{Charge (in C)} \times \frac{1 \text{ mol } e^-}{96,500 \text{ C}} \times \frac{1 \text{ mol M}}{Y \text{ mol } e^-} \times \frac{\text{Molar Mass of M}}{1 \text{ mol M}} = \text{mass M}$$

Given that silver has both the lowest charge and the highest molar mass, both factors lead to more silver metal being plated out.

9. **A** When a current is run through water, it will both reduce into hydrogen gas and oxidize into oxygen gas using the given reactions.

    Reduction: $2\,H_2O(l) + 2\,e^- \rightarrow H_2(g) + 2\,OH^-$
    Oxidation: $2\,H_2O(l) \rightarrow O_2(g) + 4\,H^+(aq) + 4\,e^-$

    The reduction reaction has to be multiplied by 2 to balance the charge, and the final reaction becomes $2\,H_2O(l) \rightarrow 2\,H_2(g) + O_2(g)$ (H^+ and OH^- combine to form H_2O, some of which cancels). Thus, twice as much H_2 is produced, meaning the H_2 is in the tube above electrode X, making electrode X the cathode as well.

10. **D** When a solid metal is placed into a solution containing aqueous cations of another metal, the SRP of the cations in solution must exceed that of the cations of the solid metal for the solution to be able to take electrons from the solid and cause a reaction to occur. The only choice in which this does not occur (i.e., in which the SRP of the cations of the solid metal > the SRP of the cations in solution) is (D).

11. **B** Aluminum's electron configuration represented by the PES is $1s^2 2s^2 2p^6 3s^2 3p^1$. The 2p peak is the third-lowest energy level that an electron can be found in, and that peak is found at about 150 eV.

12. **C** To form an ion, aluminum loses three valence electrons. Those would come from the 3s and 3p energy levels, removing those peaks from the spectrum.

13. **C** The energy from the incoming radiation has to exceed the binding energy to cause electron ejection. 200 eV is not enough energy to eject a 1s or 2s electron. Comparing the 2p and 3p orbitals, the binding energy for a 2p electron is greater, meaning there will be less kinetic energy left post-ejection.

14. **D** The necessary equation here is $\Delta G = \Delta H - T\Delta S$. Reactions are favored when ΔG is negative. Increasing the temperature in this particular reaction makes the reaction less favored, so that means the $-T\Delta S$ term is positive, making ΔS itself negative. With $-T\Delta S$ being positive, for the reaction to be favored at any temperature, ΔH must be negative.

15. **A** The configuration of an oxygen atom is $1s^2 2s^2 2p^4$. The 2p sub-level has three orbitals, and one electron must fill each of them before any of them pair up.

16. **C** All of the carbon will end up in the carbon dioxide, and all of the hydrogen will end up in the water. 44.0 g of CO_2 is equal to one mole, and with a 1:1 ratio that means there is one mole of carbon, too. 18.0 g of H_2O is equal to one mole of H_2O, but as each mole of H_2O has two moles of hydrogen, that means there are two moles of hydrogen present. The carbon-to-hydrogen ratio in the original sample is thus 1:2, which matches the ratio in option (C).

17. **C** The maximum absorbance of the copper ion occurs at approximately 640 nm. When studying a copper solution of unknown concentration, comparing it to a Beer's Law plot constructed at an absorbance of 640 nm would create the greatest possible absorbance range and give the best results.

18. **C** The tin starts at an oxidation state of 0 in its elemental state, and ends up at an oxidation state of +2 in Sn^{2+}. This requires a loss of electrons; thus tin is oxidized. To determine the oxidation state of the chromium in dichromate, realize that oxygen is –2 in the compound and the total of the oxidation states equals the charge on the ion (–2). $2(Cr) + 7(-2) = -2$. $Cr = +6$. The chromium in Cr^{3+} ends up at an oxidation state of +3, meaning it gained electrons and thus was reduced.

19. **D** Looking at the oxidation half-reaction, tin goes from 0 to +2, and does it three times in the balanced equation. That means 6 moles of electrons are transferred. You get the same answer when looking at the reduction half-reaction with chromium going from +6 to +3 twice.

20. **B** $0.033\ M = \dfrac{n}{0.100\ L}$ $\quad n = 0.0033\ \text{mol}\ K_2Cr_2O_7 \times \dfrac{3\ \text{mol Sn}}{1\ \text{mol}\ K_2Cr_2O_7} = 0.010\ \text{mol Sn}$

$0.010\ \text{mol Sn} \times \dfrac{118.71\ g}{1\ \text{mol Sn}} = 1.2\ g\ Sn$

21. **A** When a salt dissociates in solution, whether or not it is endothermic or exothermic depends on the sum of the hydration energy and the lattice energy. Hydration energy is always negative, and lattice energy is always positive. Thus, if the value for the enthalpy of solution is becoming more positive, the (positive) lattice energy is increasing at a faster rate than the (negative) hydration energy.

22. **A** An increase in the reverse reaction rate is a shift to the left. In an aqueous equilibrium, diluting the system will cause a shift to the side with more ions. In this equilibrium, there are three ions on the left and only one on the right, so adding water shifts the system left.

23. **C** $Q = \dfrac{[Cu^{2+}]}{[Ag^+]}$

K for any galvanic cell is very large, and if Q ever reaches K, the cell will be at equilibrium and have zero voltage. For this cell, increasing $[Ag^+]$ causes Q to decrease, taking it further from K, which means it is also further from equilibrium and further from zero voltage.

24. **A** $0.50\ M\ \text{NaOH} = \dfrac{n}{0.020\ L}$ $\quad n = 0.010\ \text{mol NaOH} \times \dfrac{1\ \text{mol}\ HN_3}{1\ \text{mol NaOH}} = \dfrac{0.010\ \text{mol}\ HN_3}{0.025\ L} = 0.40\ M$

25. **B** At equivalence (20.0 mL), all of the HN_3 would be converted to N^{3-}. At half-equivalence (10.0 mL), only half of the HN_3 would be converted to N^{3-}, meaning the concentration of both would be equal.

26. **C** Indicators change color at their pK_a value. The best indicator for a titration is one that changes color at the equivalence point of the titration. For this titration, that occurs at about a pH of 9, meaning the thymol blue is the best choice.

27. **B** The amount the pH changes at the very beginning of the titration is dependent on the K_a of the weak acid. The lower the K_a, the weaker the acid is and the more the addition of NaOH will change the pH of the solution. Note that (D) is wrong because a weaker acid will have a stronger conjugate base, meaning that at equivalence the solution will be more basic (higher pH).

28. **B** All three ions are isoelectric, meaning they have the same number of electrons and the same electron configuration. However, the nucleus of the aluminum ion has the most protons, and can pull the electrons in closer. This makes them harder to remove and gives the aluminum ion the greatest ionization energy.

29. **A** The negative partial charges on the water molecules are located on the oxygen end of the molecule. These partial charges would be pointing toward the sodium cation (and the positive hydrogen partial charges would point toward the chloride anion). The sodium ion is also smaller than the chloride ion, leading to the correct answer.

30. **A** The equation you need here is $D = P(MM)/RT$, where MM is the molar mass of the gas (in this case, 4.0 g/mol). At STP, the pressure is 1.0 atm and the temperature is 273 K, while the ideal gas constant is always 0.0821 atm·L/mol·K.

31. **D** Atoms with four charge clouds around the central atom have a base bond angle of 109.5°. Replacing a terminal atom with an unbonded lone pair slightly increases the repulsive forces against the remaining terminal atoms. OF_2, with two lone pairs on the central atom, would thus have the smallest bond angle. (Note: BF_3, with only three charge clouds, has a bond angle of 120°).

32. **D** When flipping the initial reaction, take the reciprocal of the equilibrium constant. $1/(5.0 \times 10^{-3}) = 2.0 \times 10^2$. Then, when the coefficients are doubled, the reaction coefficient must be squared. $(2.0 \times 10^2)^2 = 4.0 \times 10^4$.

33. **A** The substance starts at a solid, and begins to melt at 50°C. This is known because the temperature stops changing, which is always true during a phase change. While it is melting, the solid and liquid phases will exist simultaneously.

34. **C** In the gas phase, which occurs at point D, the intermolecular forces present are very weak due to the great distance between the gas molecules as well as their high velocities.

35. **B** The amount of heat needed to change the substance from a liquid to a gas (heat of vaporization) exceeds the amount of heat needed to change it from a solid to a liquid (heat of fusion). This is based on the length of the horizontal lines representing each phase change.

36. **A** The solid-to-liquid phase change (the first section of unchanging temperature) requires 200 J of energy to complete. A 1.0 sample of the substance would require half as much heat to melt, so 100 J.

37. **A** The substance is in the liquid phase starting at 50°C and ending at 150°C. During that time, 400 J of heat is added. Using $q = mc\Delta T$:   400 J = (2.0 g)c(100°C) c = 2.0 J/g·°C

38. **C** When a weak acid is diluted, the reverse reaction becomes more hindered because the H^+ and F^- ions do not collide as often due to the overwhelming number of reactants surrounding them. This has the effect of increasing the percent dissociation as the H^+ and F^- ions are more likely to remain in solution. Note that (D) is wrong because a weak acid cannot be dissociated 100%; if it were, that would make it a strong acid.

39. **D** Voltage is a function of the reaction quotient for the battery. Given that $Q = [Zn^{2+}]/[Cu^{2+}]$ in both batteries, the value for Q would be the same, making their voltages the same. However, battery Y has more metal ions both in solution and in the electrodes, meaning the battery could operate for longer.

40. **C** $C_6H_{12}O_6$ is a covalent substance, so that would have the lowest melting point. Of the three ionic substances, Mg^{2+} and O^{2-} are both smaller than and have a greater magnitude of charge than the ions present in the other salts. These factors would lead to greater lattice energy and a higher melting point.

41. **B** If all gases are at the same temperature, the average kinetic energy of their particles is the same. For this to be true, the gas with the lowest molar mass will have the fastest particles, on average.

42. **A** With number of moles and temperature constant, pressure and volume have an inverse relationship defined using $P_1V_1 = P_2V_2$. Treating each gas separately:

    Kr: $(1.0)(1.0) = P_2(3.0)$ $P_2 = 0.33$ atm

    H_2O: $(2.0)(1.0) = P_2(3.0)$ $P_2 = 0.67$ atm

    CO_2: $(3.0)(1.0) = P_2(3.0)$ $P_2 = 1.0$ atm

    $P_T = 0.33 + 0.67 + 1.0 = 2.0$ atm

43. **B** Pressure and temperature have a direct relationship which is linear in nature. As temperature increases, pressure will increase at a proportional rate.

44. **D** C_2H_5OH and PF_3 are covalent substances, which means they are poor conductors of electricity. Of the two ionic substances, NH_4Cl breaks down into two ions (NH_4^+ and Cl^-) while Na_2CO_3 breaks down into three ($2\ Na^+$ and CO_3^{2-}). More ions in solution = greater conductivity.

45. **C** The reaction is second order with respect to NO. This means doubling the concentration of NO will make the reaction go four times faster. At the same time, given that the reaction is first order with respect to O_2, cutting the concentration of O_2 in half will halve the speed of the reaction. Combining both factors yield $4 \times 0.50 = 2$.

46. **A** When determining the enthalpy of reaction using bond enthalpies, bonds broken in the reactants are assigned a positive value and bonds formed in the products are assigned a negative value prior to summing them all up. Remember to account for both the number of bonds inside a molecule and how many of that molecule there are in the balanced equation when calculating (for instance, there are 4 H-O bonds in 2 H_2O molecules).

    $$4(C\text{-}H) + (C=C) + 3(O=O) - 4(C=O) - 4(H\text{-}O) = \Delta H^\circ_{rxn}$$

    $$4(410) + 720 + 3(500) - 4(800) - 4(470) = -1{,}220 \text{ kJ/mol}$$

47.　**A**　$K_p = \dfrac{(P_{NO})^2 (P_{O_2})}{(P_{NO_2})^2}$　$K_p = \dfrac{(3 \times 10^{-1})^2 (2 \times 10^{-1})}{(6 \times 10^{-1})^2}$　$K_p = \dfrac{18 \times 10^{-3}}{36 \times 10^{-2}}$　$K_p = 0.5 \times 10^{-1} = .05$

48.　**D**　The bond angle is based on the number of charge clouds (either a lone pair or any type of bond) around the central atom in the bond. There are only three charge clouds around the carbon in the C-C=O bond, making the bond angle about 120°. There are four charge clouds that are the central atoms of the H-C-H and H-N-H bonds, meaning the bond angle is approximately 109.5°. However, one of those charge clouds on the H-N-H bond is a lone pair, which exerts slightly more repulsive force than a bonded pair, shrinking the bond angle.

49.　**B**　All single bonds are sigma bonds, but any additional shared pairs of electrons in double or triple bonds are pi bonds. There's only one double bond present, and so there is only one pi bond (since the first bonded pair is still a sigma bond).

50.　**B**　Hydrogen bonding is the strongest type of intermolecular force, and occurs when hydrogen is bonded to either nitrogen, oxygen, or fluorine. In acetamide, there are two hydrogens bonded directly to the nitrogen, creating hydrogen bonding.

51.　**C**　$Q = \dfrac{[CH_3OH]}{[CO][H_2]^2}$. When all three species are at 0.10 M, $Q = \dfrac{0.10}{(0.10)(0.10)^2}$. That simplifies to $\dfrac{1}{(0.010)}$, which is equal to 100. That is greater than the value given for the equilibrium constant, and so more reactants must be created in order for Q to decrease and eventually reach the equilibrium value. This also means the amount of product will decrease, as the additional reactants can only be created by consuming some existing product.

52.　**C**　Between trial 1 and 2, the concentration of SO_2 remains the same but the concentration of O_3 changes. However, the rate does not change, which means the O_3 concentration has no effect on rate, making O_3 zero order. Between trials 1 and 3, the concentration of O_3 remains the same while the concentration of SO_2 doubles. The rate increases by a factor of four, making the reaction second order with respect to SO_2.

53.　**B**　$K_c = [H_2S][NH_3]$　$9.0 \times 10^{-2} = (x)(x)$　$x = 3.0 \times 10^{-1} = [NH_3]$

54.　**B**　Adding the NH_3 would cause the equilibrium to shift left in order to relieve the stress. However, the NH_4HS is a solid, and solids have a set concentration which cannot be changed. That is also why adding or removing a solid (or liquid) from an equilibrium system does not cause a shift.

55.　**B**　This reaction is endothermic, meaning an increase in temperature causes a permanent shift to the right. That would increase the value for K, given that there are more H_2S and NH_3 present once equilibrium re-establishes. Notice that while the concentration of NH_4HS is a solid and cannot change, the amount can still change. In this case, creating more products means there is less reactant left.

56.　**A**　As long as the temperature remains constant, so does the value for K. If K is constant, so are the concentrations of all the products and reactants at equilibrium.

57.　**D**　The solubility reaction here is $LiOH(s) \rightleftharpoons Li^+(aq) + OH^-(aq)$. The OH^- ions in the products would react with the H^+ ions dissociated from HCl, causing $[OH^-]$ to decrease. This would cause the equilibrium to shift to the right, increasing the amount of LiOH that dissolves.

58.　**D**　The most electronegative element in a compound will have an oxidation state that is equal to the most common charge of its corresponding ion. In this case, chlorine is more electronegative than sulfur, and would thus have an oxidation state of –1. The entire molecule must have a combined oxidation state of zero, so to balance the two chlorines each sulfur must be +1.

59.　**D**　Fluorine is more electronegative than oxygen, so fluorine would have a negative partial charge and oxygen would have a partial charge dipole in OF_2. The attraction between two molecules would be of the negative (F) partial charge on one molecule to the positive (O) partial charge on the other.

60.　**C**　A negative value for $\Delta S°$ means the entropy is decreasing, so the system is becoming more ordered. The phase change from a gas to a solid in (C) demonstrates this, while all other options show entropy increasing.

Section II—Free-Response Answers and Explanations

1.　(a)　First, the solution containing the precipitate needs to be poured into a funnel that has been prepared with a piece of filter paper. The solution should be filtered as many times as needed until the filtrate is clear. The filter paper and precipitate should then be fully washed with water and dried prior to a final massing.

(b)　(i)　$1.64 \text{ g CaC}_2\text{O}_4 \times \dfrac{1 \text{ mol CaC}_2\text{O}_4}{128.10 \text{ g}} \times \dfrac{1 \text{ mol Ca}}{1 \text{ mol CaC}_2\text{O}_4} = 0.0128 \text{ mol Ca}$

(ii)　$0.0128 \text{ mol Ca} \times \dfrac{40.08 \text{ g}}{1 \text{ mol Ca}} = \dfrac{0.513 \text{ g Ca}}{1.49 \text{ g}} \times 100\% = 34.4\% \text{ Ca}$

(c)　(i)　$K_{sp} = [Ca^{2+}][C_2O_4^{2-}]$　The ions exist in a 1:1 ratio, so both unknowns can be replaced with x.

$K_{sp} = (x)(x)$　　$1.9 \times 10^{-9} = x^2$　　$x = [Ca^{2+}] = [C_2O_4^{2-}] = 4.4 \times 10^{-5} \, M$

(ii)　The concentration of both ions will remain the same. For a saturated solution, the ions are already present at their maximum possible concentration values. If some water evaporates, some of the ions that were suspended between the evaporated water molecules will precipitate out, leaving the total concentration unchanged.

(d) If the calcium oxalate is a hydrate, that means there will be some waters of hydration attached to the crystalline lattice. This will cause the mass to increase, meaning the mass that was used in the calculations is artificially high, making the moles of calcium and the calcium percentage in the tablet artificially high.

(e) (i) $5.00 \text{ g} - 4.38 \text{ g} = 0.62 \text{ g H}_2\text{O} \times \dfrac{1 \text{ mol H}_2\text{O}}{18.02 \text{ g H}_2\text{O}} = 0.034 \text{ mol H}_2\text{O}$

(ii) To determine the hydration coefficient, compare the moles of water to the moles of the anhydrous salt.

$4.38 \text{ g CaC}_2\text{O}_4 \times \dfrac{1 \text{ mol CaC}_2\text{O}_4}{128.10 \text{ g}} = 0.0342 \text{ mol CaC}_2\text{O}_4$

The moles of water and moles of salt exist in a 1:1 ratio; therefore, the hydration coefficient is 1.

2. (a) An ionic salt dissolves when the attraction of its ions to the dipoles in the water molecules exceeds the attraction of the ions to each other. In this case, the positive sodium cations will be attracted to the negative water partial charges (oxygen), and the negative fluoride ions will be attracted to positive water partial charges (hydrogen).

(b) F^- is larger. Both ions have the same number of electrons: 10. When looking at species that have the same number of electrons, the one with more protons (in this case, Na^+) will be smaller, because the nuclear charge is greater and can pull those electrons in closer.

(c) The Na^+ ions must be between the negative partial charges, which in water are on the oxygen atoms. The F^- ions must be between the positive water partial charges located on the hydrogen atoms.

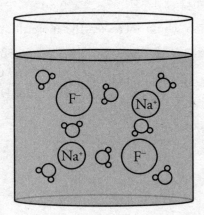

(d) (i) $q = mc\Delta T$ is the equation needed here. Note that the mass is the mass of the water plus the mass of the dissolved salt!

$q = (55.0 \text{ g})(4.18 \text{ J/g·°C})(0.52°\text{C})$

$q = 120 \text{ J}$

(ii) If the water lost 120 J of energy, that means the dissolution process required 120 J of energy. The sign flips to indicate the energy gain.

$$\Delta H = \frac{q}{n} \qquad q = 120 \text{ J} \qquad n = 5.00 \text{ g NaF} \times \frac{1 \text{ mol NaF}}{42.0 \text{ g NaF}} = 0.119 \text{ mol NaF}$$

$$\Delta H = 120 \text{ J}/0.119 \text{ mol} = 1.0 \times 10^3 \text{ J/mol} = 1.0 \text{ kJ/mol}$$

(iii) Lattice energy describes the amount of energy necessary to break the bonds in an ionic substance. Breaking bonds requires energy, so lattice energy is a positive value. Hydration energy is the amount of energy released when the dipoles of water molecules form attractions with the charged ions in the salt. Forming attractions releases energy, so hydration energy is always a negative value.

The total enthalpy of solution is the sum of the lattice and hydration energy values for a particular salt. In this case, the enthalpy of solution is positive, meaning the magnitude of the lattice energy exceeds that of the hydration energy.

(e) Hydration energy is based on Coulomb's Law. In this case, the charges of all the ions are the same — +1 on the cations and −1 on the anions. Looking at their sizes, however, Na^+ is smaller than K^+ and F^- is smaller than Cl^-. As Coulombic energy is inversely proportional with size, the substance with the smaller ions would have a greater Coulombic energy, and thus the greater hydration energy. So, the magnitude of NaF's hydration energy exceeds that of KCl.

3. (a) $K_a = \dfrac{[H^+][C_3H_5O_3^-]}{[HC_3H_5O_3]}$

(b) (i) $-\log [H^+] = 2.08 \qquad [H^+] = 10^{-2.08} \qquad [H^+] = 8.32 \times 10^{-3} \, M$

$$\% \text{ dissociation} = \frac{[H^+]}{[HC_3H_5O_3]} \times 100\% = \frac{8.32 \times 10^{-3} M}{0.50 \, M} \times 100\% = 1.7\%$$

(ii) The percent dissociation would increase. The more dilute the solution is, the less likely the reverse reaction is to occur. This means a larger percentage of ions will stay dissociated.

(c) $0.113 \, M = \dfrac{n}{0.500 \text{ L}} \qquad n = 0.0565 \text{ mol HC}_3\text{H}_5\text{O}_3 \times \dfrac{69.08 \text{ g}}{1 \text{ mol HC}_3\text{H}_5\text{O}_3} = 3.90 \text{ g HC}_3\text{H}_5\text{O}_3$

$\dfrac{3.90 \text{ g}}{500. \text{ g}} \times 100\% = 0.780\%$ lactic acid

(d) (i) Using Henderson-Hasselbalch:

$$pH = pK_a + \log\frac{[C_3H_5O_3^-]}{[HC_3H_5O_3]}$$

$$4.00 = -\log(1.38 \times 10^{-4}) + \log\frac{[C_3H_5O_3^-]}{0.10M}$$

$$4.00 = 3.86 + \log\frac{[C_3H_5O_3^-]}{0.10M}$$

$$0.14 = \log[C_3H_5O_3^-] - \log(0.1)$$

$$0.14 = \log[C_3H_5O_3^-] - (-1)$$

$$-0.86 = \log[C_3H_5O_3^-]$$

$$[C_3H_5O_3^-] = 0.14\ M$$

$$0.14\ M = \frac{n}{0.100\ L}$$

$$n = 0.014\ mol\ NaC_3H_5O_3$$

(ii) NaOH. Adding a strong base to a weak acid will cause the creation of conjugate base ions as follows:

$$HC_3H_5O_3(aq) + OH^-(aq) \rightarrow C_3H_5O_3^-(aq) + H_2O(l)$$

Those conjugate base ions will create a buffer with the remaining weak acid.

4. (a) (i) To calculate the order with respect to BB, look at the first two trials. The concentration of the ClO^- stays constant, so any change in rate is due to the change in the initial concentration of the BB.

$$\frac{3.3 \times 10^{-3} = k(1.5 \times 10^{-3})^m(1.5 \times 10^{-3})^n}{4.4 \times 10^{-3} = k(2.0 \times 10^{-3})^m(1.5 \times 10^{-3})^n}$$

$$0.75 = (0.75)^m \qquad m = 1$$

The reaction is first order with respect to BB.

(ii) The concentration of BB changes with every trial, but because you know the order, you can still determine the order with respect to ClO^- by inputting the BB order into our equation. Using trials 2 and 3,

$$\frac{4.4 \times 10^{-3} = k(2.0 \times 10^{-3})^1(1.5 \times 10^{-3})^n}{1.1 \times 10^{-2} = k(3.0 \times 10^{-3})^1(2.5 \times 10^{-3})^n}$$

$$0.40 = (0.67)(0.60)^n \qquad 0.60 = 0.60^n \qquad n = 1$$

The reaction is first order with respect to ClO^-.

(b) Data from any trial can be plugged into the rate law, along with the determined orders, to figure out the rate constant. The determined rate law is rate = $k[BB][ClO^-]$, and using data from trial 2, the magnitude of the rate constant can be determined.

$$4.4 \times 10^{-3} = k(2.0 \times 10^{-3})(1.5 \times 10^{-3})$$

$$k = 1,500$$

Using dimensional analysis, we can see that unitwise:

$$M/s = k(M)(M)$$

The units must be $M^{-1}s^{-1}$, so the final answer is $1,500 \ M^{-1}s^{-1}$.

5. (a) CH_3OCH_3 has the strongest IMFs, as shown by its low vapor pressure. The lower the vapor pressure, the less likely the molecules are to escape the liquid state, which requires a breaking of all IMFs. Stronger IMFs = less chance of molecules escaping = lower vapor pressure.

(b) The vapor pressure would increase. Adding energy to the liquids gives their molecules more kinetic energy, making them more likely to be able to overcome the IMFs holding them in the liquid state.

(c) (i) Deviations are more likely to occur at very low temperatures. This is because the gas molecules will have less energy, making them less likely to be able to overcome the IMFs between them.

In an ideal gas, IMFs are very weak, but the slower the gas molecules are moving, the more likely the IMFs are to become significant enough to cause deviations from ideal behavior.

(ii) The gas with the strongest IMFs is most likely to deviate from ideal behavior, which in this case means the CH_3OCH_3 is most likely to behave non-ideally.

6. (a) Oxygen has an oxidation state of negative two, and the total oxidation state on the chlorate must be equal to its charge of negative one. So: $Cl + 3(-2) = -1$ $Cl = +5$

    (b) Formal charge is equal to the number of valence electrons minus the number of assigned electrons drawn in the structure. Lone pairs count as two assigned electrons, while each bond attached to an atom counts as one.

	Cl	O_x	O_y	O_z
Valence e^-	7	6	6	6
Assigned e^-	7	7	6	6
Formal Charge	0	−1	0	0

    (c) They are the same length. The chlorate ion exhibits resonance, meaning the bonds are neither single bonds nor double bonds. All bonds would be somewhere in length between true single and double bonds, and all bonds are identical.

    (d) To expand an octet, there must be an empty "d" orbital available on the same principal energy level as the valence electrons for that atom. Chlorine's valence electrons are in the third principal energy level, and a $3d$ orbital does exist. Oxygen, however, has its valence electrons in the second principal energy level, and there is no $2d$ orbital, meaning there is no place for the extra electrons to go.

7. (a) (i) $\dfrac{193\ kJ}{mol} \times \dfrac{1\ mol}{6.02 \times 10^{23}\ molecules} \times \dfrac{1{,}000\ J}{1\ kJ} = 3.21 \times 10^{-19}\ J$

        (ii) $E = h\nu$ $3.21 \times 10^{-19}\ J = (6.626 \times 10^{-34}\ J \cdot s)\nu$ $\nu = 4.83 \times 10^{14}\ s^{-1}$

            $c = \nu\lambda$ $3.00 \times 10^{8}\ m/s = (4.83 \times 10^{14}\ s^{-1})\lambda$ $\lambda = 6.21 \times 10^{-7}\ m = 621\ nm$

    (b) The bond in F-F is shorter than the bond in Br-Br because a fluorine atom is smaller than a bromine atom (and they are both single bonds). A shorter bond has greater Coulombic energy. Energy and wavelength are inversely proportional, so breaking an F-F bond would require light of a shorter wavelength than the Br-Br bond.

Practice Test 5

The Exam

AP® Chemistry Exam

DO NOT OPEN THIS BOOKLET UNTIL YOU ARE TOLD TO DO SO.

At a Glance

Total Time
1 hour and 30 minutes
Number of Questions
60
Percent of Total Grade
50%
Writing Instrument
Pencil required

Instructions

Section I of this examination contains 60 multiple-choice questions. Fill in only the ovals for numbers 1 through 60 on your answer sheet.

Indicate all of your answers to the multiple-choice questions on the answer sheet. No credit will be given for anything written in this exam booklet, but you may use the booklet for notes or scratch work. After you have decided which of the suggested answers is best, completely fill in the corresponding oval on the answer sheet. Give only one answer to each question. If you change an answer, be sure that the previous mark is erased completely. Here is a sample question and answer.

Sample Question   Sample Answer

Chicago is a   Ⓐ ● Ⓒ Ⓓ
(A) state
(B) city
(C) country
(D) continent

Use your time effectively, working as quickly as you can without losing accuracy. Do not spend too much time on any one question. Go on to other questions and come back to the ones you have not answered if you have time. It is not expected that everyone will know the answers to all the multiple-choice questions.

About Guessing

Many candidates wonder whether or not to guess the answers to questions about which they are not certain. Multiple-choice scores are based on the number of questions answered correctly. Points are not deducted for incorrect answers, and no points are awarded for unanswered questions. Because points are not deducted for incorrect answers, you are encouraged to answer all multiple-choice questions. On any questions you do not know the answer to, you should eliminate as many choices as you can, and then select the best answer among the remaining choices.

GO ON TO THE NEXT PAGE.

CHEMISTRY
SECTION I
Time—1 hour and 30 minutes

INFORMATION IN THE TABLE BELOW AND ON THE FOLLOWING PAGES MAY BE USEFUL IN ANSWERING THE QUESTIONS IN THIS SECTION OF THE EXAMINATION

DO NOT DETACH FROM BOOK.

PERIODIC TABLE OF THE ELEMENTS

1	2		3	4	5	6	7	8	9	10	11	12	13	14	15	16	17	18
1 H 1.008																		2 He 4.00
3 Li 6.94	4 Be 9.01												5 B 10.81	6 C 12.01	7 N 14.01	8 O 16.00	9 F 19.00	10 Ne 20.18
11 Na 22.99	12 Mg 24.30												13 Al 26.98	14 Si 28.09	15 P 30.97	16 S 32.06	17 Cl 35.45	18 Ar 39.95
19 K 39.10	20 Ca 40.08		21 Sc 44.69	22 Ti 47.87	23 V 50.94	24 Cr 52.00	25 Mn 54.94	26 Fe 55.85	27 Co 58.93	28 Ni 58.69	29 Cu 63.55	30 Zn 65.38	31 Ga 69.72	32 Ge 72.63	33 As 74.92	34 Se 78.97	35 Br 79.90	36 Kr 83.80
37 Rb 85.47	38 Sr 87.62		39 Y 88.91	40 Zr 91.22	41 Nb 92.91	42 Mo 95.95	43 Tc	44 Ru 101.07	45 Rh 102.91	46 Pd 106.42	47 Ag 107.87	48 Cd 112.41	49 In 114.82	50 Sn 118.71	51 Sb 121.76	52 Te 127.60	53 I 126.90	54 Xe 131.29
55 Cs 132.91	56 Ba 137.33		57 57-71 *	72 Hf 178.49	73 Ta 180.95	74 W 183.94	75 Re 186.21	76 Os 190.23	77 Ir 192.22	78 Pt 195.08	79 Au 196.97	80 Hg 200.59	81 Tl 204.38	82 Pb 207.2	83 Bi 208.98	84 Po	85 At	86 Rn
87 Fr	88 Ra		89-103 †	104 Rf	105 Db	106 Sg	107 Bh	108 Hs	109 Mt	110 Ds	111 Rg	112 Cn	113 Nh	114 Fl	115 Mc	116 Lv	117 Ts	118 Og

*Lanthanoids	57 La 138.91	58 Ce 140.12	59 Pr 140.91	60 Nd 144.24	61 Pm	62 Sm 150.36	63 Eu 151.97	64 Gd 157.25	65 Tb 158.93	66 Dy 162.50	67 Ho 164.93	68 Er 167.26	69 Tm 168.93	70 Yb 173.05	71 Lu 174.97
†Actinoids	89 Ac	90 Th 232.04	91 Pa 231.04	92 U 238.03	93 Np	94 Pu	95 Am	96 Cm	97 Bk	98 Cf	99 Es	100 Fm	101 Md	102 No	103 Lr

GO ON TO THE NEXT PAGE.

AP® CHEMISTRY EQUATIONS & CONSTANTS

Throughout the exam the following symbols have the definitions specified unless otherwise noted.

L, mL	= liter(s), milliliter(s)		mm Hg	= millimeters of mercury
g	= gram(s)		J, kJ	= joule(s), kilojoule(s)
nm	= nanometer(s)		V	= volt(s)
atm	= atmosphere(s)		mol	= mole(s)

ATOMIC STRUCTURE

$E = h\nu$

$c = \lambda\nu$

E = energy

ν = frequency

λ = wavelength

Planck's constant, $h = 6.626 \times 10^{-34}$ J s

Speed of light, $c = 2.998 \times 10^8$ m s^{-1}

Avogadro's number $= 6.022 \times 10^{23}$ mol^{-1}

Electron charge, $e = -1.602 \times 10^{-19}$ coulomb

EQUILIBRIUM

$K_c = \dfrac{[C]^c[D]^d}{[A]^a[B]^b}$, where $a\,A + b\,B \rightleftarrows c\,C + d\,D$

$K_p = \dfrac{(P_C)^c(P_D)^d}{(P_A)^a(P_B)^b}$

$K_a = \dfrac{[H^+][A^-]}{[HA]}$

$K_b = \dfrac{[OH^-][HB^+]}{[B]}$

$K_w = [H^+][OH^-] = 1.0 \times 10^{-14}$ at 25°C

$\quad = K_a \times K_b$

$pH = -\log[H^+]$, $pOH = -\log[OH^-]$

$14 = pH + pOH$

$pH = pK_a + \log\dfrac{[A^-]}{[HA]}$

$pK_a = -\log K_a$, $pK_b = -\log K_b$

Equilibrium Constants

K_c (molar concentrations)

K_p (gas pressures)

K_a (weak acid)

K_b (weak base)

K_w (water)

KINETICS

$[A]_t - [A]_0 = -kt$

$\ln[A]_t - \ln[A]_0 = -kt$

$\dfrac{1}{[A]_t} - \dfrac{1}{[A]_0} = kt$

$t_{1/2} = \dfrac{0.693}{k}$

k = rate constant

t = time

$t_{1/2}$ = half-life

GO ON TO THE NEXT PAGE.

GASES, LIQUIDS, AND SOLUTIONS

$$PV = nRT$$

$$P_A = P_{total} \times X_A, \text{ where } X_A = \frac{\text{moles A}}{\text{total moles}}$$

$$P_{total} = P_A + P_B + P_C + \ldots$$

$$n = \frac{m}{M}$$

$$K = {}^\circ C + 273$$

$$D = \frac{m}{V}$$

$$KE_{molecule} = \frac{1}{2}mv^2$$

Molarity, M = moles of solute per liter of solution

$$A = \varepsilon bc$$

P = pressure
V = volume
T = temperature
n = number of moles
m = mass
M = molar mass
D = density
KE = kinetic energy
v = velocity
A = absorbance
ε = molar absorptivity
b = path length
c = concentration

Gas constant, R = 8.314 J mol^{-1} K^{-1}
$\qquad$ = 0.08206 L atm mol^{-1} K^{-1}
$\qquad$ = 62.36 L torr mol^{-1} K^{-1}
1 atm = 760 mm Hg = 760 torr
STP = 273.15 K and 1.0 atm
Ideal gas at STP = 22.4 L mol^{-1}

THERMOCHEMISTRY/ ELECTROCHEMISTRY

$$q = mc\Delta T$$

$$\Delta S^\circ = \sum S^\circ \text{ products} - \sum S^\circ \text{ reactants}$$

$$\Delta H^\circ = \sum \Delta H_f^\circ \text{ products} - \sum \Delta H_f^\circ \text{ reactants}$$

$$\Delta G^\circ = \sum \Delta G_f^\circ \text{ products} - \sum \Delta G_f^\circ \text{ reactants}$$

$$\Delta G^\circ = \Delta H^\circ - T\Delta S^\circ$$

$$= -RT \ln K$$

$$= -n F E^\circ$$

$$I = \frac{q}{t}$$

$$E_{cell} = E_{cell}^\circ - \frac{RT}{nF} \ln Q$$

q = heat
m = mass
c = specific heat capacity
T = temperature
S° = standard entropy
H° = standard enthalpy
G° = standard Gibbs free energy
n = number of moles
E° = standard reduction potential
I = current (amperes)
q = charge (coulombs)
t = time (seconds)
Q = reaction quotient

Faraday's constant, F = 96,485 coulombs per mole
$\qquad$ of electrons

$$1 \text{ volt} = \frac{1 \text{ joule}}{1 \text{ coulomb}}$$

GO ON TO THE NEXT PAGE.

$$Ag_2CO_3(s) \rightleftharpoons 2\,Ag^+(aq) + CO_3^{2-}(aq) \quad K_{sp} = 8.5 \times 10^{-12}$$

1. Which of the following is true about a beaker containing a saturated solution of silver carbonate?

    (A) The molar solubility of Ag_2CO_3 is equal to the concentration of the silver ions.

    (B) There would be no solid present at the bottom of the beaker.

    (C) The concentration of the silver ions will be twice that of the carbonate ions.

    (D) Adding more water would increase the solubility product constant.

2. Nitrous acid, HNO_2, has a pK_a value of 3.4. A solution of nitrous acid has some sodium nitrite, $NaNO_2$, added to it, creating a buffer solution. If the pH of the buffer solution is found to be 3.4, which of the following particulate representations shows the correct ratio of acid and conjugate base molecules in the buffer?

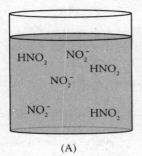

(A)

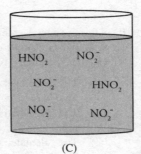

(C)

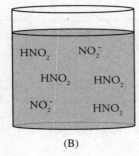

(B)

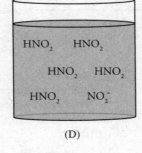

(D)

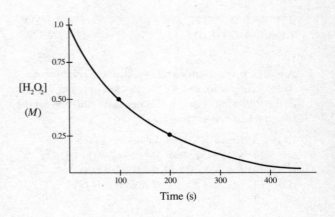

$$2\,H_2O_2(aq) \rightarrow 2\,H_2O(l) + O_2(g)$$

3. A sample of hydrogen peroxide, H_2O_2, will decompose over time. The above graph charts the concentration of an H_2O_2 sample over time. What is the rate law for the overall reaction?

    (A) Rate = k
    (B) Rate = $k[H_2O_2]$
    (C) Rate = $k[H_2O_2]^2$
    (D) Rate = $k[H_2O][O_2]$

4. The Lewis diagram of the bisulfite ion, HSO_3^-, is drawn above. Which of the following descriptions correctly compares the bond lengths of the three sulfur-oxygen bonds found in bisulfite?

    (A) All bonds are the same length.
    (B) Of the three bonds, there is one longer one and two shorter ones of identical length.
    (C) Of the three bonds, there are two longer ones of identical length and one shorter one.
    (D) All three bonds are different lengths.

GO ON TO THE NEXT PAGE.

$$2\ H^+(aq) + 2\ e^- \rightarrow H_2(g) \qquad\qquad E^\circ = +0.00\ V$$
$$K^+(aq) + e^- \rightarrow K(s) \qquad\qquad E^\circ = -2.92\ V$$

5. A piece of potassium metal is added to a solution of hydrochloric acid, and the metal starts to react. Which of the following is the correct net ionic equation for the reaction that is occurring?

(A) $2\ H^+(aq) + K^+(aq) + 3\ e^- \rightarrow H_2(g) + K(s)$
(B) $2\ H^+(aq) + 2\ K^+(aq) + 4\ e^- \rightarrow 2\ K(s) + H_2(g)$
(C) $2\ H^+(aq) + K(s) \rightarrow K^+(aq) + H_2(g)$
(D) $2\ H^+(aq) + 2\ K(s) \rightarrow 2\ K^+(aq) + H_2(g)$

6. A solution of methanol, $CH_3OH(aq)$, has a vapor pressure of 13.2 kPa at room temperature. If the solution were to be heated, what would happen to the vapor pressure, and why?

(A) Increase, because intramolecular attractions are breaking
(B) Increase, because the molecules in solution are moving faster
(C) Decrease, because stronger intermolecular attractions will form
(D) Remain unchanged, because the system is already at equilibrium

Use the following information to answer questions 7–10.

$$KNO_2(s) \rightarrow K^+(aq) + NO_2^-(aq)$$

7. A student wants to determine the enthalpy of solution for potassium nitrite, KNO_2 (molar mass = 85.1 g/mol). The student takes a sample of KNO_2 and fully dissolves it into some water that is in a Styrofoam cup, gathering the following data. The density and specific heat of the final solution are identical to that of pure water, 1.0 g/mL and 4.2 J/g°C, respectively.

Mass KNO_2	8.50 g
Volume Water	91.50 mL
Initial Water Temperature	22.5°C
Final Water Temperature	19.5°C

What is the approximate enthalpy of solution for KNO_2?
(A) –12.6 kJ/mol
(B) –1.26 kJ/mol
(C) 1.26 kJ/mol
(D) 12.6 kJ/mol

8. Which of the following statements is correct regarding the favorability of this process?

(A) The process is favored and driven by enthalpy and entropy.
(B) The process is favored and driven by enthalpy only.
(C) The process is favored and driven by entropy only.
(D) The process is not favored.

9. For the reaction to be endothermic, the magnitude of the attractive forces between the dissociated ions and the dipoles in the water molecules must be smaller than which of the following?

(A) The intermolecular forces between the water molecules
(B) The attractive forces between the K^+ and NO_3^- ions in solution
(C) The covalent bond strength within the water molecules
(D) The strength of the ionic bonds in the $KNO_3(s)$ lattice

10. After the data is gathered, the student notices that not all of the KNO_2 sample dissolved. How would that affect the calculated enthalpy of solution?

(A) The calculated enthalpy value would be artificially low.
(B) The calculated enthalpy value would be artificially high.
(C) The calculated enthalpy value would be unchanged.
(D) The calculated enthalpy value would have the incorrect sign.

$$Sr(IO_3)_2(s) \rightleftharpoons Sr^{2+}(aq) + 2\ IO_3^-(aq) \qquad K_{sp} = 1.0 \times 10^{-7}$$

11. Strontium iodate is a slightly soluble salt that dissociates in water via the above equation. What concentration of Sr^{2+} ions would be necessary to precipitate $Sr(IO_3)_2$ in a solution that contains IO_3^- ions at a concentration of 0.010 M?

(A) 0.0010 M
(B) 0.010 M
(C) 0.10 M
(D) 1.0 M

GO ON TO THE NEXT PAGE.

Acid	K_{a1}	K_{a2}	K_{a3}
Formic	1.8×10^{-4}	N/A	N/A
Sulfurous	1.5×10^{-2}	1.0×10^{-7}	N/A
Citric	8.4×10^{-4}	1.8×10^{-5}	4.0×10^{-6}

12. The K_a values for three different weak acids are given above. If the acid is polyprotic, all of the K_a values for that acid are given. If each acid were to be titrated with $0.10\ M$ NaOH, which solution would have the highest pH at the endpoint of the titration when each acid has fully reacted?

    (A) Formic
    (B) Sulfurous
    (C) Citric
    (D) All titrations would have identical pHs at the endpoint.

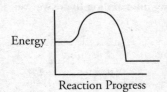

13. The above reaction coordinate shows the energy profile of an uncatalyzed reaction. If the reaction is repeated under identical collisions but a catalyst is added, which of the following coordinates is an accurate representation of what the energy profile of the catalyzed reaction could look like?

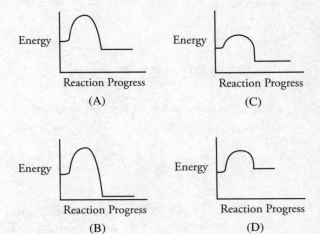

GO ON TO THE NEXT PAGE.

Use the following information to answer questions 14–18.

Name	Formula	K_a
Chlorous	$HClO_2$	1.0×10^{-2}
Carbonic	H_2CO_3	4.3×10^{-7}
Hypobromous	$HBrO$	2.0×10^{-9}

The K_a values for the first dissociation of three different weak acids are given in the above table.

14. A 0.10 M solution of which acid would have the highest concentration of H⁺ ions?

(A) $HClO_2$
(B) H_2CO_3
(C) $HBrO$
(D) [H⁺] would be equal in all solutions.

15. Identify the conjugate base of hyprobromous acid.

(A) BrO^-
(B) H_2BrO^+
(C) H_3O^+
(D) OH^-

16. Which of the following solutions would create a buffer system with chlorous acid when added in an equimolar amount?

(A) HCl
(B) NaOH
(C) $KClO_2$
(D) HOCl

17. Calculate the approximate pH of a 1.0 M solution of chlorous acid.

(A) 1
(B) 2
(C) 3
(D) 4

18. Carbonic acid is a polyprotic acid, and the second K_a value is 4.8×10^{-11}. Which of the following best explains why the K_a for the second dissociation is lower than that of the first?

(A) HCO_3^- is a more effective base than CO_3^{2-}.
(B) Acid dissociations are generally endothermic processes, and the first dissociation cools the reaction mixture, lowering the K_a value.
(C) Fewer water molecules are present to accept protons from HCO_3^- after the first dissociation has occurred.
(D) CO_3^{2-} is a more effective base than HCO_3^-.

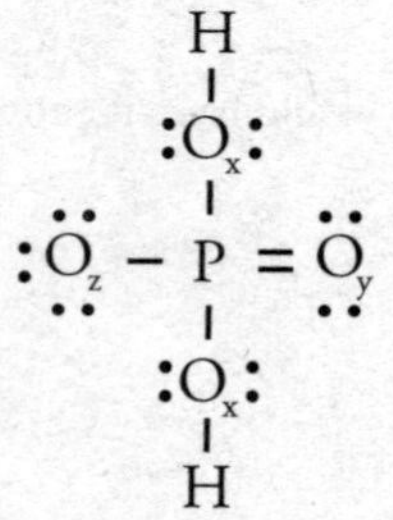

19. A Lewis diagram of the hydrogen phosphate ion is shown above. Based on this diagram, which of the atoms would have a negative formal charge?

(A) P
(B) O_x
(C) O_y
(D) O_z

$$4\,H^+(aq) + NO_3^-(aq) + 3\,e^- \rightarrow NO(g) + 2\,H_2O(l)$$

$$Zn(s) \rightarrow Zn^{2+}(aq) + 2\,e^-$$

20. Given the above half-reactions, what would the coefficient on the NO(g) be when the full reaction below is balanced?

$$\_\_ Zn(s) + \_\_ H^+(aq) + \_\_ NO_3^-(aq) \rightarrow$$
$$\_\_ NO(g) + \_\_ H_2O(l) + \_\_ Zn^{2+}(aq)$$

(A) 1
(B) 2
(C) 3
(D) 4

GO ON TO THE NEXT PAGE.

Use the following information to answer questions 21–23.

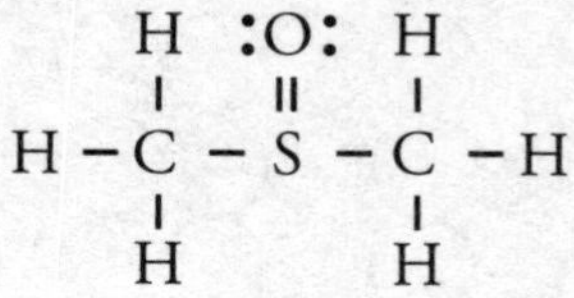

$$H \quad :O: \quad H$$
$$H-C-S-C-H$$
$$H \qquad H$$

The Lewis diagram of dimethyl sulfoxide, commonly known as DMSO, is shown above.

21. What is the orbital hybridization found around the sulfur atom?

   (A) sp
   (B) sp^2
   (C) sp^3
   (D) sp^4

22. Which of the following bonds would have the longest length?

   (A) S-O
   (B) S-C
   (C) C-H
   (D) All bonds would be equal in length.

23. Where would the strongest partial charges on a molecule of DMSO be located?

   (A) Negative partial charge on O, positive partial charge on S
   (B) Negative partial charge on S, positive partial charge on C
   (C) Negative partial charge on O, positive partial charge on C
   (D) The molecule is completely nonpolar.

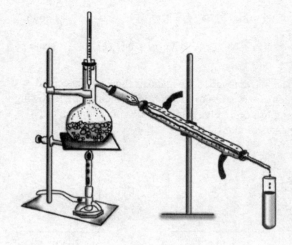

24. A student is attempting to use a distillation apparatus, shown above, to separate a mixture of propanol (boiling point: 97°C) and propionic acid (boiling point: 141°C). Which temperature should the apparatus be set at to obtain as pure a distillate of propanol as possible?

   (A)   85°C
   (B)   105°C
   (C)   125°C
   (D)   145°C

GO ON TO THE NEXT PAGE.

Step 1: $2 NO_2(g) \rightarrow NO_3(g) + NO(g)$ (slow)

Step 2: $CO(g) + NO_3(g) \rightarrow CO_2(g) + NO_2(g)$ (fast)

25. The above reaction mechanism shows the accepted elementary steps for a reaction. Which of the following graphs accurately represents the concentration of NO_2 versus time for the full reaction?

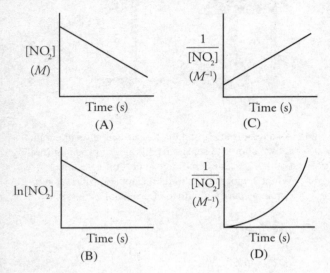

    (A)   (C)

    (B)   (D)

26. Which type of radiation would be best for studying the electron configuration of an individual atom?

    (A) Microwave
    (B) Infrared
    (C) Visible
    (D) Ultraviolet

Use the following information to answer questions 27–29.

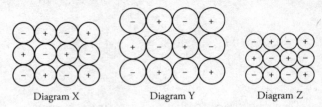

    Diagram X   Diagram Y   Diagram Z

The three diagrams above are particulate-level diagrams of three ionic solids: LiF, NaCl, and KBr. The diagrams are not organized in any particular order, and in each diagram the cations are marked with a + sign and the anions are marked with a – sign. All ions are drawn to scale.

27. Which diagram belongs to which ionic salt?

	Diagram X	Diagram Y	Diagram Z
(A)	LiF	NaCl	KBr
(B)	KBr	NaCl	LiF
(C)	KBr	LiF	NaCl
(D)	NaCl	KBr	LiF

28. Which salt would have the highest melting point?

    (A) LiF
    (B) NaCl
    (C) KBr
    (D) The melting points of all three salts would be identical.

29. Which of the following ions would have a completed *p*-subshell in their valence level?

    (A) All the cations only (Li^+, Na^+, and K^+)
    (B) All the anions only (F^-, Cl^-, and Br^-)
    (C) Only K^+ and Br^-
    (D) All ions except for Li^+

30. Methanol, CH_3OH, is heated until it boils. During the boiling process, which of the following statements is true?

    (A) The average kinetic energy of the molecules is constant.
    (B) The bonds within the molecules are breaking.
    (C) The intermolecular forces in the liquid are growing stronger.
    (D) The methanol is reacting with atmospheric gases.

GO ON TO THE NEXT PAGE.

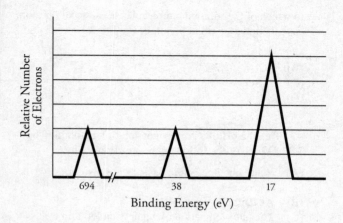

Binding Energy (eV)

Use the following information to answer questions 35–38.

$$\text{Step 1: } NO_2Cl(g) \rightarrow NO_2(g) + Cl(g)$$

$$\text{Step 2: } NO_2Cl(g) + Cl(g) \rightarrow NO_2(g) + Cl_2(g)$$

A proposed mechanism for a reaction is given above. The rate law for the overall reaction is known to be rate = $k[NO_2Cl]$.

31. The above photoelectron spectrum is for a neutral fluorine atom. Compared to that spectrum, the spectrum for a fluoride ion would have

    (A) one more peak
    (B) one less peak
    (C) the same number of peaks, but the rightmost peak would be taller
    (D) the same number of peaks, but the leftmost peak would be taller

35. What is the molecularity of the slowest elementary step?

    (A) Unimolecular
    (B) Bimolecular
    (C) Trimolecular
    (D) Tetramolecular

36. Which of the following could be the correct units on the rate constant, k?

    (A) s^{-1}
    (B) M^{-1}
    (C) $M^{-1}s^{-1}$
    (D) $M^{-1}s^{-2}$

32. Determine the bond order for the B-F bond in a molecule of boron trifluoride, BF_3.

    (A) 1
    (B) 1.33
    (C) 1.67
    (D) 2

37. Identify any catalysts and intermediates present in the proposed mechanism.

    (A) Catalyst: NO_2Cl Intermediate: Cl
    (B) Catalyst: None   Intermediate: Cl
    (C) Catalyst: Cl   Intermediate: None
    (D) There are neither catalysts nor intermediates in the proposed mechanism.

33. Copper (II) sulfate pentahydrate, $CuSO_4 \times 5\ H_2O$, has a molar mass of 250 g/mol. How many grams of oxygen are present in a 25.0 g sample of $CuSO_4 \times 5\ H_2O$?

    (A) 6.40 g
    (B) 8.00 g
    (C) 11.2 g
    (D) 14.4 g

34. A sample of a noble gas at STP with a mass of 10.0 g is found to have a volume of 5.6 L. Which of the following gases correctly identifies the gas in the flask?

    (A) Helium
    (B) Neon
    (C) Argon
    (D) Krypton

GO ON TO THE NEXT PAGE.

38. The overall reaction is known to be exothermic. Which of the following reaction coordinate diagrams represents the total reaction?

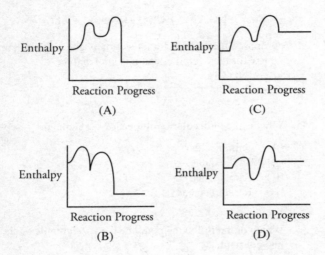

Enthalpy / Reaction Progress

(A)

Enthalpy / Reaction Progress

(C)

Enthalpy / Reaction Progress

(B)

Enthalpy / Reaction Progress

(D)

$$Fe_2O_3(s) + 3\ CO(g) \rightarrow 2\ Fe(s) + 3\ CO_2(g)$$
$$\Delta H^\circ = 26.4\ kJ/mol_{rxn}$$

39. How much heat is released or absorbed when 2.80 g of CO reacts with excess Fe_2O_3 via the above reaction?

    (A) 2.63 kJ is released.
    (B) 0.88 kJ is released.
    (C) 2.63 kJ is absorbed.
    (D) 0.88 kJ is absorbed.

$$C(s) + 2\ F_2(g) \rightleftharpoons CF_4(g)\quad \Delta H^\circ = -680\ kJ/mol_{rxn}$$

40. The above reaction is at equilibrium in a sealed flask. Which of the following actions would cause an increase in the reverse reaction rate after equilibrium has been reestablished?

    (A) Decreasing the temperature
    (B) Adding more $F_2(g)$
    (C) Removing some $F_2(g)$
    (D) Removing some $C(s)$

41. In which of the following molecules does the sulfur atom have the lowest oxidation number?

    (A) SO_2
    (B) SCl_2
    (C) H_2SO_4
    (D) SF_6

42. At 10°C, the pH of a pure water solution is found to be 7.27. Which of the following best explains why?

    (A) Water becomes more basic as temperature decreases.
    (B) At 10°C, $[OH^-] < [H^+]$.
    (C) The auto-ionization of water is endothermic.
    (D) The strength of water's intermolecular forces decreases at lower temperatures.

$$2\ HgCl_2(aq) + C_2O_4{}^{2-}(aq) \rightarrow 2\ Cl^-(aq) + 2\ CO_2(g) + Hg_2Cl_2(s)$$

43. Three trials of the above reaction are run with varying concentrations of each reactant and the initial rate of each trial is tracked. Based on the data below, what is the correct rate law for the reaction?

Trial	$[HgCl_2]$ (M)	$[C_2O_4{}^{2-}]$ (M)	Initial Rate (M/s)
1	0.10	0.10	2.5×10^{-5}
2	0.20	0.10	5.0×10^{-5}
3	0.20	0.20	2.0×10^{-4}

    (A) Rate $= k[HgCl_2][C_2O_4{}^{2-}]$
    (B) Rate $= k[C_2O_4{}^{2-}]^2$
    (C) Rate $= k[HgCl_2][C_2O_4{}^{2-}]^2$
    (D) Rate $= k[HgCl_2]^2[C_2O_4{}^{2-}]^2$

44. Which of the following ions would have the most unpaired electrons?

    (A) Na^+
    (B) S^{2-}
    (C) Co^{2+}
    (D) Cr^{2+}

GO ON TO THE NEXT PAGE.

Use the following information to answer questions 45–49.

An 0.740 g sample of an unknown alkali carbonate, identified using M_2CO_3, is dissolved in 100. mL of water. Excess aqueous $Ca(NO_3)_2$ is added to the solution, causing all of the carbonate to precipitate out as calcium carbonate ($CaCO_3$, MM = 100 g/mol). The calcium carbonate is filtered, dried, and then massed. It has a final mass of 1.00 g.

45. Identify the correct net ionic equation for the precipitation reaction.

    (A) $M_2CO_3(aq) + Ca(NO_3)_2(aq) \rightarrow CaCO_3(s) + 2\ MNO_3(aq)$
    (B) $M_2CO_3(aq) + Ca(NO_3)_2(aq) \rightarrow CaCO_3(aq) + 2\ MNO_3(s)$
    (C) $2\ M^+(aq) + CO_3^{2-}(aq) + Ca^{2+}(aq) + 2\ NO_3^-(aq) \rightarrow CaCO_3(s) + 2\ MNO_3(aq)$
    (D) $CO_3^{2-}(aq) + Ca^{2+}(aq) \rightarrow CaCO_3(s)$

46. The solution in the beaker after the two solutions are mixed can be best described as

    (A) unsaturated
    (B) saturated
    (C) supersaturated
    (D) hypersaturated

47. Which of the below diagrams correctly shows the ions present in significant amounts after the reaction has gone to completion?

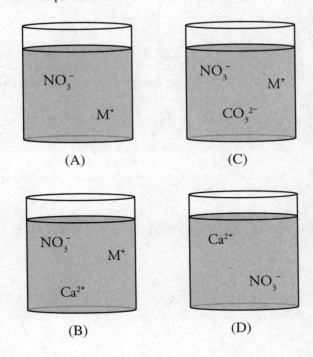

48. Which of the following is the original alkali carbonate?

    (A) Li_2CO_3
    (B) Na_2CO_3
    (C) K_2CO_3
    (D) Rb_2CO_3

49. Which of the following lab errors would NOT affect the identification of the alkali cation present in the original sample?

    (A) Spilling some of the alkali carbonate between the balance and the beaker
    (B) Not fully drying the calcium carbonate that is collected at the end
    (C) Leaving a cloudy filtrate after filtering out the precipitate
    (D) Dissolving the alkali carbonate sample in 150 mL of water instead of 100 mL

GO ON TO THE NEXT PAGE.

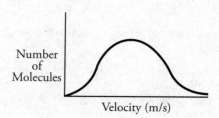

50. The Lewis diagram for nitric acid, HNO$_3$, is above. Which of the following statements is true regarding nitric acid?

(A) The nitrogen and oxygen atoms are all in the same plane.
(B) The nitrogen atom carries a negative formal charge.
(C) All N-O bonds are of identical length.
(D) There is a third equivalence resonance structure that is not shown.

Use the following information to answer questions 51–53.

$$Ni(IO_3)_2(s) \rightleftharpoons Ni^{2+}(aq) + 2\,IO_3^-(aq)$$

Nickel (II) iodate is a slightly soluble salt which dissociates in water via the above equilibrium reaction.

51. In a fully saturated solution of Ni(IO$_3$)$_2$, how do the ion concentrations compare?

(A) $[Ni^{2+}] = [IO_3^-]$
(B) $[Ni^{2+}] = 2[IO_3^-]$
(C) $2[Ni^{2+}] = [IO_3^-]$
(D) $[Ni^{2+}]^2 = [IO_3^-]$

52. In which of the following solutions would the least amount of Ni(IO$_3$)$_2$(s) dissolve?

(A) 0.5 M Ni(NO$_3$)$_2$
(B) 1.0 M KIO$_3$
(C) 2.0 M NaCl
(D) Pure water

53. If a saturated solution of Ni(IO$_3$)$_2$ is left out overnight and some of the water evaporates, how will that affect the mass of the solid present and/or the concentration of the ions in solution? Assume the temperature of the solution remains constant throughout.

	Mass Ni(IO$_3$)$_2$(s)	[Ni^{2+}] and [IO$_3^-$]
(A)	Increase	Increase
(B)	Increase	Unchanged
(C)	Unchanged	Increase
(D)	Unchanged	Unchanged

54. A compound containing only hydrogen and carbon is combusted in excess oxygen, producing 4.4 g of CO$_2$ and 1.8 g of H$_2$O. Which of the following options could be the molar mass of the compound?

(A) 18 g/mol
(B) 28 g/mol
(C) 38 g/mol
(D) 48 g/mol

55. The velocity distribution for H$_2$O(g) at a given temperature is drawn above. Which of the following distribution plots would represent the velocity distribution of an equimolar sample of CO$_2$(g) at an identical temperature? The H$_2$O distribution curve is shown as a dotted line in each option, while the CO$_2$(g) distribution curve is a solid line.

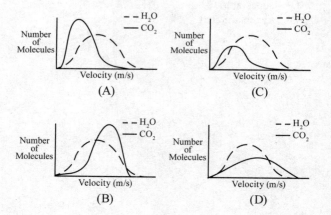

GO ON TO THE NEXT PAGE.

$$CO_2(s) \rightarrow CO_2(g)$$

56. Solid carbon dioxide is commonly known by the name dry ice. Dry ice is kept in solid form by sealing it into a high pressure container. When the valve is released, the decrease in pressure causes the CO_2 to sublimate via the above reaction. Predict the signs for $\Delta H°$ and $\Delta S°$ for the sublimation process.

	$\Delta H°$	$\Delta S°$
(A)	+	+
(B)	+	−
(C)	−	+
(D)	−	−

$$2\ N_2O_5(s) \rightleftharpoons 4\ NO(g) + 3\ O_2(g) \quad \Delta H_{rxn} = +247.4\ kJ/mol$$

57. Which of the following changes would cause the value for the equilibrium constant for the above reaction to increase?

    (A) Adding some $N_2O_5(s)$
    (B) Increasing the pressure
    (C) Adding some $O_2(g)$
    (D) Increasing the temperature

Use the following information to answer questions 58–60.

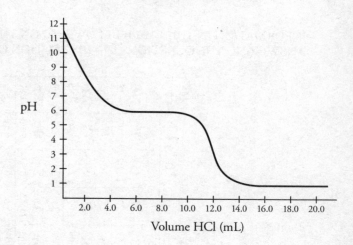

$$HONH_2(aq) + H^+(aq) \rightleftharpoons HONH^{3+}(aq)$$

Hydroxylamine, a weak base, is titrated with $1.0\ M$ hydrochloric acid, creating the above titration curve. The temperature of the solution is a constant 25°C.

58. Which of the following reactions best explains why the solution is acidic at the equivalence point of the reaction?

    (A) $HONH_3^+(aq) + H_2O(l) \rightleftharpoons HONH_2(aq) + H_3O^+(aq)$
    (B) $HONH_2(aq) + H^+(aq) \rightleftharpoons HONH_3^+(aq)$
    (C) $HONH_2(aq) + H_2O(l) \rightleftharpoons HONH_3^+(aq) + OH^-(aq)$
    (D) $HONH_3^+(aq) + OH^-(aq) \rightleftharpoons HONH_2(aq) + H_2O(l)$

59. What is the approximate pK_b for $HONH_2$?

    (A) 4
    (B) 6
    (C) 8
    (D) 10

60. Which of the following statements correctly identifies the concentration of the various ions in solution after 15.0 mL of HCl has been added?

    (A) $[H^+] > [HONH_3^+] > [HONH_2]$
    (B) $[HONH_3^+] > [HONH_2] > [H^+]$
    (C) $[HONH_3^+] > [H^+] > [HONH_2]$
    (D) $[HONH_2] > [H^+] > [HONH_3^+]$

END OF SECTION I

INFORMATION IN THE TABLE BELOW AND ON THE FOLLOWING PAGES MAY BE USEFUL IN ANSWERING THE QUESTIONS IN THIS SECTION OF THE EXAMINATION

DO NOT DETACH FROM BOOK.

PERIODIC TABLE OF THE ELEMENTS

1	2	3	4	5	6	7	8	9	10	11	12	13	14	15	16	17	18
1 **H** 1.008																	2 **He** 4.00
3 **Li** 6.94	4 **Be** 9.01											5 **B** 10.81	6 **C** 12.01	7 **N** 14.01	8 **O** 16.00	9 **F** 19.00	10 **Ne** 20.18
11 **Na** 22.99	12 **Mg** 24.30											13 **Al** 26.98	14 **Si** 28.09	15 **P** 30.97	16 **S** 32.06	17 **Cl** 35.45	18 **Ar** 39.95
19 **K** 39.10	20 **Ca** 40.08	21 **Sc** 44.69	22 **Ti** 47.87	23 **V** 50.94	24 **Cr** 52.00	25 **Mn** 54.94	26 **Fe** 55.85	27 **Co** 58.93	28 **Ni** 58.69	29 **Cu** 63.55	30 **Zn** 65.38	31 **Ga** 69.72	32 **Ge** 72.63	33 **As** 74.92	34 **Se** 78.97	35 **Br** 79.90	36 **Kr** 83.80
37 **Rb** 85.47	38 **Sr** 87.62	39 **Y** 88.91	40 **Zr** 91.22	41 **Nb** 92.91	42 **Mo** 95.95	43 **Tc**	44 **Ru** 101.07	45 **Rh** 102.91	46 **Pd** 106.42	47 **Ag** 107.87	48 **Cd** 112.41	49 **In** 114.82	50 **Sn** 118.71	51 **Sb** 121.76	52 **Te** 127.60	53 **I** 126.90	54 **Xe** 131.29
55 **Cs** 132.91	56 **Ba** 137.33	57-71 *	72 **Hf** 178.49	73 **Ta** 180.95	74 **W** 183.94	75 **Re** 186.21	76 **Os** 190.23	77 **Ir** 192.22	78 **Pt** 195.08	79 **Au** 196.97	80 **Hg** 200.59	81 **Tl** 204.38	82 **Pb** 207.2	83 **Bi** 208.98	84 **Po**	85 **At**	86 **Rn**
87 **Fr**	88 **Ra**	89-103 †	104 **Rf**	105 **Db**	106 **Sg**	107 **Bh**	108 **Hs**	109 **Mt**	110 **Ds**	111 **Rg**	112 **Cn**	113 **Nh**	114 **Fl**	115 **Mc**	116 **Lv**	117 **Ts**	118 **Og**

*Lanthanoids

57 **La** 138.91	58 **Ce** 140.12	59 **Pr** 140.91	60 **Nd** 144.24	61 **Pm**	62 **Sm** 150.36	63 **Eu** 151.97	64 **Gd** 157.25	65 **Tb** 158.93	66 **Dy** 162.50	67 **Ho** 164.93	68 **Er** 167.26	69 **Tm** 168.93	70 **Yb** 173.05	71 **Lu** 174.97

†Actinoids

89 **Ac**	90 **Th** 232.04	91 **Pa** 231.04	92 **U** 238.03	93 **Np**	94 **Pu**	95 **Am**	96 **Cm**	97 **Bk**	98 **Cf**	99 **Es**	100 **Fm**	101 **Md**	102 **No**	103 **Lr**

GO ON TO THE NEXT PAGE.

AP® CHEMISTRY EQUATIONS & CONSTANTS

Throughout the exam the following symbols have the definitions specified unless otherwise noted.

L, mL	=	liter(s), milliliter(s)	mm Hg = millimeters of mercury
g	=	gram(s)	J, kJ = joule(s), kilojoule(s)
nm	=	nanometer(s)	V = volt(s)
atm	=	atmosphere(s)	mol = mole(s)

ATOMIC STRUCTURE

$$E = h\nu$$
$$c = \lambda\nu$$

E = energy
ν = frequency
λ = wavelength

Planck's constant, $h = 6.626 \times 10^{-34}$ J s

Speed of light, $c = 2.998 \times 10^8$ m s^{-1}

Avogadro's number $= 6.022 \times 10^{23}$ mol^{-1}

Electron charge, $e = -1.602 \times 10^{-19}$ coulomb

EQUILIBRIUM

$K_c = \dfrac{[C]^c[D]^d}{[A]^a[B]^b}$, where a A $+ b$ B $\rightleftarrows c$ C $+ d$ D

$K_p = \dfrac{(P_C)^c(P_D)^d}{(P_A)^a(P_B)^b}$

$K_a = \dfrac{[H^+][A^-]}{[HA]}$

$K_b = \dfrac{[OH^-][HB^+]}{[B]}$

$K_w = [H^+][OH^-] = 1.0 \times 10^{-14}$ at 25°C

$\quad = K_a \times K_b$

pH $= -\log[H^+]$, pOH $= -\log[OH^-]$

14 = pH + pOH

pH $= pK_a + \log\dfrac{[A^-]}{[HA]}$

$pK_a = -\log K_a$, $pK_b = -\log K_b$

Equilibrium Constants

K_c (molar concentrations)

K_p (gas pressures)

K_a (weak acid)

K_b (weak base)

K_w (water)

KINETICS

$$[A]_t - [A]_0 = -kt$$

$$\ln[A]_t - \ln[A]_0 = -kt$$

$$\frac{1}{[A]_t} - \frac{1}{[A]_0} = kt$$

$$t_{1/2} = \frac{0.693}{k}$$

k = rate constant
t = time
$t_{1/2}$ = half-life

GO ON TO THE NEXT PAGE.

GASES, LIQUIDS, AND SOLUTIONS

$$PV = nRT$$

$$P_A = P_{total} \times X_A, \text{ where } X_A = \frac{\text{moles A}}{\text{total moles}}$$

$$P_{total} = P_A + P_B + P_C + \ldots$$

$$n = \frac{m}{M}$$

$$K = {}^\circ C + 273$$

$$D = \frac{m}{V}$$

$$KE_{molecule} = \frac{1}{2}mv^2$$

Molarity, M = moles of solute per liter of solution

$$A = \varepsilon bc$$

P = pressure
V = volume
T = temperature
n = number of moles
m = mass
M = molar mass
D = density
KE = kinetic energy
v = velocity
A = absorbance
ε = molar absorptivity
b = path length
c = concentration

Gas constant, R = 8.314 J mol^{-1} K^{-1}
= 0.08206 L atm mol^{-1} K^{-1}
= 62.36 L torr mol^{-1} K^{-1}
1 atm = 760 mm Hg = 760 torr
STP = 273.15 K and 1.0 atm
Ideal gas at STP = 22.4 L mol^{-1}

THERMOCHEMISTRY/ ELECTROCHEMISTRY

$$q = mc\Delta T$$

$$\Delta S^\circ = \sum S^\circ \text{ products} - \sum S^\circ \text{ reactants}$$

$$\Delta H^\circ = \sum \Delta H_f^\circ \text{ products} - \sum \Delta H_f^\circ \text{ reactants}$$

$$\Delta G^\circ = \sum \Delta G_f^\circ \text{ products} - \sum \Delta G_f^\circ \text{ reactants}$$

$$\Delta G^\circ = \Delta H^\circ - T\Delta S^\circ$$

$$= -RT \ln K$$

$$= -nFE^\circ$$

$$I = \frac{q}{t}$$

$$E_{cell} = E_{cell}^\circ - \frac{RT}{nF} \ln Q$$

q = heat
m = mass
c = specific heat capacity
T = temperature
S° = standard entropy
H° = standard enthalpy
G° = standard Gibbs free energy
n = number of moles
E° = standard reduction potential
I = current (amperes)
q = charge (coulombs)
t = time (seconds)
Q = reaction quotient

Faraday's constant, F = 96,485 coulombs per mole of electrons

$$1 \text{ volt} = \frac{1 \text{ joule}}{1 \text{ coulomb}}$$

GO ON TO THE NEXT PAGE.

CHEMISTRY
SECTION II
Time—1 hour and 45 minutes
7 Questions

Directions: Questions 1–3 are long free-response questions that require about 23 minutes each to answer and are worth 10 points each. Questions 4–7 are short free-response questions that require about 9 minutes each to answer and are worth 4 points each.

On test day, you will be asked to show your work for each part in the space provided after that part. For this practice test, you may use scrap paper. Examples and equations may be included in your responses where appropriate. For calculations, clearly show the method used and the steps involved in arriving at your answers. You must show your work to receive credit for your answer. Pay attention to significant figures.

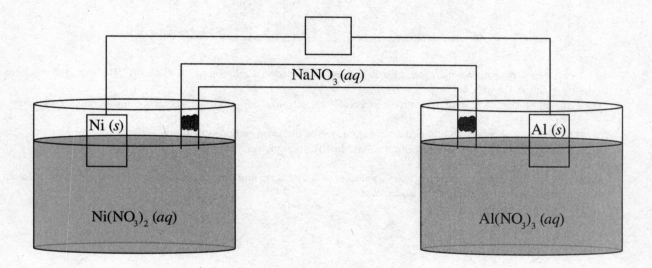

1. A nickel-aluminum voltaic cell is connected as shown above and held at 25°C. The concentration of all solutions is initially at 1.0 M. Some of the standard reduction potentials below may help you in answering the following questions.

Half-Reaction	Standard Reduction Potential (V)
$O_2(g) + 4\ H^+(aq) + 4\ e^- \rightarrow 2\ H_2O(l)$	1.23
$Ni^{2+}(aq) + 2\ e^- \rightarrow Ni(s)$	−0.25
$2\ H_2O(l) + 2\ e^- \rightarrow H_2(g) + 2\ OH^-(aq)$	−0.83
$Al^{3+}(aq) + 3\ e^- \rightarrow Al(s)$	−1.66

(a)   (i)   Write out and balance the full net ionic equation occurring in the galvanic cell.

      (ii)   Determine E^{o}_{cell} for the reaction in (a)(i).

(b)   If the concentration of the nickel (II) nitrate were to be doubled to 2.0 M while the concentration of the aluminum nitrate remained the same, how would that change the E_{cell} value, if at all? Justify your answer.

GO ON TO THE NEXT PAGE.

(c)   A student studying this cell claims that the electrons flow through the salt bridge from the anode to the cathode. Is this statement correct? Why or why not?

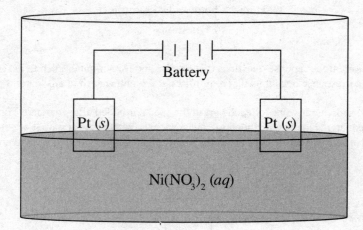

(d)   In a separate experiment, two inert platinum electrodes are placed in a solution of 1.0 M nickel (II) nitrate and a current is run through the electrodes.
   (i)   If a current of 2.0 A is run through the nickel (II) nitrate solution for 360. s, how many grams of nickel will plate out?
   (ii)   Will the nickel plate out at the platinum anode or the platinum cathode? Justify your answer.
   (iii)   Is this reaction thermodynamically favored? Justify your answer.

(e)   When the platinum electrodes are placed in a solution of 1.0 M aluminum nitrate and current is run through the solution, no aluminum plates out on either electrode. Why?

GO ON TO THE NEXT PAGE.

$$2 \, Na(s) + 2 \, H_2O(l) \rightarrow 2 \, Na^+(aq) + 2 \, OH^-(aq) + H_2(g)$$

2. Like all alkali metals, sodium reacts with water to produce hydrogen gas.

    (a)   (i)   Which element is being oxidized, and which is being reduced?

              (ii)   What property of alkali metals allows them to react so easily with water?

    (b)   (i)   If 2.00 g of sodium were to dissolve in 250. mL of pure water, and the resultant hydrogen was collected at 25°C and a pressure of 715 torr, what volume of gas would be produced?

              (ii)   What would the pH of the remaining solution be after the reaction has gone to completion? Assume the total volume remains unchanged.

    (c)   Three containers are filled completely with identical masses of three different gases, as is shown in the diagram. All containers are held at 25°C.

        (i)   In which container, if any, would the pressure be the greatest? Justify your answer.

        (ii)   In which container, if any, would the gas molecules have the highest average speed? Justify your answer.

    (d)   The three containers are cooled rapidly.

        (i)   As the containers are cooled, will the pressure inside them increase, decrease, or stay the same? Justify your answer by discussing the behavior of the gas molecules on a particulate level.

        (ii)   Which gas would condense first? Justify your answer.

GO ON TO THE NEXT PAGE.

3.

$$2\ KClO_3(s) \rightleftharpoons 2\ KCl(s) + 3\ O_2(g)$$

Potassium chlorate, $KClO_3$, will partially decompose at any temperature. 1.00 g of solid potassium chlorate is placed in a 250 mL flask at 23.0°C. The flask is capped, and the pressure inside is monitored until it stabilizes.

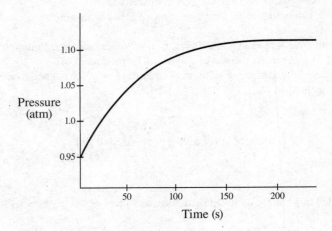

(a) (i) Calculate the partial pressure of the oxygen gas produced by the decomposition of the $KClO_3$.
   (ii) Calculate the value for the equilibrium constant expression, K_p, at 23.0°C.

(b) Would the mass of the $KClO_3$ increase, decrease, or remain the same between the beginning of the experiment and the point at which equilibrium is established? Justify your answer.

(c) If the experiment were repeated at the same temperature, but using a 125 mL flask, how would that change the total pressure inside the flask at the end, if at all? Justify your answer.

(d) Catalysts are often used to lower the activation energy for a given reaction. The experiment is repeated, this time with an MnO_2 catalyst added.
   (i) How do catalysts lower a reaction's activation energy?
   (ii) Using a dashed line, sketch a line on the graph that would indicate any changes that would occur in the data gathered during this trial.

(e) Calculate the value for ΔG for this reaction at 23.0°C.

(f) Is the entropy change that occurs during this reaction positive or negative? Justify your answer.

4. The primary acid in lemonade is citric acid ($H_3C_6H_5O_7$, $pK_a = 3.08$). To control the pH of many commercially available lemonade drinks, some sodium dihydrogen citrate ($NaH_2C_6H_5O_7$) is added to create a buffer system.

(a) Write out the reaction which would occur if a strong base were added to a buffered lemonade solution such as the one described above.

(b) What is the mole ratio of $H_2C_6H_5O_7^- : H_3C_6H_5O_7$ in a buffered solution with a pH of 3.25?

(c) If a buffered solution of lemonade were to be diluted, would that increase the pH, decrease it, or leave it the same? Justify your answer.

GO ON TO THE NEXT PAGE.

5.

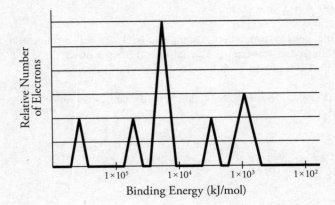

Binding Energy (kJ/mol)

Peak 1	Peak 2	Peak 3	Peak 4	Peak 5
2.7×10^5 kJ/mol	5.4×10^4 kJ/mol	2.1×10^4 kJ/mol	4.9×10^3 kJ/mol	1.0×10^3 kJ/mol

The photoelectron spectrum for a neutral atom is displayed above.

(a)   What is the total number of valence electrons shown on the above spectrum?

A separate sample of the element is bombarded with radiation that has a frequency of 6.3×10^{15} s^{-1}. The energy level is sufficient to only remove the outermost electron.

(b)   How much energy does a single photon of the radiation have?

(c)   How much kinetic energy would the ejected electron have?

6. Glyoxal is an organic compound commonly used in many synthesis reactions.

(a)   In the box below, complete the Lewis electron-dot diagram for the glyoxal molecule by drawing in all electron pairs.

(b)   Identify all types of intermolecular forces that are present in a solution of glyoxal.

(c)   Would you expect glyoxal to be miscible with water? Justify your answer.

GO ON TO THE NEXT PAGE.

7.

$$2 \, N_2O_5(g) \rightarrow 4 \, NO_2(g) + O_2(g)$$

The decomposition of N_2O_5 is a first-order reaction that goes to completion. A sample of N_2O_5 gas with a concentration of 1.0 M is placed in a sealed container at 50°C and allowed to decompose. The half-life of the reaction at 50°C is 1,200 seconds.

(a) On the graph below, draw a graph showing how the concentration of the N_2O_5 sample will change over time.

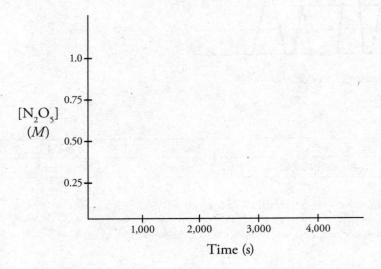

(b) Calculate the rate constant for this reaction at 50°C.

(c) Using your graph, determine the instantaneous rate of disappearance for N_2O_5 at $t = 2,000$ s.

STOP

END OF EXAM

Practice Test 5:
Answers and
Explanations

PRACTICE TEST 5: MULTIPLE-CHOICE ANSWER KEY

1.	C	21.	B	41.	B
2.	A	22.	B	42.	C
3.	B	23.	A	43.	C
4.	B	24.	B	44.	D
5.	D	25.	C	45.	D
6.	B	26.	D	46.	B
7.	D	27.	D	47.	B
8.	C	28.	A	48.	A
9.	D	29.	D	49.	D
10.	A	30.	A	50.	A
11.	A	31.	C	51.	C
12.	B	32.	A	52.	B
13.	C	33.	D	53.	B
14.	A	34.	C	54.	B
15.	A	35.	A	55.	A
16.	C	36.	A	56.	A
17.	A	37.	B	57.	D
18.	D	38.	A	58.	A
19.	D	39.	D	59.	C
20.	B	40.	B	60.	C

Section I—Multiple-Choice Answers and Explanations

1. **C** When the Ag_2CO_3 dissociates, it produces two Ag^+ ions for every one CO_3^{2-} ion.

2. **A** Henderson-Hasselbalch is used here, so pH = pK_a + log $[NO_2^-]/[HNO_2]$. For the pH to be the same as the pK_a of the acid, the logarithm term must cancel out. That happens when $[NO_2^-]$ = $[HNO_2]$.

3. **B** The graph indicates the decomposition reaction has a constant half-life. This means the reaction must be first order.

4. **B** There is resonance between the two S-O bonds where the H is not attached to the other side of the oxygen atom. These bonds are identical in length, but would be shorter than the true single S-O bond with the oxygen that does have the hydrogen bonded to it.

5. **D** The H^+ ions are being reduced, but the K(s) is being oxidized, so the potassium half-reaction has to be flipped. It also has to be multiplied by two so the electrons will cancel out when the two half-reactions are combined.

6. **B** Vapor pressure is a function of the strength of the intermolecular forces. When the methanol is heated, the methanol molecules will be moving faster, meaning they are more likely to have enough energy to break free of the IMFs, meaning more of them will end up in a gaseous state, increasing the vapor pressure.

7. **D** First, solve for the heat lost by the water using $q = mc\Delta T$. In this case, the mass is the mass of the dissolved compound plus that of the water, so about 100 g.

    $q = (100 \text{ g})(4.2 \text{ J/g°C})(-3.0°C)$ $q = -1,260 \text{ J}$

    To determine the enthalpy of reaction, flip the sign and then divide by the number of moles.

    $\Delta H = q_{rxn}/n$ $\Delta H = 1,260 \text{ J}/0.10 \text{ mol} = 12,600 \text{ J/mol} = 12.6 \text{ kJ/mol}$

8. **C** The reaction occurs, so it must be favored. The surrounding water lost energy, meaning the reaction required energy to occur. This means the reaction is endothermic, which is not a favorable enthalpy change. The KNO_2 starts as a solid and ends up as dissociated ions. Those ions are much more dispersed than the solid, meaning dispersion increases, which is a favorable entropy change.

9. **D** When an ionic salt dissolves in water, the energy required is that to break up the ionic lattice, whereas the energy released comes from the attraction of the ions to the dipoles of the surrounding water molecules. The reaction was endothermic, which indicates the lattice energy is greater than the ion-dipole attractions. (Note that the reaction still proceeds due to its favorable entropy change!)

10. **A** If not all of the salt dissolved, the moles of salt in the denominator of the enthalpy of reaction expression would be artificially high, making the enthalpy of reaction artificially low.

11. **A** For the salt to precipitate, the value for the reaction quotient has to be equal to or greater than the equilibrium constant for the reaction.

$$Q_{sp} = [Sr^{2+}][IO_3^-]^2 \qquad 1.0 \times 10^{-7} = [Sr^{2+}](1.0 \times 10^{-2})^2 \qquad [Sr^{2+}] = 1.0 \times 10^{-3} \, M$$

12. **B** After an acid has completely reacted during a titration, all that is present in the solutions is the conjugate base of that acid. In a polyprotic titration, the endpoint occurs after all of the protons have been neutralized. In each case, the lower the K_a is for the final dissociation of the weak acid, the higher the K_b will be of its conjugate base (because for any conjugate pair, $K_a \times K_b = 1.0 \times 10^{-14}$). A higher K_b means a stronger conjugate base, leading to a higher pH at the endpoint.

13. **C** A catalyst speeds up a reaction by lowering the activation energy of that reaction. The activation energy is shown on the graph as the height of the "hump," so a catalyzed reaction will have a lower hump. Note that the energy change of the reaction (the distance between the energy level of the reactants and the products) will remain unchanged.

14. **A** The higher the K_a value, the greater the percent dissociation for the acid is, and the higher the concentration of H^+ ions will be.

15. **A** To determine the conjugate base of a weak acid, simply remove the proton (H^+).

16. **C** To create a buffer, similar amounts of a weak acid and its conjugate must be present. The conjugate base of $HClO_2$ is ClO_2^-, so adding a soluble salt with the ClO_2^- ion present would create a buffer.

17. **A** $HClO_2(aq) \rightleftharpoons H^+(aq) + ClO_2^-(aq) \qquad K_a = [H^+][ClO_2^-]/[HClO_2^-]$

$$1.0 \times 10^{-2} = x^2/1.0 - x \qquad x^2 = 1.0 \times 10^{-2} \, M \qquad x = 1.0 \times 10^{-1} \, M = [H^+]$$

$$-\log[H^+] = pH \qquad -\log(1.0 \times 10^{-1}) = 1$$

Note that the above calculations ignore the "$-x$" term in the denominator. While $HClO_2$ has a high enough K_a that the dissociation will be significant and affect the pH, it can still be ignored for the purposes of approximation.

18. **D** The first dissociation is $H_2CO_3 \rightleftharpoons H^+ + HCO_3^-$. The second is $HCO_3^- \rightleftharpoons H^+ + CO_3^{2-}$. If CO_3^{2-} is a more effective, or stronger, base than HCO_3^-, then the second dissociation will be less favored because stronger acids have weaker conjugate bases. That means the second K_a should be lower than the first.

19. **D** Formal charge is calculated by subtracting the number of assigned electrons on an atom in a Lewis diagram from the number of its valence electrons. When looking at an atom, lone pairs are considered to be two assigned electrons, and any type of bond is considered to be one (so a single bond is one, a double bond would be two, etc). Constructing a formal charge chart for the identified atoms shows that, of the identified atoms, only O_z would have a negative formal charge.

	P	O_x	O_y	O_z
Valence	5	6	6	6
Assigned	5	6	6	7
Formal Charge	0	0	0	−1

20. **B** To fully balance the reaction, the charge must be balanced. This required multiplying the zinc oxidation half-reaction by three, and the nitrate reduction half-reaction by two. Doing that and combining the reactions yields:

$$3 \text{ Zn}(s) + 8 \text{ H}^+(aq) + 2 \text{ NO}_3^-(aq) \rightarrow 2 \text{ NO}(g) + 4 \text{ H}_2\text{O}(l) + 3 \text{ Zn}^{2+}(aq)$$

21. **B** Orbital hybridization can be determined by looking at the number of charge clouds around the atom in question. The sulfur has three charge clouds—two lone pair, and one bond. That is consistent with sp^2 hybridization.

22. **B** Single bonds are longer than double bonds, and when comparing the S–C and C–H bonds, a sulfur atom is bigger than a hydrogen atom, which would mean the S–C bond is longer.

23. **A** Oxygen is very electronegative, and would be the most effective at attracting electrons toward it. That would give it a negative partial charge, and since oxygen is directly bonded to sulfur, sulfur would have a positive partial charge.

24. **B** In order to separate out the propanol, it must be boiled, so the temperature must be above the boiling point of propanol. However, the temperature should also be below the boiling point of propionic acid, because if both substances boil, they will not separate effectively. There are two options in the temperature range, and the lower one is better because the propionic acid will have a lower vapor pressure at 105°C, meaning less of it will evaporate and contaminate the distillate.

25. **C** For any reaction, the coefficients on the reactants in the slowest elementary step are equal to the order of those reactants in the rate law. So, the rate law for the given mechanism is rate = $k[\text{NO}_2]^2$. A second-order reactant produces a straight line graph when graphing the reciprocal of that reactant vs. time.

26. **D** The type of radiation that is used to study sub-atomic particles depends on the size of the particles. The higher the frequency of the radiation, the smaller the particles that can be studied. For studying electrons, ultraviolet radiation is often used. Studying bonds (shared pairs of electrons) generally uses infrared radiation, while studying entire molecules uses microwave radiation.

27. **D** The diagrams can be diagnosed according to the relative atomic sizes of the ions. Li^+ and F^- would be the smallest ions, which are present in diagram Z. Na^+ and Cl^- would be the next biggest, so diagram X. Finally, K^+ and Br^- would be the largest, which corresponds with diagram Y.

28. **A** All three compounds contain ions of identical charge, and the only other factor that can affect the lattice energy (a type of Coulombic energy) is size, which is inversely proportional. Li^+ and F^- are the smallest ions, meaning they have the greatest amount of lattice energy and the highest melting point.

29. **D** When the cations lose their only valence electron, the valence shell becomes the one beneath it. For instance, Na^+ would have a configuration of $1s^2 2s^2 2p^6$. When the halide anions gain an electron, they too complete their outmost p-subshell. Li^+ is the exception because after lithium loses an electron, all that is left is the first energy level, which has no p-subshell.

30. **A** During a liquid-to-gas phase change, any energy added goes into breaking the intermolecular forces between molecules. The temperature of a boiling solution remains constant, and thus, so does the average kinetic energy of the molecules.

31. **C** A fluoride atom is $[He]2s^2 2p^5$. A fluorine ion would be $[He]2s^2 2p^6$. The $2p$ peak is the rightmost peak, and in the ion it would contain one more electron and thus be higher than the $2p$ peak from the neutral atom.

32. **A** Boron has a very low electronegativity value for a nonmetal, and within a molecule it is considered stable with just six valence electrons. Thus, a molecule of BF_3 would be stable with three single bonds and would look like this:

33. **D** There are nine oxygen atoms present in the hydrate: four in the anhydrous $CuSO_4$ salt, and five more from the five water molecules. In one mole of the hydrate, the mass of those oxygen atoms would be $16(9) = 144$ g. The oxygen thus represents 144 g out of a 250 g sample. It would maintain the same mass ratio in a 25.0 g sample, which would be 14.4 g.

34. **C** At STP, 1 mol of any gas has a volume of 22.4 L. If the gas in question has a volume of 5.6 L, that represents $5.6/22.4 = 0.25$ mole of the gas. Multiplying the mass of the sample by four would then yield the mass of 1.0 mole of the gas. $10.0 \text{ g} \times 4 = 40.0$ g/mol, which is the molar mass of argon.

35. **A** The coefficients on the reactants present in the slowest elementary step become the order of those reactants in the overall reaction. Therefore, the first step must be the slow/rate-determining step. Molecularity is dependent on how many reactant molecules there are, which in this case is just one.

36. **A** The units of rate are M/s. With that in mind, it's just a matter of using dimensional analysis on the rate law. Rate = $k[NO_2Cl]$ $M/s = k(M)$ $k = s^{-1}$

37. **B** Catalysts are present in the reactants of the first step and the products of the final step. No species in this reactant fits that profile. Intermediates are present in the products of an early step and the reactants of a later step. Cl fits the bill.

38. **A** The overall reaction is exothermic, meaning the enthalpy of the products must be lower than the enthalpy of the reactants. Additionally, step 1 of the process must be endothermic, since bonds are broken in step 1 without any new bonds being formed. The first hump is for the first step, so after that hump the energy level must be higher than where it started.

39. **D** $2.80 \text{ g CO} \times \dfrac{1 \text{ mol CO}}{28.0 \text{ g CO}} \times \dfrac{1 \text{ mol}_{rxn}}{3 \text{ mol CO}} \times \dfrac{26.4 \text{ kJ}}{1 \text{ mol}_{rxn}} = 0.88 \text{ kJ}$.

The positive sign means the reaction is endothermic, and thus heat was absorbed.

40. **B** Adding the $F_2(g)$ would cause an immediate shift to the right, which would increase the forward reaction rate. However, the key is that after equilibrium is reestablished there is more matter in the flask, meaning both the forward and reverse reaction rates will be higher than they were initially (and still equal to each other).

41. **B** SO_2: $S + (-2)2 = 0$ $S = +4$ SCl_2: $S + (-1)(2) = 0$ $S = +2$

H_2SO_4: $(+1)2 + S + (-2)4 = 0$ $S = +6$ SF_6: $S + (-1)(6) = 0$ $S = +6$

42. **C** Pure water has a neutral pH of 7 at 25°C only. When the temperature changes, so does the pH of water. The auto-ionization of water is $H_2O(l) \rightleftharpoons H^+(aq) + OH^-(aq)$. When temperature decreases, that equilibrium shifts to the left, as it would in any endothermic process. This decreases the value for the equilibrium constant, meaning that both $[H^+]$ and $[OH^-]$ decrease. If $[H^+]$ and $[OH^-]$ decrease, the pH of the water increases, as $[H^+]$ and pH are inversely proportional.

43. **C** Between trials 1 and 2, the $[HgCl_2]$ doubles while $[C_2O_4^{2-}]$ stays constant. The rate of the reaction also doubles, which demonstrates a first-order relationship with regard to $[HgCl_2]$. Between trials 2 and 3, $[C_2O_4^{2-}]$ doubles while $[HgCl_2]$ remains the same. The rate increases by a factor of four, demonstrating a second-order relationship with regard to $[C_2O_4^{2-}]$.

44. **D** Choices (A) and (B) have no unpaired electrons. Comparing (C) and (D), both are transition metals, which always lose their s electrons first when forming ions. Cr^{2+} would thus have four unpaired d electrons, while Co^{2+} would only have three.

45. **D** In a net ionic reaction, the only ions that appear are those that undergo a chemical change. In this case, that is the Ca^{2+} and CO_3^{2-} ions. The M^+ and NO_3^- are spectator ions and would not appear in the net ionic equation.

46. **B** Any solution that has a solid precipitate at the bottom is considered to be saturated at the very least. For it to be supersaturated, the solution would need to be heated to dissolve more precipitate and then cooled back to its original temperature.

47. **B** The calcium nitrate is described as being in excess, so there will definitely be Ca^{2+} and NO_3^- ions in solution. In addition, the M^+ alkali cation does not react, so it will also be present in the final solution.

48. **A** $1.00 \text{ g CaCO}_3 \times \dfrac{1 \text{ mol CaCO}_3}{100. \text{ g CaCO}_3} = 0.010 \text{ mol CaCO}_3 \times \dfrac{1 \text{ mol CO}_3^{2-}}{1 \text{ mol CaCO}_3} = 0.010 \text{ mol CO}_3^{2-}$

$0.010 \text{ mol CO}_3^{2-} \times \dfrac{1 \text{ mol M}_2\text{CO}_3}{1 \text{ mol CO}_3^{2-}} = 0.010 \text{ mol M}_2\text{CO}_3 \qquad \dfrac{0.74 \text{ g}}{0.010 \text{ mol}} \doteq 74 \text{ g/mol}$

74 g/mol is the molar mass of lithium carbonate.

49. **D** If the alkali carbonate dissolves fully in 100. mL of water, it will also dissolve fully in 150. mL of water. The same amount of carbonate will be in solution, leading to no difference in the final mass of the $CaCO_3$ precipitate.

50. **A** The resonance shown here is for the N–O bonds that do not have hydrogen attached on the other end of the oxygen. This means that two of the N–O bonds are identical (bond order = 1.5), but the third one (from N–O–H) is a true single bond. The three O atoms are all coplanar with the N, which has three charge clouds and is trigonal planar.

51. **C** Based on the balanced equilibrium reaction, in a saturated solution there will be twice as many IO_3^- ions as there are Ni^{2+} ions.

52. **B** The common ion effect states that if the salt that is being dissolved shares an ion with the solution that it is being dissolved in, less of that salt will dissolve. This is because the presence of that ion will push the solubility equilibrium to the left. Both (A) and (B) share ions with $Ni(IO_3)_2$, but there are more of those ions present in (B) due to the higher concentration.

53. **B** As long as temperature stays unchanged, so does the K_{sp} value for the salt, meaning the ion concentration will not change. However, if some water evaporates, that means the volume decreases. For the ion concentration to remain unchanged, a decrease in volume must be accompanied by a decrease in mass of the ions dissolved in solution. Those ions will recombine and precipitate out, increasing the mass of solid present at the bottom.

54. **B** For any hydrocarbon combustion, all of the carbon in the original compound ends up in CO_2, and all of the hydrogen ends up in H_2O. Some stoichiometry gives us the mole ratio of carbon to hydrogen in the original compound.

$$4.4 \text{ g } CO_2 \times \frac{1 \text{ mol } CO_2}{44.0 \text{ g } CO_2} \times \frac{1 \text{ mol C}}{1 \text{ mol } CO_2} = 0.10 \text{ mol C}$$

$$1.8 \text{ g } H_2O \times \frac{1 \text{ mol } H_2O}{18.0 \text{ g } H_2O} \times \frac{2 \text{ mol H}}{1 \text{ mol } H_2O} = 0.20 \text{ mol H}$$

Thus, the ratio of C:H in the compound must be 1:2. A single CH_2 unit would have a molar mass around 14 g/mol and two of those units (C_2H_4) would have a molar mass around 28 g/mol.

55. **A** Given that both samples are at the same temperature, their particles have the same average kinetic energy. As $KE = \frac{1}{2} mv^2$, the sample with the lower molar mass (H_2O) must have a higher average velocity. Additionally, as the samples are equimolar, the area under the curves must be the same. Thus, the apex on the H_2O curve must be lower to compensate for the wider potential velocity distribution.

56. **A** The solid turning into a gas is an increase in dispersion, making $\Delta S°$ positive. In order for the solid to turn into a gas, IMFs must be broken, which requires the input of energy from the environment. So, $\Delta H°$ is also positive.

57. **D** The only way to change the equilibrium constant for any reaction is by changing the temperature. Increasing the temperature on an endothermic reaction permanently shifts it right, creating more products and increasing the value of the equilibrium constant.

58. **A** At the equivalence point, all of the $HONH_2$ has been protonated by the H^+ because equal amounts have been added. The only ion present at this point is the $HONH_3^+$, which reacts with water to create an acidic environment.

59. **C** At the half-equivalence point of this reaction, the pOH of the solution is equal to the pK_b of hydroxylamine. This occurs at about 6.0 mL, at which point the pH is 6. That means the pOH is 8 (because at 25°C pH + pOH = 14), as is the pK_b for hydroxylamine.

60. **C** 15.0 mL is past the equivalence point, so all of the original $HONH_2$ has been protonated and there is none left in solution. It took 12.0 mL of H^+ to create the $HONH_3^+$ in solution, and after that, only 3.0 mL of H^+ have been added. So, the amount of $HONH_3^+$ is still greater than the amount of unreacted H^+ in solution.

Section II—Free-Response Answers and Explanations

1. (a) (i) The trick here is to make sure the electrons are balanced in the two half-reactions. This requires multiplying the nickel half-reaction by 3, and the aluminum half-reaction by 2. When you combine them, 6 moles of electrons will then cancel.

$$3\ Ni^{2+}(aq) + 2\ Al(s) \rightarrow 2\ Al^{3+}(aq) + 3\ Ni(s)$$

(ii) $-0.25\ V - (-1.66\ V) = 1.41\ V$

(b) Whether or not the voltage increases or decreases depends on if the change described brings the reaction closer to or further from equilibrium. To determine that, look at Q.

$$Q = \frac{[Al^{3+}]^2}{[Ni^{2+}]^3}$$

If the nickel concentration increases while the aluminum concentration remains the same, that means Q will decrease. As this cell is voltaic, that means K is much, much bigger than one. If Q is getting smaller, that is taking the cell further from equilibrium. For any cell, at equilibrium the voltage is zero, so taking the cell further from equilibrium takes the voltage further from zero. Thus, E_{cell} would increase.

(c) This statement is incorrect. Electrons do not flow through the salt bridge. Instead, they flow through the wires.

(d) (i) Electroplating problems are all about dimensional analysis/unit conversions. Remember, an ampere is equivalent to a coulomb per second.

$$360.\ s \times \frac{2.0\ C}{s} \times \frac{1\ mol\ e^-}{96{,}500\ C} \times \frac{1\ mol\ Ni}{2\ mol\ e^-} \times \frac{58.69\ g}{1\ mol\ Ni} = 0.22\ g\ Ni$$

(ii) It will plate out at the cathode, because the cathode is where the reduction reaction occurs.

$$Ni^{2+} + 2\ e^- \rightarrow Ni(s)$$

(iii) This reaction is not favored, as an outside source of energy (in this case, electrical current) is required for the reaction to occur.

(e) The reduction potential of the Al^{3+} ion, $-1.66\ V$, is lower than that of water, $-0.83\ V$. Thus, when current is run through the solution, the reduction of water will occur. This will produce hydrogen gas as water is reduced, but the aluminum ions will stay in solution.

2. (a) (i) Sodium is being oxidized, going from an oxidation state of 0 to +1. Hydrogen is being reduced, going from a state of +1 in water to 0 in hydrogen gas.

(ii) Alkali metals all have very low ionization energies, and so they can be readily oxidized by water.

(b) (i) First, figure out how many moles of hydrogen gas are being produced.

$$2.00 \text{ g Na} \times \frac{1 \text{ mol Na}}{22.99 \text{ g}} \times \frac{1 \text{ mol H}_2}{2 \text{ mol Na}} = 0.0435 \text{ mol H}_2$$

Then, use the Ideal Gas Law to convert that to volume.

$$P = \frac{715 \text{ torr}}{760 \text{ torr}} = 0.941 \text{ atm} \qquad T = 25°C + 273 = 298 \text{ K}$$

$$PV = nRT$$

$$(0.941 \text{ atm})(V) = (0.0435 \text{ mol})(0.08206 \text{ atm·L/mol·K})(298 \text{ K})$$

$$V = 1.13 \text{ L}$$

(ii) The pH will be dependent on the concentration of the OH^- ions, as they are a strong base.

$$2.00 \text{ g Na} \times \frac{1 \text{ mol Na}}{22.99 \text{ g}} \times \frac{2 \text{ mol OH}^-}{2 \text{ mol Na}} = 0.0870 \text{ mol OH}^-$$

$$\frac{0.0870 \text{ mol OH}^-}{0.250 \text{ L}} = 0.348 \ M$$

$$-\log[OH^-] = pOH \quad -\log(0.348) \quad = 0.458$$

$$pH + pOH = 14 \quad pH + 0.458 = 14$$

$$pH = 13.542$$

(c) (i) If the same mass of gas is present in each container, the gas with the lowest molar mass would have the greatest number of moles present. Moles and pressure are directly proportional, so in this case, the container with the hydrogen gas would have the greatest pressure.

(ii) As all containers are at the same temperature, the gases inside have the same amount of kinetic energy. Both the mass and velocity of the individual gas molecules contribute to their total kinetic energy. If the kinetic energy is the same, the gas molecules with the lowest mass would also have the highest velocity to compensate. Thus, the hydrogen molecules would have the greatest average velocity.

(d) (i) The pressure in each container will decrease. As the temperature drops, the molecules in each container will slow down. They will thus hit the sides of the container less often and with less force, meaning the total pressure decreases.

(ii) A phase change from gas to liquid will occur once the gas molecules are moving slowly enough that the intermolecular forces between them become significant enough to affect their behavior. The gas with the strongest IMFs will condense first. As all three gases are non-polar, the only IMFs present are London dispersion forces. The CO_2 gas has the most polarizable electron cloud, and so it has the strongest LDFs and would condense first.

3.   (a)   (i)   The initial pressure of the flask at $t = 0$ s is the pressure of just the atmosphere. The total pressure at the end consists of both the atmosphere and the produced oxygen gas. Subtracting the two values would give you the partial pressure of the oxygen gas.

1.11 atm − 0.95 atm = 0.16 atm = P_{O_2}

(ii)   Remember, solids do not appear in equilibrium expressions.

$K_p = (P_{O_2})^3$

$K_p = (0.16)^3 = 4.1 \times 10^{-3}$

(b)   The mass of the $KClO_3$ would decrease. The reason that solids do not appear in the equilibrium expression is because their concentration is unchanging. However, the mass of the $KClO_3$ can still decrease without changing the concentration. For any products to be produced, at least some of the $KClO_3$ must decompose, reducing its mass.

(c)   The pressure in the flask would be the same. The initial atmospheric pressure would be the same, and as long as temperature is constant, the value for K_p would not change. If the K_p value does not change, neither does the partial pressure of the oxygen.

(d)   (i)   Catalysts lower activation energy by stabilizing an existing transition state or by providing an alternative reaction pathway (a different set of elementary steps by which the reaction occurs).

(ii)   The catalyst would not change the value for K_p, so the pressure would be the same. However, equilibrium would be established faster.

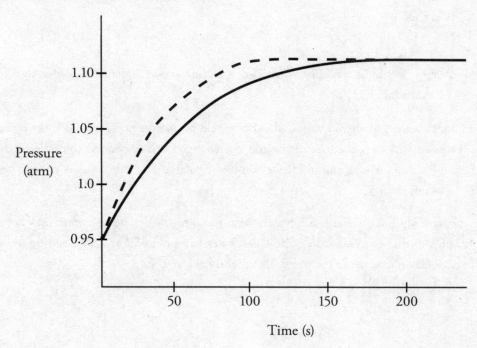

(e) $\Delta G = -RT \ln K$ $\Delta G = (-8.31 \text{ J/mol·K})(296 \text{ K})(\ln 4.1 \times 10^{-3})$ $\Delta G = +14{,}000 \text{ J}$

(f)   The entropy change is positive. The reaction begins with a solid, and ends with a solid and a gas. The gas is much more dispersed than the solid, leading to an increase in entropy.

4.   (a)   The exact reaction depends on how much strong base is added. Any of the following three answers would be acceptable.

$$H_3C_6H_5O_7(aq) + 3\ OH^-(aq) \rightarrow C_6H_5O_7^{3-}(aq) + 3\ H_2O(l)$$

OR

$$H_3C_6H_5O_7(aq) + 2\ OH^-(aq) \rightarrow HC_6H_5O_7^{2-}(aq) + 2\ H_2O(l)$$

OR

$$H_3C_6H_5O_7(aq) + OH^-(aq) \rightarrow H_2C_6H_5O_7^-(aq) + H_2O(l)$$

(b)   Using Henderson-Hasselbalch:

$$pH = pK_a + \log \frac{[C_6H_5O_7^{3-}]}{[H_3C_6H_5O_7]}$$

$$3.25 = 3.08 + \log \frac{[C_6H_5O_7^{3-}]}{[H_3C_6H_5O_7]}$$

$$0.17 = \log \frac{[C_6H_5O_7^{3-}]}{[H_3C_6H_5O_7]}$$

$$1.5 = \frac{[C_6H_5O_7^{3-}]}{[H_3C_6H_5O_7]}$$

The mole ratio would be the same as the concentration ratio calculated above, as the volume of the solution is constant.

(c)   There are two ways to answer this, and either would be acceptable as long as it is justified correctly. The textbook answer is that the pH would not change. This is because the ratio of the conjugate base:acid does not change, and in Henderson-Hasselbalch if that ratio doesn't change, the pH of the buffer does not change.

That being said, if the buffer is diluted enough, the auto-dissociation of water may become a larger contributor to the solution acidity, and since water has a pH of 7 at 25°C, adding more water to a buffered solution may cause the pH to rise slightly as a result.

5. (a) The peaks, from left to right, are $1s$, $2s$, $2p$, $3s$, and $3p$. Valence electrons are those on the outmost principal energy level, which in this case is the third energy level. Looking at the $3s$ and $3p$ peaks, they represent a total of five electrons.

    (b) $E = h\nu = (6.626 \times 10^{-34} \text{ J} \cdot \text{s})(6.3 \times 10^{15} \text{ s}^{-1}) = 4.2 \times 10^{-18} \text{ J}$

    (c) First, convert the energy from a single photon into kJ/mol.

    $$\frac{4.2 \times 10^{-18} \text{ J}}{\text{photon}} \times \frac{6.02 \times 10^{23} \text{ photons}}{1 \text{ mol}} \times \frac{1 \text{ kJ}}{1,000 \text{ J}} = 2,500 \text{ kJ/mol}$$

    For photoelectron spectroscopy, the incoming energy goes into two places. First, it overcomes the binding energy of the electron. Any energy left over after that goes into the kinetic energy of the electron once it is ejected. You can see from the data table that the valence ($3p$) electrons from phosphorus have a binding energy of 1,000 kJ/mol. So:

    Incoming Energy = Binding Energy + Kinetic Energy

    2,500 kJ/mol = 1,000 kJ/mol + KE

    KE = 1,500 kJ/mol

6. (a) Glyoxal has a total of 22 valence electrons to place. If all single bonds are used, 26 electrons are placed. So, two double bonds are needed. Hydrogen cannot form double bonds, so that's out. If you double bond the two carbons and one carbon-oxygen pair, that would give the carbon ten electrons, which is impossible because carbon cannot expand its octet. Thus, the only workable diagram has both carbon-oxygen pairs double bonded.

    (b) London dispersion forces are present, as are dipole-dipole attractions. Note that there are no hydrogen bonds, as for that to occur, the hydrogen would need to be directly bonded to the oxygen, which it is not.

    (c) Miscibility occurs when two liquids have similar intermolecular forces. Water is very polar due to hydrogen bonding, and even though glyoxal does not have hydrogen bonding, it is still polar, meaning it should be miscible in water.

7. (a) A first-order reaction has a constant half-life. So, every 1,200 s, half of the sample will have decayed. At 1,200 s, the concentration would be 0.50 M. At 2,400 s, the concentration would be 0.25 M, and so on.

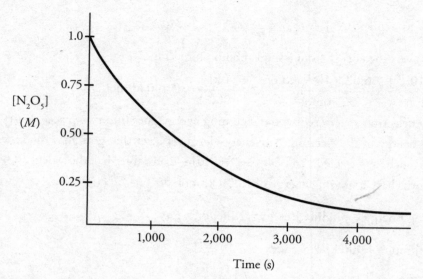

(b) Given that this is a first-order reaction, we can use the half-life equation to solve for the rate constant.

$$t_{\frac{1}{2}} = \frac{0.693}{k} \qquad\qquad 2{,}000\ \text{s} = \frac{0.693}{k} \qquad\qquad k = 5.8 \times 20^{-4}\ \text{s}^{-1}$$

(c) The integrated rate law for a first-order reaction is $\ln[N_2O_5]_t = -kt + \ln[N_2O_5]_0$. Plugging in what you know,

$$\ln[N_2O_5]_{2{,}000} = -(5.8 \times 10^{-4}\ \text{s}^{-1})(2{,}000\ \text{s}) + \ln(1.0\ M)$$
$$\ln[N_2O_5]_{2{,}000} = -1.2\ M$$
$$[N_2O_5]_{2{,}000} = 0.30\ M$$

Once the concentration at $t = 2{,}000$ s is known, then it can be plugged into the rate law to determine the rate at that point.

$$\text{Rate} = k[N_2O_5]$$
$$\text{Rate} = (5.8 \times 10^{-4}\ \text{s}^{-1})(0.30\ M)$$
$$\text{Rate} = 1.7 \times 10^{-4}\ M/s$$

Completely darken bubbles with a No. 2 pencil. If you make a mistake, be sure to erase mark completely. Erase all stray marks.

1.

YOUR NAME: _____
(Print)
                Last   First   M.I.

SIGNATURE: _____   DATE: _____ / _____ / _____

HOME ADDRESS: _____
(Print)
                                   Number and Street

_____
          City   State   Zip Code

PHONE NO.: _____

IMPORTANT: Please fill in these boxes exactly as shown on the back cover of your test book.

2. TEST FORM

3. TEST CODE

4. REGISTRATION NUMBER

5. YOUR NAME

First 4 letters of last name				FIRST INIT	MID INIT
Ⓐ	Ⓐ	Ⓐ	Ⓐ	Ⓐ	Ⓐ
Ⓑ	Ⓑ	Ⓑ	Ⓑ	Ⓑ	Ⓑ
Ⓒ	Ⓒ	Ⓒ	Ⓒ	Ⓒ	Ⓒ
Ⓓ	Ⓓ	Ⓓ	Ⓓ	Ⓓ	Ⓓ
Ⓔ	Ⓔ	Ⓔ	Ⓔ	Ⓔ	Ⓔ
Ⓕ	Ⓕ	Ⓕ	Ⓕ	Ⓕ	Ⓕ
Ⓖ	Ⓖ	Ⓖ	Ⓖ	Ⓖ	Ⓖ
Ⓗ	Ⓗ	Ⓗ	Ⓗ	Ⓗ	Ⓗ
Ⓘ	Ⓘ	Ⓘ	Ⓘ	Ⓘ	Ⓘ
Ⓙ	Ⓙ	Ⓙ	Ⓙ	Ⓙ	Ⓙ
Ⓚ	Ⓚ	Ⓚ	Ⓚ	Ⓚ	Ⓚ
Ⓛ	Ⓛ	Ⓛ	Ⓛ	Ⓛ	Ⓛ
Ⓜ	Ⓜ	Ⓜ	Ⓜ	Ⓜ	Ⓜ
Ⓝ	Ⓝ	Ⓝ	Ⓝ	Ⓝ	Ⓝ
Ⓞ	Ⓞ	Ⓞ	Ⓞ	Ⓞ	Ⓞ
Ⓟ	Ⓟ	Ⓟ	Ⓟ	Ⓟ	Ⓟ
Ⓠ	Ⓠ	Ⓠ	Ⓠ	Ⓠ	Ⓠ
Ⓡ	Ⓡ	Ⓡ	Ⓡ	Ⓡ	Ⓡ
Ⓢ	Ⓢ	Ⓢ	Ⓢ	Ⓢ	Ⓢ
Ⓣ	Ⓣ	Ⓣ	Ⓣ	Ⓣ	Ⓣ
Ⓤ	Ⓤ	Ⓤ	Ⓤ	Ⓤ	Ⓤ
Ⓥ	Ⓥ	Ⓥ	Ⓥ	Ⓥ	Ⓥ
Ⓦ	Ⓦ	Ⓦ	Ⓦ	Ⓦ	Ⓦ
Ⓧ	Ⓧ	Ⓧ	Ⓧ	Ⓧ	Ⓧ
Ⓨ	Ⓨ	Ⓨ	Ⓨ	Ⓨ	Ⓨ
Ⓩ	Ⓩ	Ⓩ	Ⓩ	Ⓩ	Ⓩ

TEST CODE:

Ⓐ Ⓙ, Ⓑ Ⓚ, Ⓒ Ⓛ, Ⓓ Ⓜ, Ⓔ Ⓝ, Ⓕ Ⓞ, Ⓖ Ⓟ, Ⓗ Ⓠ, Ⓘ Ⓡ

Digit columns (Test Code / Registration Number): 0 1 2 3 4 5 6 7 8 9

6. DATE OF BIRTH

Month	Day		Year	
◯ JAN				
◯ FEB	⓪	⓪	⓪	⓪
◯ MAR	①	①	①	①
◯ APR	②	②	②	②
◯ MAY	③	③	③	③
◯ JUN		④	④	④
◯ JUL		⑤	⑤	⑤
◯ AUG		⑥	⑥	⑥
◯ SEP		⑦	⑦	⑦
◯ OCT		⑧	⑧	⑧
◯ NOV		⑨	⑨	⑨
◯ DEC				

The **Princeton Review**®

1. Ⓐ Ⓑ Ⓒ Ⓓ
2. Ⓐ Ⓑ Ⓒ Ⓓ
3. Ⓐ Ⓑ Ⓒ Ⓓ
4. Ⓐ Ⓑ Ⓒ Ⓓ
5. Ⓐ Ⓑ Ⓒ Ⓓ
6. Ⓐ Ⓑ Ⓒ Ⓓ
7. Ⓐ Ⓑ Ⓒ Ⓓ
8. Ⓐ Ⓑ Ⓒ Ⓓ
9. Ⓐ Ⓑ Ⓒ Ⓓ
10. Ⓐ Ⓑ Ⓒ Ⓓ
11. Ⓐ Ⓑ Ⓒ Ⓓ
12. Ⓐ Ⓑ Ⓒ Ⓓ
13. Ⓐ Ⓑ Ⓒ Ⓓ
14. Ⓐ Ⓑ Ⓒ Ⓓ
15. Ⓐ Ⓑ Ⓒ Ⓓ
16. Ⓐ Ⓑ Ⓒ Ⓓ
17. Ⓐ Ⓑ Ⓒ Ⓓ
18. Ⓐ Ⓑ Ⓒ Ⓓ
19. Ⓐ Ⓑ Ⓒ Ⓓ
20. Ⓐ Ⓑ Ⓒ Ⓓ

21. Ⓐ Ⓑ Ⓒ Ⓓ
22. Ⓐ Ⓑ Ⓒ Ⓓ
23. Ⓐ Ⓑ Ⓒ Ⓓ
24. Ⓐ Ⓑ Ⓒ Ⓓ
25. Ⓐ Ⓑ Ⓒ Ⓓ
26. Ⓐ Ⓑ Ⓒ Ⓓ
27. Ⓐ Ⓑ Ⓒ Ⓓ
28. Ⓐ Ⓑ Ⓒ Ⓓ
29. Ⓐ Ⓑ Ⓒ Ⓓ
30. Ⓐ Ⓑ Ⓒ Ⓓ
31. Ⓐ Ⓑ Ⓒ Ⓓ
32. Ⓐ Ⓑ Ⓒ Ⓓ
33. Ⓐ Ⓑ Ⓒ Ⓓ
34. Ⓐ Ⓑ Ⓒ Ⓓ
35. Ⓐ Ⓑ Ⓒ Ⓓ
36. Ⓐ Ⓑ Ⓒ Ⓓ
37. Ⓐ Ⓑ Ⓒ Ⓓ
38. Ⓐ Ⓑ Ⓒ Ⓓ
39. Ⓐ Ⓑ Ⓒ Ⓓ
40. Ⓐ Ⓑ Ⓒ Ⓓ

41. Ⓐ Ⓑ Ⓒ Ⓓ
42. Ⓐ Ⓑ Ⓒ Ⓓ
43. Ⓐ Ⓑ Ⓒ Ⓓ
44. Ⓐ Ⓑ Ⓒ Ⓓ
45. Ⓐ Ⓑ Ⓒ Ⓓ
46. Ⓐ Ⓑ Ⓒ Ⓓ
47. Ⓐ Ⓑ Ⓒ Ⓓ
48. Ⓐ Ⓑ Ⓒ Ⓓ
49. Ⓐ Ⓑ Ⓒ Ⓓ
50. Ⓐ Ⓑ Ⓒ Ⓓ
51. Ⓐ Ⓑ Ⓒ Ⓓ
52. Ⓐ Ⓑ Ⓒ Ⓓ
53. Ⓐ Ⓑ Ⓒ Ⓓ
54. Ⓐ Ⓑ Ⓒ Ⓓ
55. Ⓐ Ⓑ Ⓒ Ⓓ
56. Ⓐ Ⓑ Ⓒ Ⓓ
57. Ⓐ Ⓑ Ⓒ Ⓓ
58. Ⓐ Ⓑ Ⓒ Ⓓ
59. Ⓐ Ⓑ Ⓒ Ⓓ
60 Ⓐ Ⓑ Ⓒ Ⓓ

Completely darken bubbles with a No. 2 pencil. If you make a mistake, be sure to erase mark completely. Erase all stray marks.

1.

YOUR NAME: _____
(Print)   Last   First   M.I.

SIGNATURE: _____ DATE: ____ / ____ / ____

HOME ADDRESS: _____
(Print)   Number and Street

_____
City   State   Zip Code

PHONE NO.: _____

IMPORTANT: Please fill in these boxes exactly as shown on the back cover of your test book.

2. TEST FORM

_____
_____

6. DATE OF BIRTH

Month	Day		Year	
○ JAN				
○ FEB	⓪	⓪	⓪	⓪
○ MAR	①	①	①	①
○ APR	②	②	②	②
○ MAY	③	③	③	③
○ JUN		④	④	④
○ JUL		⑤	⑤	⑤
○ AUG		⑥	⑥	⑥
○ SEP		⑦	⑦	⑦
○ OCT		⑧	⑧	⑧
○ NOV		⑨	⑨	⑨
○ DEC				

3. TEST CODE   **4. REGISTRATION NUMBER**

⓪	Ⓐ	Ⓙ	⓪	⓪	⓪	⓪	⓪	⓪	⓪	⓪
①	Ⓑ	Ⓚ	①	①	①	①	①	①	①	①
②	Ⓒ	Ⓛ	②	②	②	②	②	②	②	②
③	Ⓓ	Ⓜ	③	③	③	③	③	③	③	③
④	Ⓔ	Ⓝ	④	④	④	④	④	④	④	④
⑤	Ⓕ	Ⓞ	⑤	⑤	⑤	⑤	⑤	⑤	⑤	⑤
⑥	Ⓖ	Ⓟ	⑥	⑥	⑥	⑥	⑥	⑥	⑥	⑥
⑦	Ⓗ	Ⓠ	⑦	⑦	⑦	⑦	⑦	⑦	⑦	⑦
⑧	Ⓘ	Ⓡ	⑧	⑧	⑧	⑧	⑧	⑧	⑧	⑧
⑨			⑨	⑨	⑨	⑨	⑨	⑨	⑨	⑨

The Princeton Review®

5. YOUR NAME

First 4 letters of last name				FIRST INIT	MID INIT
Ⓐ	Ⓐ	Ⓐ	Ⓐ	Ⓐ	Ⓐ
Ⓑ	Ⓑ	Ⓑ	Ⓑ	Ⓑ	Ⓑ
Ⓒ	Ⓒ	Ⓒ	Ⓒ	Ⓒ	Ⓒ
Ⓓ	Ⓓ	Ⓓ	Ⓓ	Ⓓ	Ⓓ
Ⓔ	Ⓔ	Ⓔ	Ⓔ	Ⓔ	Ⓔ
Ⓕ	Ⓕ	Ⓕ	Ⓕ	Ⓕ	Ⓕ
Ⓖ	Ⓖ	Ⓖ	Ⓖ	Ⓖ	Ⓖ
Ⓗ	Ⓗ	Ⓗ	Ⓗ	Ⓗ	Ⓗ
Ⓘ	Ⓘ	Ⓘ	Ⓘ	Ⓘ	Ⓘ
Ⓙ	Ⓙ	Ⓙ	Ⓙ	Ⓙ	Ⓙ
Ⓚ	Ⓚ	Ⓚ	Ⓚ	Ⓚ	Ⓚ
Ⓛ	Ⓛ	Ⓛ	Ⓛ	Ⓛ	Ⓛ
Ⓜ	Ⓜ	Ⓜ	Ⓜ	Ⓜ	Ⓜ
Ⓝ	Ⓝ	Ⓝ	Ⓝ	Ⓝ	Ⓝ
Ⓞ	Ⓞ	Ⓞ	Ⓞ	Ⓞ	Ⓞ
Ⓟ	Ⓟ	Ⓟ	Ⓟ	Ⓟ	Ⓟ
Ⓠ	Ⓠ	Ⓠ	Ⓠ	Ⓠ	Ⓠ
Ⓡ	Ⓡ	Ⓡ	Ⓡ	Ⓡ	Ⓡ
Ⓢ	Ⓢ	Ⓢ	Ⓢ	Ⓢ	Ⓢ
Ⓣ	Ⓣ	Ⓣ	Ⓣ	Ⓣ	Ⓣ
Ⓤ	Ⓤ	Ⓤ	Ⓤ	Ⓤ	Ⓤ
Ⓥ	Ⓥ	Ⓥ	Ⓥ	Ⓥ	Ⓥ
Ⓦ	Ⓦ	Ⓦ	Ⓦ	Ⓦ	Ⓦ
Ⓧ	Ⓧ	Ⓧ	Ⓧ	Ⓧ	Ⓧ
Ⓨ	Ⓨ	Ⓨ	Ⓨ	Ⓨ	Ⓨ
Ⓩ	Ⓩ	Ⓩ	Ⓩ	Ⓩ	Ⓩ

1. Ⓐ Ⓑ Ⓒ Ⓓ
2. Ⓐ Ⓑ Ⓒ Ⓓ
3. Ⓐ Ⓑ Ⓒ Ⓓ
4. Ⓐ Ⓑ Ⓒ Ⓓ
5. Ⓐ Ⓑ Ⓒ Ⓓ
6. Ⓐ Ⓑ Ⓒ Ⓓ
7. Ⓐ Ⓑ Ⓒ Ⓓ
8. Ⓐ Ⓑ Ⓒ Ⓓ
9. Ⓐ Ⓑ Ⓒ Ⓓ
10. Ⓐ Ⓑ Ⓒ Ⓓ
11. Ⓐ Ⓑ Ⓒ Ⓓ
12. Ⓐ Ⓑ Ⓒ Ⓓ
13. Ⓐ Ⓑ Ⓒ Ⓓ
14. Ⓐ Ⓑ Ⓒ Ⓓ
15. Ⓐ Ⓑ Ⓒ Ⓓ
16. Ⓐ Ⓑ Ⓒ Ⓓ
17. Ⓐ Ⓑ Ⓒ Ⓓ
18. Ⓐ Ⓑ Ⓒ Ⓓ
19. Ⓐ Ⓑ Ⓒ Ⓓ
20. Ⓐ Ⓑ Ⓒ Ⓓ

21. Ⓐ Ⓑ Ⓒ Ⓓ
22. Ⓐ Ⓑ Ⓒ Ⓓ
23. Ⓐ Ⓑ Ⓒ Ⓓ
24. Ⓐ Ⓑ Ⓒ Ⓓ
25. Ⓐ Ⓑ Ⓒ Ⓓ
26. Ⓐ Ⓑ Ⓒ Ⓓ
27. Ⓐ Ⓑ Ⓒ Ⓓ
28. Ⓐ Ⓑ Ⓒ Ⓓ
29. Ⓐ Ⓑ Ⓒ Ⓓ
30. Ⓐ Ⓑ Ⓒ Ⓓ
31. Ⓐ Ⓑ Ⓒ Ⓓ
32. Ⓐ Ⓑ Ⓒ Ⓓ
33. Ⓐ Ⓑ Ⓒ Ⓓ
34. Ⓐ Ⓑ Ⓒ Ⓓ
35. Ⓐ Ⓑ Ⓒ Ⓓ
36. Ⓐ Ⓑ Ⓒ Ⓓ
37. Ⓐ Ⓑ Ⓒ Ⓓ
38. Ⓐ Ⓑ Ⓒ Ⓓ
39. Ⓐ Ⓑ Ⓒ Ⓓ
40. Ⓐ Ⓑ Ⓒ Ⓓ

41. Ⓐ Ⓑ Ⓒ Ⓓ
42. Ⓐ Ⓑ Ⓒ Ⓓ
43. Ⓐ Ⓑ Ⓒ Ⓓ
44. Ⓐ Ⓑ Ⓒ Ⓓ
45. Ⓐ Ⓑ Ⓒ Ⓓ
46. Ⓐ Ⓑ Ⓒ Ⓓ
47. Ⓐ Ⓑ Ⓒ Ⓓ
48. Ⓐ Ⓑ Ⓒ Ⓓ
49. Ⓐ Ⓑ Ⓒ Ⓓ
50. Ⓐ Ⓑ Ⓒ Ⓓ
51. Ⓐ Ⓑ Ⓒ Ⓓ
52. Ⓐ Ⓑ Ⓒ Ⓓ
53. Ⓐ Ⓑ Ⓒ Ⓓ
54. Ⓐ Ⓑ Ⓒ Ⓓ
55. Ⓐ Ⓑ Ⓒ Ⓓ
56. Ⓐ Ⓑ Ⓒ Ⓓ
57. Ⓐ Ⓑ Ⓒ Ⓓ
58. Ⓐ Ⓑ Ⓒ Ⓓ
59. Ⓐ Ⓑ Ⓒ Ⓓ
60. Ⓐ Ⓑ Ⓒ Ⓓ

Completely darken bubbles with a No. 2 pencil. If you make a mistake, be sure to erase mark completely. Erase all stray marks.

1.

YOUR NAME: _____
(Print)   Last   First   M.I.

SIGNATURE: _____   DATE: __ / __ / __

HOME ADDRESS: _____
(Print)   Number and Street

_____
City   State   Zip Code

PHONE NO.: _____

IMPORTANT: Please fill in these boxes exactly as shown on the back cover of your test book.

2. TEST FORM

_____

3. TEST CODE

⓪	Ⓐ	Ⓙ	⓪	⓪	
①	Ⓑ	Ⓚ	①	①	
②	Ⓒ	Ⓛ	②	②	
③	Ⓓ	Ⓜ	③	③	
④	Ⓔ	Ⓝ	④	④	
⑤	Ⓕ	Ⓞ	⑤	⑤	
⑥	Ⓖ	Ⓟ	⑥	⑥	
⑦	Ⓗ	Ⓠ	⑦	⑦	
⑧	Ⓘ	Ⓡ	⑧	⑧	
⑨			⑨	⑨	

4. REGISTRATION NUMBER

⓪	⓪	⓪	⓪	⓪	⓪	⓪
①	①	①	①	①	①	①
②	②	②	②	②	②	②
③	③	③	③	③	③	③
④	④	④	④	④	④	④
⑤	⑤	⑤	⑤	⑤	⑤	⑤
⑥	⑥	⑥	⑥	⑥	⑥	⑥
⑦	⑦	⑦	⑦	⑦	⑦	⑦
⑧	⑧	⑧	⑧	⑧	⑧	⑧
⑨	⑨	⑨	⑨	⑨	⑨	⑨

5. YOUR NAME

First 4 letters of last name				FIRST INIT	MID INIT
Ⓐ	Ⓐ	Ⓐ	Ⓐ	Ⓐ	Ⓐ
Ⓑ	Ⓑ	Ⓑ	Ⓑ	Ⓑ	Ⓑ
Ⓒ	Ⓒ	Ⓒ	Ⓒ	Ⓒ	Ⓒ
Ⓓ	Ⓓ	Ⓓ	Ⓓ	Ⓓ	Ⓓ
Ⓔ	Ⓔ	Ⓔ	Ⓔ	Ⓔ	Ⓔ
Ⓕ	Ⓕ	Ⓕ	Ⓕ	Ⓕ	Ⓕ
Ⓖ	Ⓖ	Ⓖ	Ⓖ	Ⓖ	Ⓖ
Ⓗ	Ⓗ	Ⓗ	Ⓗ	Ⓗ	Ⓗ
Ⓘ	Ⓘ	Ⓘ	Ⓘ	Ⓘ	Ⓘ
Ⓙ	Ⓙ	Ⓙ	Ⓙ	Ⓙ	Ⓙ
Ⓚ	Ⓚ	Ⓚ	Ⓚ	Ⓚ	Ⓚ
Ⓛ	Ⓛ	Ⓛ	Ⓛ	Ⓛ	Ⓛ
Ⓜ	Ⓜ	Ⓜ	Ⓜ	Ⓜ	Ⓜ
Ⓝ	Ⓝ	Ⓝ	Ⓝ	Ⓝ	Ⓝ
Ⓞ	Ⓞ	Ⓞ	Ⓞ	Ⓞ	Ⓞ
Ⓟ	Ⓟ	Ⓟ	Ⓟ	Ⓟ	Ⓟ
Ⓠ	Ⓠ	Ⓠ	Ⓠ	Ⓠ	Ⓠ
Ⓡ	Ⓡ	Ⓡ	Ⓡ	Ⓡ	Ⓡ
Ⓢ	Ⓢ	Ⓢ	Ⓢ	Ⓢ	Ⓢ
Ⓣ	Ⓣ	Ⓣ	Ⓣ	Ⓣ	Ⓣ
Ⓤ	Ⓤ	Ⓤ	Ⓤ	Ⓤ	Ⓤ
Ⓥ	Ⓥ	Ⓥ	Ⓥ	Ⓥ	Ⓥ
Ⓦ	Ⓦ	Ⓦ	Ⓦ	Ⓦ	Ⓦ
Ⓧ	Ⓧ	Ⓧ	Ⓧ	Ⓧ	Ⓧ
Ⓨ	Ⓨ	Ⓨ	Ⓨ	Ⓨ	Ⓨ
Ⓩ	Ⓩ	Ⓩ	Ⓩ	Ⓩ	Ⓩ

6. DATE OF BIRTH

Month		Day		Year	
◯ JAN					
◯ FEB	⓪	⓪	⓪	⓪	
◯ MAR	①	①	①	①	
◯ APR	②	②	②	②	
◯ MAY	③	③	③	③	
◯ JUN		④	④	④	
◯ JUL		⑤	⑤	⑤	
◯ AUG		⑥	⑥	⑥	
◯ SEP		⑦	⑦	⑦	
◯ OCT		⑧	⑧	⑧	
◯ NOV		⑨	⑨	⑨	
◯ DEC					

The **Princeton Review**®

1. Ⓐ Ⓑ Ⓒ Ⓓ
2. Ⓐ Ⓑ Ⓒ Ⓓ
3. Ⓐ Ⓑ Ⓒ Ⓓ
4. Ⓐ Ⓑ Ⓒ Ⓓ
5. Ⓐ Ⓑ Ⓒ Ⓓ
6. Ⓐ Ⓑ Ⓒ Ⓓ
7. Ⓐ Ⓑ Ⓒ Ⓓ
8. Ⓐ Ⓑ Ⓒ Ⓓ
9. Ⓐ Ⓑ Ⓒ Ⓓ
10. Ⓐ Ⓑ Ⓒ Ⓓ
11. Ⓐ Ⓑ Ⓒ Ⓓ
12. Ⓐ Ⓑ Ⓒ Ⓓ
13. Ⓐ Ⓑ Ⓒ Ⓓ
14. Ⓐ Ⓑ Ⓒ Ⓓ
15. Ⓐ Ⓑ Ⓒ Ⓓ
16. Ⓐ Ⓑ Ⓒ Ⓓ
17. Ⓐ Ⓑ Ⓒ Ⓓ
18. Ⓐ Ⓑ Ⓒ Ⓓ
19. Ⓐ Ⓑ Ⓒ Ⓓ
20. Ⓐ Ⓑ Ⓒ Ⓓ

21. Ⓐ Ⓑ Ⓒ Ⓓ
22. Ⓐ Ⓑ Ⓒ Ⓓ
23. Ⓐ Ⓑ Ⓒ Ⓓ
24. Ⓐ Ⓑ Ⓒ Ⓓ
25. Ⓐ Ⓑ Ⓒ Ⓓ
26. Ⓐ Ⓑ Ⓒ Ⓓ
27. Ⓐ Ⓑ Ⓒ Ⓓ
28. Ⓐ Ⓑ Ⓒ Ⓓ
29. Ⓐ Ⓑ Ⓒ Ⓓ
30. Ⓐ Ⓑ Ⓒ Ⓓ
31. Ⓐ Ⓑ Ⓒ Ⓓ
32. Ⓐ Ⓑ Ⓒ Ⓓ
33. Ⓐ Ⓑ Ⓒ Ⓓ
34. Ⓐ Ⓑ Ⓒ Ⓓ
35. Ⓐ Ⓑ Ⓒ Ⓓ
36. Ⓐ Ⓑ Ⓒ Ⓓ
37. Ⓐ Ⓑ Ⓒ Ⓓ
38. Ⓐ Ⓑ Ⓒ Ⓓ
39. Ⓐ Ⓑ Ⓒ Ⓓ
40. Ⓐ Ⓑ Ⓒ Ⓓ

41. Ⓐ Ⓑ Ⓒ Ⓓ
42. Ⓐ Ⓑ Ⓒ Ⓓ
43. Ⓐ Ⓑ Ⓒ Ⓓ
44. Ⓐ Ⓑ Ⓒ Ⓓ
45. Ⓐ Ⓑ Ⓒ Ⓓ
46. Ⓐ Ⓑ Ⓒ Ⓓ
47. Ⓐ Ⓑ Ⓒ Ⓓ
48. Ⓐ Ⓑ Ⓒ Ⓓ
49. Ⓐ Ⓑ Ⓒ Ⓓ
50. Ⓐ Ⓑ Ⓒ Ⓓ
51. Ⓐ Ⓑ Ⓒ Ⓓ
52. Ⓐ Ⓑ Ⓒ Ⓓ
53. Ⓐ Ⓑ Ⓒ Ⓓ
54. Ⓐ Ⓑ Ⓒ Ⓓ
55. Ⓐ Ⓑ Ⓒ Ⓓ
56. Ⓐ Ⓑ Ⓒ Ⓓ
57. Ⓐ Ⓑ Ⓒ Ⓓ
58. Ⓐ Ⓑ Ⓒ Ⓓ
59. Ⓐ Ⓑ Ⓒ Ⓓ
60 Ⓐ Ⓑ Ⓒ Ⓓ

Completely darken bubbles with a No. 2 pencil. If you make a mistake, be sure to erase mark completely. Erase all stray marks.

1.

YOUR NAME: _____
(Print)   Last   First   M.I.

SIGNATURE: _____ DATE: ___/___/___

HOME ADDRESS: _____
(Print)   Number and Street

_____
City   State   Zip Code

PHONE NO.: _____

IMPORTANT: Please fill in these boxes exactly as shown on the back cover of your test book.

2. TEST FORM

3. TEST CODE

4. REGISTRATION NUMBER

⓪	Ⓐ	Ⓙ	⓪	⓪	⓪	⓪	⓪	⓪	⓪	⓪
①	Ⓑ	Ⓚ	①	①	①	①	①	①	①	①
②	Ⓒ	Ⓛ	②	②	②	②	②	②	②	②
③	Ⓓ	Ⓜ	③	③	③	③	③	③	③	③
④	Ⓔ	Ⓝ	④	④	④	④	④	④	④	④
⑤	Ⓕ	Ⓞ	⑤	⑤	⑤	⑤	⑤	⑤	⑤	⑤
⑥	Ⓖ	Ⓟ	⑥	⑥	⑥	⑥	⑥	⑥	⑥	⑥
⑦	Ⓗ	Ⓠ	⑦	⑦	⑦	⑦	⑦	⑦	⑦	⑦
⑧	Ⓘ	Ⓡ	⑧	⑧	⑧	⑧	⑧	⑧	⑧	⑧
⑨			⑨	⑨	⑨	⑨	⑨	⑨	⑨	⑨

6. DATE OF BIRTH

Month	Day		Year	
◯ JAN				
◯ FEB	⓪	⓪	⓪	⓪
◯ MAR	①	①	①	①
◯ APR	②	②	②	②
◯ MAY	③	③	③	③
◯ JUN		④	④	④
◯ JUL		⑤	⑤	⑤
◯ AUG		⑥	⑥	⑥
◯ SEP		⑦	⑦	⑦
◯ OCT		⑧	⑧	⑧
◯ NOV		⑨	⑨	⑨
◯ DEC				

The Princeton Review®

5. YOUR NAME

First 4 letters of last name				FIRST INIT	MID INIT
Ⓐ	Ⓐ	Ⓐ	Ⓐ	Ⓐ	Ⓐ
Ⓑ	Ⓑ	Ⓑ	Ⓑ	Ⓑ	Ⓑ
Ⓒ	Ⓒ	Ⓒ	Ⓒ	Ⓒ	Ⓒ
Ⓓ	Ⓓ	Ⓓ	Ⓓ	Ⓓ	Ⓓ
Ⓔ	Ⓔ	Ⓔ	Ⓔ	Ⓔ	Ⓔ
Ⓕ	Ⓕ	Ⓕ	Ⓕ	Ⓕ	Ⓕ
Ⓖ	Ⓖ	Ⓖ	Ⓖ	Ⓖ	Ⓖ
Ⓗ	Ⓗ	Ⓗ	Ⓗ	Ⓗ	Ⓗ
Ⓘ	Ⓘ	Ⓘ	Ⓘ	Ⓘ	Ⓘ
Ⓙ	Ⓙ	Ⓙ	Ⓙ	Ⓙ	Ⓙ
Ⓚ	Ⓚ	Ⓚ	Ⓚ	Ⓚ	Ⓚ
Ⓛ	Ⓛ	Ⓛ	Ⓛ	Ⓛ	Ⓛ
Ⓜ	Ⓜ	Ⓜ	Ⓜ	Ⓜ	Ⓜ
Ⓝ	Ⓝ	Ⓝ	Ⓝ	Ⓝ	Ⓝ
Ⓞ	Ⓞ	Ⓞ	Ⓞ	Ⓞ	Ⓞ
Ⓟ	Ⓟ	Ⓟ	Ⓟ	Ⓟ	Ⓟ
Ⓠ	Ⓠ	Ⓠ	Ⓠ	Ⓠ	Ⓠ
Ⓡ	Ⓡ	Ⓡ	Ⓡ	Ⓡ	Ⓡ
Ⓢ	Ⓢ	Ⓢ	Ⓢ	Ⓢ	Ⓢ
Ⓣ	Ⓣ	Ⓣ	Ⓣ	Ⓣ	Ⓣ
Ⓤ	Ⓤ	Ⓤ	Ⓤ	Ⓤ	Ⓤ
Ⓥ	Ⓥ	Ⓥ	Ⓥ	Ⓥ	Ⓥ
Ⓦ	Ⓦ	Ⓦ	Ⓦ	Ⓦ	Ⓦ
Ⓧ	Ⓧ	Ⓧ	Ⓧ	Ⓧ	Ⓧ
Ⓨ	Ⓨ	Ⓨ	Ⓨ	Ⓨ	Ⓨ
Ⓩ	Ⓩ	Ⓩ	Ⓩ	Ⓩ	Ⓩ

1. Ⓐ Ⓑ Ⓒ Ⓓ
2. Ⓐ Ⓑ Ⓒ Ⓓ
3. Ⓐ Ⓑ Ⓒ Ⓓ
4. Ⓐ Ⓑ Ⓒ Ⓓ
5. Ⓐ Ⓑ Ⓒ Ⓓ
6. Ⓐ Ⓑ Ⓒ Ⓓ
7. Ⓐ Ⓑ Ⓒ Ⓓ
8. Ⓐ Ⓑ Ⓒ Ⓓ
9. Ⓐ Ⓑ Ⓒ Ⓓ
10. Ⓐ Ⓑ Ⓒ Ⓓ
11. Ⓐ Ⓑ Ⓒ Ⓓ
12. Ⓐ Ⓑ Ⓒ Ⓓ
13. Ⓐ Ⓑ Ⓒ Ⓓ
14. Ⓐ Ⓑ Ⓒ Ⓓ
15. Ⓐ Ⓑ Ⓒ Ⓓ
16. Ⓐ Ⓑ Ⓒ Ⓓ
17. Ⓐ Ⓑ Ⓒ Ⓓ
18. Ⓐ Ⓑ Ⓒ Ⓓ
19. Ⓐ Ⓑ Ⓒ Ⓓ
20. Ⓐ Ⓑ Ⓒ Ⓓ

21. Ⓐ Ⓑ Ⓒ Ⓓ
22. Ⓐ Ⓑ Ⓒ Ⓓ
23. Ⓐ Ⓑ Ⓒ Ⓓ
24. Ⓐ Ⓑ Ⓒ Ⓓ
25. Ⓐ Ⓑ Ⓒ Ⓓ
26. Ⓐ Ⓑ Ⓒ Ⓓ
27. Ⓐ Ⓑ Ⓒ Ⓓ
28. Ⓐ Ⓑ Ⓒ Ⓓ
29. Ⓐ Ⓑ Ⓒ Ⓓ
30. Ⓐ Ⓑ Ⓒ Ⓓ
31. Ⓐ Ⓑ Ⓒ Ⓓ
32. Ⓐ Ⓑ Ⓒ Ⓓ
33. Ⓐ Ⓑ Ⓒ Ⓓ
34. Ⓐ Ⓑ Ⓒ Ⓓ
35. Ⓐ Ⓑ Ⓒ Ⓓ
36. Ⓐ Ⓑ Ⓒ Ⓓ
37. Ⓐ Ⓑ Ⓒ Ⓓ
38. Ⓐ Ⓑ Ⓒ Ⓓ
39. Ⓐ Ⓑ Ⓒ Ⓓ
40. Ⓐ Ⓑ Ⓒ Ⓓ

41. Ⓐ Ⓑ Ⓒ Ⓓ
42. Ⓐ Ⓑ Ⓒ Ⓓ
43. Ⓐ Ⓑ Ⓒ Ⓓ
44. Ⓐ Ⓑ Ⓒ Ⓓ
45. Ⓐ Ⓑ Ⓒ Ⓓ
46. Ⓐ Ⓑ Ⓒ Ⓓ
47. Ⓐ Ⓑ Ⓒ Ⓓ
48. Ⓐ Ⓑ Ⓒ Ⓓ
49. Ⓐ Ⓑ Ⓒ Ⓓ
50. Ⓐ Ⓑ Ⓒ Ⓓ
51. Ⓐ Ⓑ Ⓒ Ⓓ
52. Ⓐ Ⓑ Ⓒ Ⓓ
53. Ⓐ Ⓑ Ⓒ Ⓓ
54. Ⓐ Ⓑ Ⓒ Ⓓ
55. Ⓐ Ⓑ Ⓒ Ⓓ
56. Ⓐ Ⓑ Ⓒ Ⓓ
57. Ⓐ Ⓑ Ⓒ Ⓓ
58. Ⓐ Ⓑ Ⓒ Ⓓ
59. Ⓐ Ⓑ Ⓒ Ⓓ
60. Ⓐ Ⓑ Ⓒ Ⓓ

1.

YOUR NAME:
(Print)

Last   First   M.I.

SIGNATURE: _____ DATE: ___/___/___

HOME ADDRESS:
(Print) _____
Number and Street

_____
City   State   Zip Code

PHONE NO.: _____

5. YOUR NAME

First 4 letters of last name				FIRST INIT	MID INIT
Ⓐ	Ⓐ	Ⓐ	Ⓐ	Ⓐ	Ⓐ
Ⓑ	Ⓑ	Ⓑ	Ⓑ	Ⓑ	Ⓑ
Ⓒ	Ⓒ	Ⓒ	Ⓒ	Ⓒ	Ⓒ
Ⓓ	Ⓓ	Ⓓ	Ⓓ	Ⓓ	Ⓓ
Ⓔ	Ⓔ	Ⓔ	Ⓔ	Ⓔ	Ⓔ
Ⓕ	Ⓕ	Ⓕ	Ⓕ	Ⓕ	Ⓕ
Ⓖ	Ⓖ	Ⓖ	Ⓖ	Ⓖ	Ⓖ
Ⓗ	Ⓗ	Ⓗ	Ⓗ	Ⓗ	Ⓗ
Ⓘ	Ⓘ	Ⓘ	Ⓘ	Ⓘ	Ⓘ
Ⓙ	Ⓙ	Ⓙ	Ⓙ	Ⓙ	Ⓙ
Ⓚ	Ⓚ	Ⓚ	Ⓚ	Ⓚ	Ⓚ
Ⓛ	Ⓛ	Ⓛ	Ⓛ	Ⓛ	Ⓛ
Ⓜ	Ⓜ	Ⓜ	Ⓜ	Ⓜ	Ⓜ
Ⓝ	Ⓝ	Ⓝ	Ⓝ	Ⓝ	Ⓝ
Ⓞ	Ⓞ	Ⓞ	Ⓞ	Ⓞ	Ⓞ
Ⓟ	Ⓟ	Ⓟ	Ⓟ	Ⓟ	Ⓟ
Ⓠ	Ⓠ	Ⓠ	Ⓠ	Ⓠ	Ⓠ
Ⓡ	Ⓡ	Ⓡ	Ⓡ	Ⓡ	Ⓡ
Ⓢ	Ⓢ	Ⓢ	Ⓢ	Ⓢ	Ⓢ
Ⓣ	Ⓣ	Ⓣ	Ⓣ	Ⓣ	Ⓣ
Ⓤ	Ⓤ	Ⓤ	Ⓤ	Ⓤ	Ⓤ
Ⓥ	Ⓥ	Ⓥ	Ⓥ	Ⓥ	Ⓥ
Ⓦ	Ⓦ	Ⓦ	Ⓦ	Ⓦ	Ⓦ
Ⓧ	Ⓧ	Ⓧ	Ⓧ	Ⓧ	Ⓧ
Ⓨ	Ⓨ	Ⓨ	Ⓨ	Ⓨ	Ⓨ
Ⓩ	Ⓩ	Ⓩ	Ⓩ	Ⓩ	Ⓩ

IMPORTANT: Please fill in these boxes exactly as shown on the back cover of your test book.

2. TEST FORM

_____

6. DATE OF BIRTH

Month		Day		Year	
◯ JAN					
◯ FEB	⓪	⓪	⓪	⓪	
◯ MAR	①	①	①	①	
◯ APR	②	②	②	②	
◯ MAY	③	③	③	③	
◯ JUN		④	④	④	
◯ JUL		⑤	⑤	⑤	
◯ AUG		⑥	⑥	⑥	
◯ SEP		⑦	⑦	⑦	
◯ OCT		⑧	⑧	⑧	
◯ NOV		⑨	⑨	⑨	
◯ DEC					

3. TEST CODE

⓪	Ⓐ	Ⓙ	⓪	⓪
①	Ⓑ	Ⓚ	①	①
②	Ⓒ	Ⓛ	②	②
③	Ⓓ	Ⓜ	③	③
④	Ⓔ	Ⓝ	④	④
⑤	Ⓕ	Ⓞ	⑤	⑤
⑥	Ⓖ	Ⓟ	⑥	⑥
⑦	Ⓗ	Ⓠ	⑦	⑦
⑧	Ⓘ	Ⓡ	⑧	⑧
⑨			⑨	⑨

4. REGISTRATION NUMBER

⓪	⓪	⓪	⓪	⓪	⓪	⓪
①	①	①	①	①	①	①
②	②	②	②	②	②	②
③	③	③	③	③	③	③
④	④	④	④	④	④	④
⑤	⑤	⑤	⑤	⑤	⑤	⑤
⑥	⑥	⑥	⑥	⑥	⑥	⑥
⑦	⑦	⑦	⑦	⑦	⑦	⑦
⑧	⑧	⑧	⑧	⑧	⑧	⑧
⑨	⑨	⑨	⑨	⑨	⑨	⑨

The **Princeton Review**®

1. Ⓐ Ⓑ Ⓒ Ⓓ
2. Ⓐ Ⓑ Ⓒ Ⓓ
3. Ⓐ Ⓑ Ⓒ Ⓓ
4. Ⓐ Ⓑ Ⓒ Ⓓ
5. Ⓐ Ⓑ Ⓒ Ⓓ
6. Ⓐ Ⓑ Ⓒ Ⓓ
7. Ⓐ Ⓑ Ⓒ Ⓓ
8. Ⓐ Ⓑ Ⓒ Ⓓ
9. Ⓐ Ⓑ Ⓒ Ⓓ
10. Ⓐ Ⓑ Ⓒ Ⓓ
11. Ⓐ Ⓑ Ⓒ Ⓓ
12. Ⓐ Ⓑ Ⓒ Ⓓ
13. Ⓐ Ⓑ Ⓒ Ⓓ
14. Ⓐ Ⓑ Ⓒ Ⓓ
15. Ⓐ Ⓑ Ⓒ Ⓓ
16. Ⓐ Ⓑ Ⓒ Ⓓ
17. Ⓐ Ⓑ Ⓒ Ⓓ
18. Ⓐ Ⓑ Ⓒ Ⓓ
19. Ⓐ Ⓑ Ⓒ Ⓓ
20. Ⓐ Ⓑ Ⓒ Ⓓ

21. Ⓐ Ⓑ Ⓒ Ⓓ
22. Ⓐ Ⓑ Ⓒ Ⓓ
23. Ⓐ Ⓑ Ⓒ Ⓓ
24. Ⓐ Ⓑ Ⓒ Ⓓ
25. Ⓐ Ⓑ Ⓒ Ⓓ
26. Ⓐ Ⓑ Ⓒ Ⓓ
27. Ⓐ Ⓑ Ⓒ Ⓓ
28. Ⓐ Ⓑ Ⓒ Ⓓ
29. Ⓐ Ⓑ Ⓒ Ⓓ
30. Ⓐ Ⓑ Ⓒ Ⓓ
31. Ⓐ Ⓑ Ⓒ Ⓓ
32. Ⓐ Ⓑ Ⓒ Ⓓ
33. Ⓐ Ⓑ Ⓒ Ⓓ
34. Ⓐ Ⓑ Ⓒ Ⓓ
35. Ⓐ Ⓑ Ⓒ Ⓓ
36. Ⓐ Ⓑ Ⓒ Ⓓ
37. Ⓐ Ⓑ Ⓒ Ⓓ
38. Ⓐ Ⓑ Ⓒ Ⓓ
39. Ⓐ Ⓑ Ⓒ Ⓓ
40. Ⓐ Ⓑ Ⓒ Ⓓ

41. Ⓐ Ⓑ Ⓒ Ⓓ
42. Ⓐ Ⓑ Ⓒ Ⓓ
43. Ⓐ Ⓑ Ⓒ Ⓓ
44. Ⓐ Ⓑ Ⓒ Ⓓ
45. Ⓐ Ⓑ Ⓒ Ⓓ
46. Ⓐ Ⓑ Ⓒ Ⓓ
47. Ⓐ Ⓑ Ⓒ Ⓓ
48. Ⓐ Ⓑ Ⓒ Ⓓ
49. Ⓐ Ⓑ Ⓒ Ⓓ
50. Ⓐ Ⓑ Ⓒ Ⓓ
51. Ⓐ Ⓑ Ⓒ Ⓓ
52. Ⓐ Ⓑ Ⓒ Ⓓ
53. Ⓐ Ⓑ Ⓒ Ⓓ
54. Ⓐ Ⓑ Ⓒ Ⓓ
55. Ⓐ Ⓑ Ⓒ Ⓓ
56. Ⓐ Ⓑ Ⓒ Ⓓ
57. Ⓐ Ⓑ Ⓒ Ⓓ
58. Ⓐ Ⓑ Ⓒ Ⓓ
59. Ⓐ Ⓑ Ⓒ Ⓓ
60 Ⓐ Ⓑ Ⓒ Ⓓ

NOTES

NOTES

NOTES